BIOLOGY
The World of Life

Robert A. Wallace
University of California, Santa Barbara

Goodyear Publishing Company, Inc., Pacific Palisades, California

Copyright © 1975 Goodyear Publishing Company, Inc.
Pacific Palisades, California

All rights reserved. No part of this book
may be reproduced in any form or by any
means without permission in writing from
the publisher.

Current printing (last digit):
10 9 8 7 6 5 4 3 2

ISBN: 0-87620-114-1

Library of Congress Catalog Number: 74-31026

Y-1141-4

Library of Congress Cataloging in Publication Data

Wallace, Robert Ardell, 1938-
  Biology, the world of life.

  Includes index.
  1. Biology.   I. Title.
QH308.2.W35   574   74-31026
ISBN 0-87620-114-1

Designer/Art Director: Don McQuiston
Supervising Editor: Gail Weingart
Scientific Consultant: Gerald Sanders
Illustrator: John Dawson
Cartoons: Howard Saunders
Photos Courtesy: Salt Marsh, Harry Crosby/Photophile;
Eastern Sierras, Arthur Bilsten/Photophile;
Monument Valley, Gail Weingart;
Anacapa Island, Mike Muckley/Photophile;
Earth, NASA

Printed in the United States of America

This book is for Ruth Ann, Michael, Rip,
Ronald, Sancho Spannza, Uncle Scott, Zsolt,
one fat man in North Carolina, and the
La Playa Rummies Track and Discussion Club.

Charles Darwin

Contents

SODIUM ION (Na⁺)

Preface

I really don't think they were crouching. I think it's just that the windows were high. After all, why would they crouch? It probably only seemed that way, because they held a particular air of mystery for me on those early mornings—and perhaps partly because I could never see them very well, at least not without being obvious. And then there were those rumors about the Mafia living out there! But, the few times I was able to glimpse them, they seemed innocuous enough—reading the paper or just sitting there, in their limousines, resolutely looking ahead, never noticing the other cars, or the people, or the land, on either side.

I often wondered about those people in their sumptuous automobiles as I slowly made my way from my house on Atlantic Beach into the city where I was a visiting professor at Queens College. I suppose my fascination with what I imagined to be the rich or powerful belied my middle-class upbringing, but crawling along on a rough road between the marshy fields on those cold mornings one might search for anything to occupy his mind.

Another thought that came to me from time to time was, why do they keep filling in these marshes? Great machines stood scattered over the land on either side, squat and hard-looking or standing with their long necks disappearing into the early morning fog. As a biologist I was aware of the importance of the marshes in the ocean's food chain, and I resented their destruction.

It dawned on me one morning that no one else seemed to pay any attention to the building activities. Least of all, not those reclining with a kind of heavy presence in their chauffered limousines. Their windows were up and they were removed from the carnage. Whatever was going on out there was of no interest to them. Not them. And then I realized that, in a sense, they *were* being insulted. All of us were. But, of course, *I* was the only one who knew it.

I'm sure everyone on the road on those mornings had his own problems. We were all going to work and had things to do and who had the time or inclination to worry about marshes? The people in the limousines, it seemed to me, were the most removed from it all. Their car temperature was regulated and they weren't even involved in the physics of dodging chuckholes. So how could those ungainly machines constitute an insult to them?

Being stuck in traffic, even before coffee, breeds philosophy, and it seemed to me that I could see a larger question here. The incessantly irritated and impatient commuters around me had taken no offense at the construction because they simply didn't realize that something

more than land was being exploited. But those marshes were part of a critical and fast-disappearing link in a life-chain—and it was a life-chain in which we were all unalterably a part. Rolling up windows, or failing to see, changed nothing. What was being destroyed out there was an important part of our ecosystem—our environment, and, in a sense, part of us.

It seemed to me then, and it seems now, that the most important lesson we can learn is that we are a *part* of the world out there. It is not "our home," nor can we think in terms of its being "them or us." Any effort to "have dominion" over nature can only be self-defeating since we are a *part* of nature. We may be a highly intelligent part, but our newly developed abilities cannot negate our heritage. Rather, what we must do, if we are wise as well as intelligent, is to use our abilities to insure that the rest of the natural world is sustained. Our reward will be our own sustenance.

And this will be the point of this book—to help to show that while we are a special case in many ways, our species is but one of many in a complex and mysterious world, and that our own best interest lies in coming to understand that world and to live in harmony with it.

About the Book

The format of this book, as you can see, is traditional. We go from the history of biology to the establishment of an elementary molecular background which will be used primarily in understanding energy and the mechanisms of DNA activity. Then from cellular, through systemic and organismic biology, ending with discussions of behavior, populations and ecology. The underlying theme, or the unifying thread, if you will, involves the twin concepts of evolution and adaptation: how each life form has come to fit into its particular part of the world.

Adherence to tradition, of course, does not justify the appearance of yet another book in an already crowded field which a few truly excellent books dominate. So what particular or unusual features does this book have to offer to justify its existence? First, and I believe this is important, I think that students might enjoy reading it. I had fun writing it and I'm counting on some of that same spirit to be absorbed by the reader. Life is fascinating, and biology is the study of life, but many of the biology textbooks I have encountered in my years of teaching have somehow managed to be dull. Above all, I hope this book is not dull.

Perhaps a faint sense of honesty, but probably more of what I consider to be a literary flair, compels me to say something else: this book is biased. Throughout the text I have expressed my own opinion on a number of topics. This was done partly to get it off my chest, but primarily as a teaching technique—a means of stimulating discussion in these critical times. Hopefully, the opinion is clearly

distinguishable from what we like to call "fact," so that the student will be stimulated and not misled at such points. My own training is in behavior and ecology so perhaps the major part of my opinion is interjected in related discussions (if one may consider reproductive and population biology as areas of ecology). Anyway, I'm sure the professor who uses this book will have his or her own biases and will be quick to point out mine to the student.

Primarily, the book is written for students who have little or no training (or perhaps interest) in the sciences. I assume most of the readers will be non-majors, although several reviewers have suggested the book could be used as the framework for more rigorous courses with the material expanded by the instructor. It was my intention (1) to describe the material as simply as possible, (2) to interject biological viewpoint into questions of philosophy and values— questions to which the student can relate in a very personal sense, and, (3) in the event that some students are turned on by the material and should wish to pursue biology further, to present them with a good, solid background in biology upon which they can base future work.

Acknowledgements

I owe a debt of gratitude to a great many people for the development of this book. First, I would like to express my appreciation to the authors of the 37 reviews from various biology departments across the country. These were invaluable, content-wise, and held my tangents to a minimum. Secondly, I would like to say it has again been a great pleasure to work with Goodyear Publishing Company. I am grateful for the encouragement and various assistances of Al Goodyear, Jim Levy, Gail Weingart. Colette Conboy, Sharon Jones, Gene Schwartz, Bob Hollander, Don McQuiston, Debra McQuiston, John Dawson, Max Neufeldt, Judith Johnson, and Nancy Clark. In particular I thank my editor, Clay J. Stratton, who seemed to have the uncanny ability to be there at the right time with whatever was needed to keep things rolling. It seems that such competence in an initially so intangible, deferred, and elusive a field as publishing must be a rarity. Vernon Avila of San Diego State University served notably as final review consultant. Gerald Sanders of San Diego State University assisted the editorial staff and myself in the accuracy of both content and art during the publication process. The months he has contributed to this have been uniquely important to the book. Finally I would thank my good friend and wife, Bonni, for more-or-less putting up with me through the months of writing.

1
Biology Yesterday and Today

(Courtesy: The Granger Collection, New York and Stanley N. Botwinik from National Audubon Society)

The door closed quietly. Someone had left the room. Behind, a child, without benefit of anesthesia, lay on the operating table as huddled surgeons prepared to make the initial incision. A group of beginning medical students watched intently from just outside the bright circle of overhead light. As the operation progressed the eyes of one sixteen-year-old had faltered. Finally, no longer able to watch, Charles Darwin had hurried out of the room.

He had failed again, this time at medicine. What was he to do, this younger son of an English gentleman? Perhaps he would be better suited for the clergy, which was, after all, an admirable calling for one of his social position. So the following year his father enrolled him at Cambridge to study theology—with equally disastrous results. Although young Charles spent three years as a theology student, he seemed to spend more time with beetles than beatitudes. He was not alone in his interests, however. At that time the English countryside was alive with amateur butterfly collectors, rockhounds, and plant fanciers whose position and wealth permitted them to indulge in such hobbies. Even so, Darwin's academic prowess had been so thoroughly unremarkable that at one point his father had told his trifling son, "You care for nothing but shooting, dogs, and rat-catching, and you will be a disgrace to yourself and all your family."

THE VOYAGE OF THE BEAGLE

It must be admitted that at this point in Charles Darwin's career there was little to suggest a mind that was to be regarded as one of the most brilliant and inquisitive in history. In a short time, however, this diffident young man was to hear, through his friend the Reverend John Henslow, of an offer of free passage on a survey ship called the *Beagle*. A naturalist was needed for a voyage which was to last five years. There would be no pay, and the person chosen would have to sleep in a hammock in the cramped chartroom (although he would be permitted to share the captain's table). Armed with Henslow's recommendation, Darwin eagerly applied

Figure 1.1 Young Charles Darwin, on his way to becoming a dismal failure in life, or so his family believed. His hobbies included collecting rocks and insects. He couldn't interest himself much in medicine or the ministry. Then an event happened that changed his life. A botany professor, John S. Henslow, was offered a job as a naturalist on the research ship, the *Beagle* but his wife wouldn't let him go. Henslow then offered the job to Darwin. (Courtesy: The Granger Collection, New York)

Figure 1.2 Copernicus in 1543 proclaimed that the sun, not the earth, was at the center of the solar system. This was a shocking notion for authoritarian and religious society of the time. Some of his ideas were erroneous (he miscalculated the cycles of Mercury and Venus and implied the sun was at the center of the universe) but his basic contributions were great. Although he empirically "proved" his system, he died without his ideas ever being accepted. (Courtesy: The Bettmann Archive)

for the position, and was nearly rejected because of the shape of his nose. Captain Fitzroy, himself only twenty-three, believed that the nose reflected the character of its bearer, and Darwin's nose just didn't show much character.

Darwin's family required some persuasion to accept this "madcap scheme," which they considered scarcely in keeping for a prospective member of the clergy. And evidently young Charles had his own trepidations about such a momentous decision. In a letter to his sister Susan he wrote:

. . . *Fitzroy says the stormy sea is exaggerated; that if I do not choose to remain with them, I can at any time get home to England; that and that if I like, I shall be left in some healthy, safe and nice country; that I shall always have assistance; that he has many books, all instruments, guns, at my service. . . . There is indeed a tide in the affairs of men, and I have experienced it. Dearest Susan, Goodbye.*

Ultimately all the arrangements were made, and in 1831 the *H.M.S. Beagle* set out from Devonport, with Charles Darwin aboard as the official ship's naturalist. As the heavy wooden vessel, about the size of a tugboat, creaked and groaned its way across the Atlantic toward South America, young Darwin's worst fears were realized. As if things weren't hard enough on such a voyage, Darwin tended to get seasick. At long last the *Beagle* reached South America and then headed down the coast on the first leg of its voyage—past the coasts of Brazil and Argentina, through the pounding gales of Cape Horn, and finally northward along the desolate coasts of Chile and Peru. Fortunately there were periods of respite when the *Beagle* dropped anchor to put foraging parties ashore. At these times a grateful Darwin scrambled off the ship to explore the new and fascinating countryside, returning with specimens of all sorts, to the amusement and the occasional dismay of the crew. He also used this time to make copious notes on everything he saw. Some of the stops lasted for months, and on these occasions Darwin ventured far into the wild terrain of South America with the rugged local gauchos, for whose horsemanship he soon developed the greatest respect.

To understand just what this adventure meant to Charles Darwin, and would later mean to the world, we must stop at this point to consider the state of scientific thinking at that time. What was known of the biological world? What prejudices or beliefs did Darwin have? More important, what seeds of ideas? What hunches?

THE HISTORY OF AN IDEA

New scientific ideas have appeared throughout man's history. Many of these ideas have been erroneous, of course, but they often provided at least a framework for the expansion of our knowledge. Nevertheless, new ideas—and even new facts—have not always been met with enthusiasm. In the best of times, when an assumption did not fit what was known about the real world, the assumption was discarded. The notions which remained, then, were likely to be based to some extent on the available facts, and as new information came to light, our body of scientific knowledge expanded. It would be satisfying, somehow, if it could be said that scientific knowledge progressed stepwise from ancient times until the present, knowledge and understanding accumulating until we reach a crescendo of scientific expansion, all to the good of man. Alas, this is not the case.

As far back as the early civilizations of Babylonia, Egypt, and Greece, scholars were actively engaged in trying to figure out the nature of the world in which they lived. These early scientists were looking for some unifying concept to encompass the order of the universe and the relative positions of the species within it. However, after the fall of these ancient civilizations there was a steady decline in scientific curiosity. Then for a period of more than ten centuries, between 200 and 1200 A.D., there were virtually no scientific advances at all. In fact, much of what had been known was forgotten! Although the Greek mathematician Eratosthenes had calculated the circumference of the earth to within fifty miles more than 200 years before the birth of Christ, in 1492 Columbus had trouble convincing anyone except a few Moorish astronomers that the world was even round.

During these centuries the writings of a few ancient scholars were preserved in dusty monasteries, and some of the ideas they contained were dutifully studied and transcribed from one parchment to another. But instead of active investigation to verify statements or ideas, men turned to religious authority for absolute answers as the only kernel of order in a world that was beyond their understanding. As a result, science and religion became hopelessly intertwined, so that "scientific" statements which were otherwise testable became religious doctrine and therefore not to be questioned. Thus, by a twist of fate and politics, the church became on one hand the seat of higher learning and on the other hand a formidable opponent of new ideas. The Polish astronomer Copernicus (1473–1543), who voiced the heretical theory that the earth was not the center of the universe, escaped retribution by dying shortly after his work was published (from fumes he had inhaled when he peered over the rim of Mt. Vesuvius). However, when the great Galileo (1564–1642) produced detailed evidence that the earth did indeed revolve around the sun, his writings were banned

Figure 1.3 Galileo Galilei embraced the findings of Copernicus which stated that the sun stands at the center of the universe with all other bodies moving around it. But this implied, at least to religious minds, that man did not stand at the center of all creation. The old scholar of 69 was forced to recant under dire threats of the Church. He knelt before ten cardinals, robed in their scarlet, and placed the earth back in its immovable position before all the world. It is said that as he arose from his knees he muttered, "Eppur si Muove"—"It moves nevertheless." (Courtesy: The Granger Collection, New York)

Figure 1.4 Isaac Newton, one of the truly important figures in the history of science, analyzing light rays in his laboratory. Newtonian physics was to dominate the world of the physical sciences until the 20th century. It was Newton who, in 1687, provided final and irrefutable evidence for the Copernican doctrine of the solar system. (Courtesy: The Granger Collection, New York)

in Rome and he was summoned before the Inquisition and forced to recant his belief in the Copernican theory.

Nevertheless, as the world became more predictable, and men began to feel less at the mercy of magical forces and unseen beings, there was a great surge of scientific thinking. The year Galileo died marked the birth of one of the greatest scientists of all time—Isaac Newton (1642–1727). By the age of twenty-four Newton had already formulated the idea of universal gravitation, and in 1685 he presented a set of carefully proven laws of motion which were to revolutionize the physical sciences.

THE BEGINNINGS OF BIOLOGY

Until the eighteenth century, science was limited for the most part to topics dealing with inanimate matter—mathematics, astronomy, physics. It was one thing to apply physical laws to the motion of a falling rock or the stars, but quite another to consider *life* in those terms. After all, man was a living thing, and therefore life must surely have some special purpose, some grand design. We just weren't quite ready to view ourselves as merely another physical phenomenon.

In the 1700s it was universally believed that all species, or "kinds," of living things were created in their present form—in other words, that they had not changed in any way over the millennia. This was certainly the view of the Swedish botanist Carl von Linné (1707–1778), who devised a system of classification for all living organisms and, in his fondness for Latin, even signed his own name as Carolus Linnaeus. Many argued that Linnaeus' designations, which involved lumping the species together on the basis of their similarities, were artificial and based on personal whim. Few, however, considered criticizing his assumption of special creation or his concept that each species was fixed in its original form.

One small departure was suggested by a French contemporary, George-Louis Leclerc de Buffon (1707–1788). In 1753 Buffon proposed that in addition to those animals which had originated in the Creation, there were also lesser families "conceived by Nature and produced by Time." He explained that changes of this kind were the result of imperfections in the Creator's expression of the ideal—a philosophical, rather than a biological, point of view.

A decade later, however, another doubter appeared on the scene in England—Erasmus Darwin (1731–1802), the grandfather of Charles. Erasmus was a physician and an amateur naturalist who wrote on the subjects of botany and zoology, often in verse form. He referred almost incidentally to the relationships among animals, the importance of competition in the formation of species, the effect of environment on changes in species, and the heritability of these changes. There has, of course, been much speculation on the extent to which young Charles may have been influenced by Erasmus. The effect may not have been as great as one might imagine because Charles apparently had little respect for the musings of his grandfather.

Other men were also beginning to toy with the notion of the heritability of change. In France Jean Baptiste de Lamarck (1744–1829), a protégé of Buffon's, boldly suggested that not only had one species given rise to another, but man himself had arisen from other species. A passionate classifier, Lamarck believed that every organism had a relative position on the "scale of nature"—with man, of course, as the highest form of life. He pointed out, however, that the fossil animals found in older layers of rock were simpler forms and contended that they had become higher animals through a gradual progression. In his view there was a "force of life" which caused an organism to generate new structures and organs to meet its biological needs. Such structures then continued to develop through use, with the change acquired in each generation transmitted to succeeding generations. Lamarck cited as an example the long neck of the giraffe, which he maintained had evolved as each generation of giraffes had stretched their necks in an effort to reach the topmost branches of trees and then transmitted genes for this longer neck to their offspring.

Although Lamarck's concept of evolution is contradicted by genetic evidence, certain aspects of his theory were given serious consideration a number of years later. At the time, however, the arguments he put forth did little to persuade his lecture audiences. With the foment of the French Revolution, there was some lively discussion in intellectual circles, but in society at large the firm conviction that each form of life had arisen through special creation ruled out serious consideration of any other concept.

Figure 1.5 The Swedish naturalist, Carolus Linnaeus, the founder of the scientific method of naming living things. He apparently conceived of species as unchangeable entities, the products of divine creation. He lived and worked in the century before Darwin and had no idea of the concept of evolution. (Courtesy: The Granger Collection, New York)

TIME AND THE INTELLECTUAL MILIEU

The intellectual climate into which Darwin was born was far more conservative than that in France. The English had been horrified by the French Revolution, and ideas such as those put forth by the "French atheists" were not only dismissed as out of hand, but viewed with suspicion. Partly for this reason, the church held strong sway over the sciences, and biologists in England continued to adhere rigidly to the traditional tenets.

Even so, men such as Erasmus Darwin had begun to consider the idea of heritability of change, at least from a philosophical standpoint. It has been suggested that the similarity between these

Longer
neck passed
on to
offspring

Neck
stretched
longer
in each
generation

Stretches
neck to
reach higher
leaves

Original
short-necked
ancestor

LAMARCK'S GIRAFFE

'DARWIN'S GIRAFFE

Original
neck length
varies

Longer-necked
animals can
reach higher
leaves

Longer-necked
animals
favored
in each
generation

Figure 1.6 Opposing explanations of how the giraffe got a long neck.

ideas and the concepts formulated later by Charles Darwin merely reflect an inescapable convergence of thinking, given the state of knowledge and a changing intellectual milieu. In other words, perhaps the time was right.

It must be stressed, however, that such views were by no means universal throughout the scientific community. Half a century later, after Darwin had published his *Origin of Species*, he observed:

It has sometimes been said that the success of the "Origin" proved "that the subject was in the air," or "that men's minds were prepared for it." I do not think that this is strictly true, for I occasionally sounded out a few naturalists, and never happened to come across a single one who seemed to doubt about the permanence of species. . . . I tried once or twice to explain to able men what I meant by Natural Selection, but signally failed.

DARWIN'S THEORY OF EVOLUTION

When Charles Darwin set out on his voyage aboard the *Beagle*, it was with the same conviction shared by others of his time that life had originated through special creation and all species were fixed in form. The physical sciences, however, were less hampered by religious tenets, and important changes were stirring in other disciplines. Charles Lyell (1797–1875), who was only a few years older than Darwin, had published the first volume of his work *Principles of Geology* before the *Beagle* set sail, and Darwin had asked to have the second volume sent to him en route. Lyell had rejected the Biblical thesis that the earth had been created in the year 4004 B.C. and contended that it was actually much older. On this basis he theorized that the earth's features had not appeared suddenly through catastrophic upheavals, as had been thought, but were instead the result of slow, steady processes over an exceedingly long period of time. Lyell had not extended his theory to any speculation about the origin of living organisms, but Darwin immediately seized on the implications. He had dismissed the idea that all species had sprung from simpler forms through spontaneous generation. The idea that mud would give rise to frogs or that dog food left in a corner would turn into cockroaches was no longer seriously considered. But the dimension of time opened up possibilities for some other kind of evolutionary process. Nevertheless, this was only the seed of an idea. Darwin had been thoroughly schooled in beliefs of long tradition, and rejection of those beliefs was to be a very slow process.

As Darwin continued his investigation of the plant and animal life and fossil beds of South America, he was struck by the differences in living things from one place to the next. The shells awash on the Pacific shores were not at all like those he had found on the Atlantic side. He also noticed differences in the birds and mammals as he traveled from one place to another. In some cases the change was marked. In other cases there were gradual changes in form as one type of terrain gave rise to another. Had each of these forms been created separately, Darwin wondered, each with subtle characteristics which distinguished it as a species?

THE GALÁPAGOS ISLANDS

When the *Beagle* finally left the shores of South America, an impatient Captain Fitzroy set the sails of his sturdy boat for a straight run to the Galápagos, a chain of islands which lie about 580 miles off the coast of Ecuador. Little did he or Darwin realize the turmoil which would be precipitated, or perhaps only hastened, by this decision.

On this group of islands Darwin encountered a strange assortment of animals. "A little world in itself," he wrote, "with inhabitants such as are found nowhere else." There were lizards three feet long which grazed beneath the turbulent sea—in Darwin's words, "imps of darkness, black as the porous rocks over which they crawl." And there were the incredible giant tortoises for which the islands are named. Even more remarkable, each island had its own type of tortoise, distinguished by characteristics that enabled local fishermen to tell immediately from which island any one of these creatures had come since tortoises are strictly land animals.

Darwin also noticed a group of small birds—the Galápagos finches, sometimes called Darwin's finches. All these birds have the distinguishing characteristics of finches, but there are thirteen species differentiated by the size and shape of their beaks. Each species uses its specialized beak differently and picks up its food in a specific way. Darwin surmised that these birds must have come originally from the South American mainland, since the volcanic islands of the Galápagos would have been formed later than the continent. But how could these finches differ so markedly from those on the mainland, and why was it that the species on the islands differed so much from each other? It seemed almost as if one type of finch had been modified into a number of new types. Darwin was unable to dismiss this thought. But it was another twenty years before he was to resolve it in the context of any comprehensive theory.

THE IMPACT OF MALTHUS

After his return to England Darwin had occasion to read an earlier

The Galapagos Islands

A. Vistas such as this at Bartolome Islet led Darwin to comment that the Galapagos were a most inhospitable place. There are 18 major islands but almost all are uninhabitable by humans on a year 'round basis. (Courtesy: Photophile) B. Point Espinosa, Fernandina Island. The first permanent resident of the islands was an Irishman, Patrick Watkins, who was described as ragged, covered with vermin, wild, savage, and occasionally cannibalistic. There were about 250 people, mostly of doubtful repute, on the islands when Darwin arrived there. (Courtesy: Richard Howmiller) C. The volcanic origin of the islands is revealed in the great craters which remain. There are those who believe the islands were once connected to the mainland, but probably most scientists believe they arose from the sea through volcanic action and were never part of the mainland. The great craters often enclose huge lakes which periodically dry up. (Courtesy: Ron Garrison, San Diego Zoo) D. Marine iguanas at Point Espinosa. Generally, islands which are flatter have fewer species than those with more altitudinal variation. The taller islands usually have more "vegetation zones." (Courtesy: Richard Howmiller)

A

A. Marine iguanas at Point Espinosa. Darwin called them "imps of darkness." In the non-breeding season they are gregarious, but early in the breeding season the males fight for a space on the breeding ground. (Courtesy: Richard Howmiller) B. The cormorants, which no longer have any need to fly, now have reduced wings that are not used for flight but they always spread them after a dive. The reason for this is not known, but it may be a behavioral relic of a time when it was necessary to have dry wings. (Courtesy: Richard Howmiller) C. Penguins normally breed at or near Antarctica. The Galapagos penguin, however, is able to breed as far north as the equator because the frigid Humboldt current sweeps past the Galapagos through much of the year. (Courtesy: Photophile) D. The Galapagos flamingo has characteristics of both the West Indian and South American species. This is the Galapagos bird species most in danger of extinction.

(Courtesy: Ron Garrison, San Diego Zoo) E. The tameness of many island species such as this frigate bird has led to great slaughters. Earlier, hundreds of thousands of doves were killed by sticks and U.S. servicemen obliterated the species on Baltra

Island. (Courtesy: Richard Howmiller) F. Land iguanas are very tame and eat almost any kind of plant matter. They are now extinct on several islands where they once thrived. (Courtesy: Photophile)

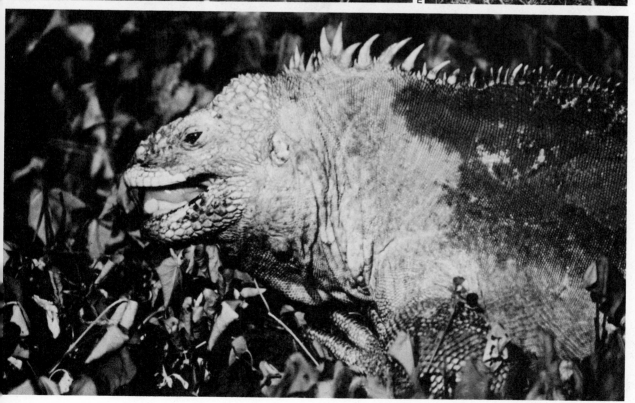

A. Saddle-back tortoises are usually found under harsh conditions. Their shell, flared upward in front, permits them to feed higher on the prickly-pear cactus. (Courtesy: A. W. Ambler from National Audubon Society)
B. Round-shelled tortoises are usually found on those islands with the most lush vegetation. The Galapagos tortoises were often taken aboard sailing vessels and tied upside down until they were slaughtered. (Courtesy: Stanley N. Botwinik from National Audubon Society) C. The prickly-pear cactus grows into tall treelike forms. Its height protects it against browsing tortoises. (Courtesy: Ron Garrison, San Diego Zoo) D. A colony of *Brachycereus nesioticus*, a small cactus which is one of the first colonizers of fresh lava flow. Since it is the same on all the islands, it is thought to be a recent colonizer.

(Courtesy: Ron Garrison, San Diego Zoo) E. The lava which formed the islands can be so rough as to make foot-travel almost impossible. Because the volcanic rock is porous, rain is absorbed and does not form pools. Thus the lava does not weather much, and little soil has formed in most places. (Courtesy: Ron Garrison, San Diego Zoo) F. The Galapagos vegetation is often tough and thorny, a reflection of the harsh conditions under which the islands evolved. The finch doesn't seem to mind. (Courtesy: Ron Garrison, San Diego Zoo) G. A female Galapagos sea lion. Unlike the fur seal, which shares the islands, the sea lion is quite tame. In breeding season, however, the large males may rush ashore and charge inquisitive biologists. (Courtesy: Ron Garrison, San Diego Zoo)

G

Figure 1.7 A dairy cow is a prodigious producer of milk. The breed was developed through generations of artificial selection; continually mating the offspring of high milk producers. (Courtesy: The Crescent Beauty Farm, Ft. Atkinson, Wisc.)

essay by the Reverend Thomas Malthus (1766–1834), which was probably our first clear warning of the danger of overpopulation. In this essay, which appeared in 1798, Malthus pointed out that populations increased in a geometric progression, and if man continued to reproduce at the same rate, he would inevitably outstrip his food supply and create a teeming world full of misery and vice. Darwin gleaned one important idea from this treatise. The reproductive potential of animals is high, but not all animals survive long enough to reproduce; hence it is this difference in survival among members of populations which has kept those populations in check. Darwin calculated that a pair of slow-breeding elephants, for example, could produce 19 million progeny in only 750 years. Yet through the years the total number of elephants seemed to stay about the same. If some members of a population were unable to survive, what factors determined which among them were to be the successful ones? Darwin could only account for the phenomena by the operation of some system of natural selection.

NATURAL SELECTION

Darwin had long been familiar with the principles of *artificial selection* from his experiences as the son of a country gentleman. He knew that breeders were able, through careful selection of animals for mating, to accentuate desired characteristics in the offspring. By mating only the offspring of the greatest milk producers, they could develop high-yield dairy cattle. And by permitting only the offspring of good laying hens to breed, they could eventually produce hens which were veritable egg-laying machines.

The results of artificial selection could be seen in only a few generations, since individuals with undesirable qualities were kept from transmitting these qualities at all.

Darwin saw natural selection as an analogous process in which the animals which would reproduce were selected, not by the breeder, but by the environment. The process in this case was far less efficient, since individuals with less-desirable characteristics were often able to produce some offspring and could thus maintain their traits in the population for a time. Those individuals with characteristics totally out of keeping with their environments would leave no offspring at all; hence their characteristics would disappear from the population. And since those individuals with characteristics which provided some advantage in reproducing could be expected to outreproduce the rest of the population, they would eventually come to predominate.

DESCENT WITH MODIFICATION

Since this theory of natural selection depends on inequalities in members of a population, the source of the variation had to be accounted for. Darwin proposed that variations occur at random—that no driving force, no direction, and no design are necessary. A heritable trait, if it provides some advantage, would be expected to increase the probability that the animal will leave offspring. If the long neck of the giraffe is inherited and is helpful in acquiring food, then the giraffes with longer necks will be better nourished, and hence more likely to have the energy to leave offspring. Some of these offspring will have longer necks than others, and so the long-necked ones will continually be favored through the generations, as long as the species specializes in eating leaves of trees. The result is a tendency for any generation of such animals to have longer necks than any generation of their predecessors.

It is important to realize that Darwin had no hard proof for this idea. There was no experimental evidence he could offer, since he knew nothing of the field we now call genetics. He was therefore reluctant to present his theory without being able to explain the mechanism by which the process of natural selection actually worked. However, in 1858, when Alfred Russell Wallace (1823–1913) outlined a similar theory of natural selection, Darwin enthusiastically pulled together his voluminous notes and prepared a paper to be presented to the Linnaean Society of London. In spite of the overwhelming data he had gathered during the twenty years since his voyage, Darwin asked that Wallace be permitted to present his paper first and receive credit for the idea, rather than have any man think he had behaved in a "paltry spirit." Wallace, just as much a gentleman, declined. The outcome was that they presented their papers together, with Darwin speaking first, in keeping with his much more substantial evidence.

Darwin's carefully formulated ideas were greeted with enthusiasm in some quarters, but needless to say, the response was not universal. He was forced to defend his suggestion of "descent with modification" not only against scientists who demanded hard evidence, but also against the attacks of philosophers, theologians, and a general public who were reluctant to part with the idea that they were members of a unique and chosen species.

Darwin himself was poorly equipped for such a battle. When he had returned to England after his five-year odyssey, he was ailing, and he had never recovered his health. Plagued by anxiety and self-doubt, he had remained in seclusion since his return, rarely venturing outside his house and avoiding everyone but his family and a few close friends. His defense however, was by no means single-handed. Many of the best minds of the time leaped to Darwin's defense. Some of these seemed to be imbued much more than Darwin with a taste for battle. Here, after all, was a unifying concept, one that made sense of it all. It accounted for the observations. It was not to be rejected on any basis other than a better explanation. And there was none.

We will discuss the principles of evolution throughout this book, but for now let us note that the concept of evolutionary descent is generally accepted by the scientific community on the basis that it accounts for the data. Today the major arguments concern the mechanism of its operation. For example, are new species formed through an accumulation of small changes over a long period of time, or are such changes produced only by major upheavals?

Darwin's theory of natural selection has also given rise to a number of other questions. If the species we see around us are the products of evolutionary change, what are their histories? More important, how are they related to each other? And the question which hangs over it all: where does man fit in? Are we really the cousins of the animals in our zoos? Can we really be related to them?

MECHANISM AND VITALISM
One of the basic questions which arose is where it all began. If one animal is descended from another, what did the *earliest* animals look like? Did they have much the same general form we see in animals today, or have they changed radically from some simple ancestor? In other words, how did life get started?

Figure 1.8 As the notion of natural selection became publicized, the political cartoonists had a field day with Darwin. This is one of the more complimentary likenesses. (Courtesy: The Granger Collection, New York)

One idea was that life might have originated in some sort of chance combination of chemicals. If even the possibility of such a beginning could be demonstrated, then perhaps the quality of life could be defined in terms of simple chemical interactions. According to this view, called *mechanism*, a living thing might conceivably be no more than the sum total of a number of chemical reactions.

Taken to its extreme, this mechanistic view implied that if everything were known about the chemical make-up of the members of any group, their behavior could be predicted under any given circumstances. We could know when any individual would feel hungry or sexy, whom he would favor in a football game, and how he would feel about entering the European Common Market. Such thinking was a bit much for those who believed that living things have certain special qualities which are lacking in nonliving things—qualities which are not reducible to mere chemical reactions. According to their view, called *vitalism*, living things uniquely possess some mysterious sort of "life force."

In the absence of any direct evidence of such a life force, the mechanistic approach at least offered a framework for continuing the search for the possible chemical beginnings of life. The working hypothesis was that life could have arisen from the interactions of nonliving chemicals as a form of "spontaneous generation," in which, at some minute point, matter crossed the dividing line between nonlife and LIFE.

2
The Beginnings

(Courtesy: National Aeronautics and Space Administration)

To understand how life could have started on earth, we must consider what the earth must have been like in those days. Of course, no one can give us a firsthand description of conditions on the early earth. We have to make many assumptions, but these assumptions must be based on certain facts that we do have available. First of all, it has been determined that the atmosphere was different then. Today we have an oxygen-laden atmosphere. Our air contains about 78 percent molecular nitrogen (N_2), 21 percent molecular oxygen (O_2), 0.03 percent carbon dioxide (CO_2), and traces of rare gases such as argon and helium. The Russian scientist A. I. Oparin proposed in 1936 that the atmosphere of the early earth must have contained much more hydrogen (H_2) than it does now. The result was that much of the free oxygen, nitrogen, and carbon dioxide could combine with the hydrogen to form new molecules. So instead of free nitrogen, for example, there was ammonia (NH_3), oxygen joined with hydrogen to form water vapor (H_2O), and of particular importance for the beginnings of life, carbon joined with hydrogen to form methane (CH_4). Methane is a simple hydrocarbon, an organic molecule, and organic molecules are associated with living systems, as we shall see.

Only our imaginations can tell us what the earth was like in those days. Perhaps the most difficult thing to imagine is that the planet—this planet, our earth—was lifeless. There were not, nor had there ever been, ears to hear or eyes to see. There were no crusty lichens or slimy molds. Nothing lurked anywhere. The globe was a lifeless unchronicled ball of matter covered by a very thin layer of hot swirling gas. The earth itself was at first hot, molten, volcanic. Then the surface began to cool and solidify. The piercing shrieks and groans and deep rumblings from below the crust startled no one. Heavy billowing clouds, miles thick, surrounded the darkened sphere. The blackness was split by spewing gapes in the earth and thunderous lightning from above. When the water vapor condensed and fell to the sterile earth, it exploded with the heat and was lofted skyward again with a crackling hiss. What could

have been happening to the molecules of the earth's substance at such a time? How might they have become rearranged to form the progenitors of life?

THE BEGINNINGS OF LIFE

Molecules associated with life are, for the most part, highly complex structures. Often they are long branching chains of atoms, far more complicated than the simple water, ammonia, and methane which comprised the early earth's atmosphere. For life to appear, somehow these simpler molecules had to be joined together. The trouble was that the molecules, by themselves, had no tendency to join. They could coexist quite nicely side by side for an indefinite period. As time went by, however, changes took place which increased the likelihood of their interacting.

With the cooling of the earth, pools of water condensed on its surface. More important, as the atmosphere itself cooled there must have been torrential rains of incredible proportion. As the water rushed over the exposed surface it swept away the salts and minerals of the topmost rock. These mineral-laden waters poured into the cracks and crevices of a parched earth, turning pattering streams into pounding rivers. The harsh waters fell into the lowest reaches of the planet's surface, and the oceans were born. It was here, in these violent briny waters, that the earth's myriad chemicals were forced into immediate proximity for long periods.

Still, there was little likelihood of any of the molecules joining together on their own. If they were to join, some outside source would have to provide the energy to make it happen. And there are, in fact, a number of possible sources of such energy. One is the sun. In addition to visible light rays, the sun produces ultraviolet rays, x rays, gamma rays, and various other radiations. Another possible source of the critical energy is heat. The earth's surface was still far from placid. Violent eruptions were occurring, and molten and cooling rock covered much of the earth's surface. There was also lightning streaking earthward continually from the dense, heavy clouds above, so the critical energy could have been electrical.

But was this enough? Could a mixture of methane, water, and ammonia form complex molecules by the addition of some outside source of energy?

MILLER'S EXPERIMENTS

In 1953 a graduate student at the University of Chicago provided an answer. Stanley L. Miller built an airtight apparatus through which four gases—methane, ammonia, water, and hydrogen—could

Figure 2.2 Miller's apparatus is an airtight system through which methane, ammonia, water and hydrogen are circulated past electrodes.

Figure 2.1 (left) Artist's concept of the primitive earth.

be circulated past electrodes. He permitted the gases to circulate together in his chamber for a week, energizing the mixture electrically, and then he analyzed the contents of the chamber. What he found surpassed his expectations. There were an astounding number of organic compounds. Among these were certain important amino acids, the components of protein, those magical molecules of life.

So that no one would question his experimental method, Miller ran two controls. In one case he sterilized the gas mixture at 130°C for eighteen hours before he subjected it to the electrical power source. The yield of complex molecules was the same as in his first test. In the second he ran the test exactly as before, except for the step of adding electric energy. In this case there was no significant yield of the new molecules. He then felt safe in concluding that the formation of the organic molecules was not due to the presence of living contaminants, but was the result of the fusing of smaller molecules which had been subjected to a source of energy.

Since then, Miller's experiment has been duplicated many times, and even extended. Researchers have subjected methane in hydrogen-laden atmospheres to energy sources and have produced a great variety of organic compounds. Among these are important molecules of life, such as purines, pyrimidines, and sugars, which we will consider later. Other workers, using other gases and a number of types of energy sources, have added to the long list of synthesized organic compounds.

In the face of so much evidence, it now appears that the synthesis of complex organic compounds in the early days of the earth was not only possible, but perhaps *inevitable*. Any large molecules so produced would probably survive for a considerable time, and thus have greater likelihood of entering into other reactions. Why should these large molecules have survived for long periods? Their stability is due to two phenomena: the relative lack of oxygen and the absence of microorganisms. Oxygen can break down complex substances through the processes of oxidation. This is why both iron and a half-eaten apple turn rust colored in air. Their constituents are breaking down in the presence of oxygen. And microorganisms, such as bacteria, can act to degrade, or rot, complex substances. Of course, in earth's present atmosphere, which is high in oxygen and filled with all kinds of microorganisms, no free complex organic molecule could survive for long. Thus life could not again arise on earth through such a pathway.

THE COACERVATE DROPLET

So here we are. We now have a system whereby large, complex

organic molecules could have been produced and maintained when the earth was young. It is apparent however, that the problem of the origin of life is not solved, not even theoretically. It is a long step from acknowledging the existence of organic molecules to accounting for even the simplest forms of life. Any living body is composed of a fantastically ordered arrangement of molecules. The precision of their relationships and interactions is even more astounding than the early biologists dreamed. How, then, could a living system arise from the simple hot soup of chemicals we have been talking about?

Again we have no firm answers, but we have some ideas. Oparin pointed out that *colloidal* protein molecules, in which the particles are suspended in a gel-like state, tend to clump together into increasingly complex masses. These are called *coacervate droplets*. Each droplet consists of an inner cluster of colloidal molecules surrounded by a shell of water. The molecules of this water shell are arranged in a specific manner in relation to the colloid center, so that there is a clear demarcation between the colloidal protein mass and the water in which it is suspended.

Figure 2.3 Artist's concept of the rich primordial soup of the early earth, filled with myriad forms of coacervate droplets. Some are doomed to disintegrate but others grow and fragment.

THE GROWTH OF COACERVATE DROPLETS

The peculiar orientation of the water molecules around the colloidal mass gives this water shell special physical properties. In effect, it acts as a sort of membrane; that is, the water layer functions, to a slight degree, as a screen for the kinds of molecules that can pass through it. Some kinds pass through with ease; others meet some resistance. The colloidal mass inside has its own peculiar qualities as well, since the complex molecules of which it is composed also tend to arrange themselves in an orderly manner. This comes about through the interaction of a number of forces in the mass, such as positive and negative charges along the molecules. (We will discuss these positive and negative charges in more detail later.) The colloid, as a result of its peculiar arrangement, has a tendency to absorb other molecules from the medium in which it is suspended. However, it does not absorb all molecules with equal vigor. In fact the mechanism is so selective that the colloid may take almost all of certain chemicals from the medium and practically none of others. As a result of this continual absorption of certain kinds of molecules, the droplet grows. In some cases the arrangement of the colloid may become further differentiated or specialized. The molecules at the surface may become more active and rearrange themselves into a membranelike structure just under the water layer. This new membrane is even more selective in what molecules it will allow to pass than was the water "membrane."

Thus, with increasing selectivity in the kinds of molecules the mass will accept, the structure becomes more and more specific. At this stage the complex droplets have many of the properties of living things and under a microscope may be mistaken for living things. (Even experienced biologists have, on occasion, attempted to classify such structures as species.)

INTERNAL STRUCTURE

It is important to realize that the specialized coacervate droplet provides a special environment for any molecule absorbed into its structure. First of all, because of the selectivity of the droplet surface, only certain types of molecules are brought together. Then, once these molecules have been incorporated into a droplet, they are brought into increasingly closer association as a result of the compactness (or high density) of the droplet. Thus certain kinds of reactions become more likely inside the droplet than outside it, and each of these reactions is likely to influence other reactions in new ways. In addition, the catalytic activity of substances such as metal compounds is facilitated by the regular arrangement of the colloidal molecules.

So what we have is a rather highly structured mass of matter which is susceptible to increased orderliness as it grows. And as a result of its orderliness, the likelihood of certain molecular events is increased. In a sense the droplet "regulates," or at least influences, the types of chemical reactions which can take place inside its mass. Now we begin to see that, whereas coacervate droplets are not "alive" as we usually understand the term, they do have many of the qualities we associate with living things.

REPRODUCING DROPLETS

If we are going to accept these droplets as precursors of life (or as distant relatives), we must somehow account for their reproduction. Again we rely on a strong imagination—held in check by circumstantial evidence. To begin with, *most* of these primordial droplets obviously did not reproduce. They must have formed in an astounding array of sizes and compositions, each with its own peculiar structure. Some were more stable than others, and these, of course, had the greatest chance of finally reproducing. Of course, here we are using the term "reproduction" in the broadest sense. Consider one example of how reproduction might have occurred. As certain of the droplets continued to survive, they would have kept on growing as new molecules entered their masses. At some point they would have probably fragmented as a direct or an indirect result of their having reached a critical size. Now, if their structures were highly regular and ordered, the resulting fragments would have each been very similar to the original droplet, and each fragment could then begin its own growth.

Reproduction of the droplets could also have occurred in another way. Among the many types of molecules comprising the colloid there were *nucleic acids*. We will discuss the chemical makeup and behavior of these remarkable molecules later. For now, however, bear in mind that they are the critical components in the mechanisms of heredity. It is possible that nucleic acids in a pool of their component parts, called *nucleotides*, could arrange those free nucleotides to correspond to their own images—another form of "reproduction." It is even possible that nucleic acids could form some kind of "code" for molecules other than replicas of themselves, and thus produce such critical molecules as proteins. Just how this might have come about is not at all clear, but if you are interested, you might read the June 1, 1973 volume of *Science*. The point is that there are ways in which certain highly complex coacervate droplets *could* have come to reproduce themselves. The most "successful" of these droplets, of course, would be those which came closest to producing *replicas* of themselves, since they had effectively demonstrated the stability of their own particular structure by their very presence.

SUCCESSFUL DROPLETS

Through eons of slow accumulation, as the less successful ones disappeared, the more successful droplets could have developed into the first cell-like bodies. With the continual disruption of less satisfactory bodies, as time passed the percentage of more stable forms increased, and the components of the lost droplets were recovered by increasingly efficient early cells. These cells, then, were the result of the most fortuitous of the molecular combinations that had taken place in an untold number of trials through millions of years. The point at which a droplet had become what we can call a cell is arbitrary. There was no sudden and dramatic appearance of life on the earth. No trumpets blared; no flags waved. There were only a few peculiar droplets which were perhaps a little more organized than the rest.

This picture of how life was first generated on earth is by no means universally accepted. Some scientists contend that coacervate droplets probably did not nurture reproducing nucleic acids. Instead, they consider it more likely that free nucleic acids came together and then built a cell around themselves. Others, of course, feel that there is an essential distinction between life and nonlife, and that all life—or at least that of man and the other vertebrates—

was created by some higher being. And then there are a host of lesser theories which need not concern us here. The whole question is an academically fascinating one, however, so let us briefly consider how these early cells might have made their living.

THE EARLY CELLS

As you probably already know, there are only two basic ways an organism can get its energy, or "food." Plants utilize carbon dioxide, water, and sunlight to make their food. They are called *autotrophs*, meaning "self-feeders." Animals do not normally make their own food, but must derive it from other organisms that have made it. They are called *heterotrophs*, meaning "other-feeders."

The prevailing notion is that the earliest cells were heterotrophs—that they utilized nutrients from their environment. Such a life style enabled them to survive without having to develop a food-producing mechanism. But it presented other problems. At first, with so few cells in the rich, warm primordial soup, there was no problem in getting food. As time went by, however, the cell population began to expand drastically, since the cells were actively reproducing themselves generation after generation. And eventually the nutrients on which they subsisted must have fallen into short supply.

There must also have been many different types of cells in those days, since they had sprung from a wide array of coacervate droplets which had formed more or less through various chance combinations. Some types of cells had advantages over the others. Those which were more adept at obtaining food from the environment were, naturally, able to reproduce at a more rapid rate, and it follows that in time these cells would come to be more prevalent. In this ever-changing and complex early world it is also likely that further chance changes took place within certain cells. When such changes impaired their food-absorbing ability, these altered cells perished in the competitive environment. But when the changes produced some slight advantage in competing for food, these altered cells increased in number over the rest. Again, the net result as time went by was that these early cells became increasingly efficient.

AUTOTROPHY AND HETEROTROPHY

The incredibly complex biochemical reactions which take place in cells today are probably a far cry from what went on in the early cells. So we might ask how they could have ever developed such complex systems. What was the impetus? Perhaps such complexity was spawned by the need to find new sources of supply for specific

Figure 2.4 Cheetahs are heterotrophs which eat other heterotrophs. The amoeba is also a heterotroph. It engulfs tiny autotrophs and heterotrophs and obtains the energy it needs by breaking down prey material and turning it into amoeba material. Here it has just captured two elongated paramecia. One is already partially digested. (Courtesy: Mark Boulton from National Audubon Society and Carolina Biological Supply House)

nutrients which had fallen into short supply. We know that certain food molecules were critical for cell reproduction. Thus a cell unable to find such nutrients had to either develop some alternative or perish. Among the small chance changes which were continually arising in the cells, then, there could have been some change which enabled a cell to manufacture critical nutrient A from a more available nutrient B. These cells, which had been held in check by scarce nutrient A, were then able to reproduce wildly—that is, until nutrient B fell into short supply. At this point perhaps there appeared, by chance, some other slight change which enabled a cell to manufacture B from nutrient C and thereby to make nutrient A. These cells were the ones which survived. Finally, the most viable cells would have been those with the ability to carry out long chains of biochemical reactions. Such biochemical complexity would have set the stage for existing in an increasingly complex and variable world. Any such system, of course, would demand an efficient energy system to drive the mechanism. It is interesting that the chemical ATP, which is a prime source of energy in present cells, is one of the molecules that is synthesized in experiments simulating the conditions of the early earth.

Today all animals are heterotrophs. They no longer absorb their food from a surrounding hot soup, it is true. Instead they eat autotrophs—or other heterotrophs which originally got their food from autotrophs they had consumed. All plants are autotrophs; they are able to make their own food, and as a result they have played a unique role in the history of life. Let's briefly consider how they might have arisen and how they might have affected other life.

AN AUTOTROPHY CALLED PHOTOSYNTHESIS

Basically, autotrophy involves the manufacture of complex molecules out of simpler materials. It is likely that as some of the early cells developed complex chemical pathways for making the nutrients they needed, they were finally able to make carbohydrates from carbon dioxide and water. Carbon dioxide would have been abundant as a by-product of heterotrophic cycles. We can only speculate about how early plants might have manufactured their food, but the process by which plants now do this is called *photosynthesis*. The simplified chemical description of this process is written as

$$6CO_2 + 12H_2O \xrightarrow[\text{chlorophyll}]{\text{sunlight}} C_6H_{12}O_6 + 6O_2 + 6H_2O$$

That is, six molecules of carbon dioxide and twelve molecules of water, in the presence of sunlight for energy and a pigment called chlorophyll, form one molecule of glucose (a sugar), six molecules of oxygen, and six molecules of water. We will discuss the pathway in more specific terms later.

OXYGEN: A BANE AND BLESSING

It is important to note that oxygen is released in the photosynthetic reaction. This is good news for us, since we need oxygen, but it sounded the death knell for many of the early heterotrophs. The release of more and more oxygen gradually changed the atmosphere from one full of hydrogen to one rich in oxygen. Now, it is logical to assume that some cells developed a means of utilizing the free oxygen, perhaps in energy-producing reactions. Oxidation, as we shall see, is a much more efficient means of producing energy than the fermentation-like processes the cells had been using. New types of cells with such abilities would have had a great advantage over the old types, and in time they would have come largely to replace them. However, the oxygen which proved a boon for the new kinds of cells probably acted as a poison for many of the old heterotrophic organisms and hastened their disappearance from the earth.

Not all scientists believe that the increase in the earth's oxygen is primarily the result of photosynthesis by plants. Some believe that most of this oxygen has come from the sun's action on the water of the upper atmosphere. We know that as water (H_2O) is split into its component parts oxygen is released. This knowledge may provide some consolation as we watch the destruction of the world's great forests and the slow poisoning of the algae-laden seas.

Whatever its source, the appearance of high levels of oxygen had important effects. As oxygen collected in the upper atmosphere in the form of ozone (O_3), it formed a shield against certain types of the sun's rays. The ultraviolet rays in sunlight had probably provided energy for the activities of early cells, but the disruptive effects of these rays would also have been a threat to their continued survival. In fact, the early cells probably survived only by staying five to ten meters beneath the surface of the water, where the force of the ultraviolet rays was diminished. The development of a protective layer of ozone meant that dry land was now safe to inhabit, and soon the shorelines were invaded by strange primitive life forms. An important effect of the oxygen-rich atmosphere, however, was that life could no longer arise in the way it once had. In other words, as life developed, it destroyed the very mechanism that had produced it.

THE IDEA IN SCIENCE

The foregoing description of the possible beginnings of life is, of course, pure conjecture. Almost every point can be argued, and some key points defy conceptualization, let alone proof. However, the scientific mind has never been marked by a reluctance to launch expansive theories from small scraps of information—and in a way that's good. It provides a rather loose intellectual milieu in which to rigorously test small facets of great theories. And it is only the accumulation of hard data that can ever bring an idea past the state of hypothesis, or even of theory.

THEORY

And by the way, what is a *theory*? There are some who discount the concept of evolution by saying, "It's only a theory." However, bear in mind that the status of theory in science is a relatively lofty one. To illustrate, suppose someone comes up with a good idea, one which explains a host of observed phenomena in nature. At first it is just that, an idea. After it has been carefully described and its premises precisely defined, it may then be a hypothesis which can be subjected to testing. If rigorous, carefully controlled testing substantiates the hypothesis, more and more confidence will be placed in it. If it continues to account for incoming data from the real world, it may finally gain the status of a theory. The theory itself, however, may remain unproven and unprovable.

The reception of a theory—or, for that matter, of a fact—may depend on the intellectual milieu into which it is introduced. The theory of evolution, for example, caused an initial shock in the Victorian world but was assimilated by the intellectual community

Figure 2.5 It is possible to learn about chimpanzees by watching them in their natural habitat. Other animals, such as whales, do not offer this opportunity for close observation. Jane van Lawick-Goodall has learned much through her observation of chimpanzees. (Courtesy: San Diego Zoo)

surprisingly rapidly. One possible reason for its widespread acceptance is that the deeply religious citizens of that era saw in evolution a preexisting *program* for change—a program authored by the Deity. Of course, the theory also had strong opposition at the time and is still met with strong resistance in some quarters today. It is interesting to note, however, that the arguments against evolution are not universally applied. For example, no one has ever seen a hydrogen atom, but detailed descriptions of its behavior may be found in any chemistry book. Moreover, we have been able to use the description of this theoretical atom to make water! In spite of the same kind of abundant evidence for evolutionary theory, people somehow have much stronger feelings about evolution than they do about chemistry.

In science any idea or hypothesis must (ideally) be testable to gain full acceptance. This does not mean that scientists are a breed of eminently rational people, unfettered by personal prejudice. They aren't, but fortunately such sterling character is not actually necessary. Important contributions have stemmed from hunches or from ideas that simply appealed to the prejudices of individual scientists. The fact is that the "scientific method" works in the face of all bias simply because almost any idea is sure to be attacked by someone and if it has weaknesses they are likely to be revealed.

THE USUAL PROCESS

When a scientist reports the results of an experiment, which is only one way of gaining information, several hurdles confront their acceptance. First, the general method comes into question. Other scientists working in that area of research have some idea whether the data could have been gotten in the way the experiment was conducted. Jane van Lawick-Goodall followed chimpanzees through African forests for several years and through observation and a few simple experiments brought back important new information regarding their behavior. Her findings were enthusiastically accepted. However, any reports based on extended periods of following the great blue whale through its daily activity might be greeted with some skepticism. Whale following is presently impossible.

Second, the scientific audience will question the experimental technique. Could the results have been produced only by the experimental conditions as described? Third, the readers will ask whether the conclusion is, in fact, supported by the data. Finally, the experiment will be repeated to see whether the results correspond to those originally reported. Not only do experiments face such rigorous tests, but even simple observations may be subjected to the same skeptical analysis.

Of course, this isn't the only way information gets into circulation. Someone who happened to witness the birth of a volcano might provide us with some observations which it would be difficult, or at least inadvisable, to duplicate. Nor can it be said that all good scientists formulate theories only after a painstaking analysis of voluminous and detailed evidence. Some of the best science is often dreamed up from the armchair, as it were. The actual testing may be done by more pedestrian scientists. In fact, it has been said that the inspired guess comes first, and that the data are only a dodge to give "proof to the disbelievers." The scientific world may be full of "disbelievers," it is true, but they, in their role of constant skeptic, have undoubtedly saved us from many "great ideas" which just couldn't be made to fit a real and very complex world.

WHAT KINDS OF ANSWERS CAN SCIENCE PROVIDE?

Because science is limited to testable evidence, it can provide answers only to certain kinds of questions. Unfortunately the confusion between science and religion has often blurred the issues, but science cannot, by definition, resolve questions either of faith or of "right" and "wrong." Thus science *cannot* tell us whether there is a God since any such Being, by definition, could choose not to be examined. Science also cannot tell us whether war is good or bad

Figure 2.6 The emblem of Science for the People, an organization of scientists devoted to focusing attention on the responsibility of science to humanity. (Courtesy: Science for the People)

or whether capital punishment or abortion is immoral, since such statements would have to be based on some system of values about what is good or bad. However, science *can* measure certain physiological responses and arbitrarily tell us at some point whether an organism is dead or alive. And science *can* measure the effects of certain conditions on living organisms and tell us the effects of birth and death rates on population size. And scientists *can* tell us what bombs do to children.

SCIENCE AND SOCIAL RESPONSIBILITY

For some time now there has been a movement to make science more responsible to the needs of man and society. Many have wondered, for example, whether a scientist has the moral right to develop a destructive device and at the same time disclaim responsibility for the use to which it is put. However, this question leads to a larger issue. Should scientists determine for themselves what areas of knowledge mankind should or should not have? Should they refuse to develop anything that could cause the wholesale destruction of people or the mindless poisoning of our environment? Or is it the responsibility of the scientist to seek knowledge in any form, so that an informed society is better able to make its own decisions?

One argument for greater personal involvement on the part of scientists is that they are in the best position to realize the full implications of their work. This argument is bolstered by the obvious lack of information among the general public. An all-too-common lament is, "But why haven't they said something? Why haven't they told us?" The truth of the matter is, prominent biologists have covered the country on speaking tours and have written bestselling books on the state of our environment. Television talk shows have been inundated with alarmed scientists from every discipline. Activist student groups and concerned-citizen groups have sprung up all over the country. Each day the news media bring us stories of environmental disasters. Yet somehow the general populace remains either uninformed or unconcerned.

FADS AND TRENDS

Part of the problem is that *ecology* has become a household word—and like any household word, it has become so familiar that people are sure they know exactly what it means. In fact, the general public has been bombarded to the saturation point with "eco-messages" from every quarter—often in contexts that had little to do with the environment. To make matters worse, the worst offenders among us took up the ecology cry, generally in hope of selling us something.

Figure 2.7 Bobcats killed in Texas on display. The reasons for hunting this endangered species are varied. Some ranchers cite bobcat attacks on small livestock, such as chickens. In such reasoning, the cat's important role in rodent control is often overlooked. (Courtesy: H. Charles Laun from National Audubon Society)

Paper companies touted their tree-planting programs while they stripped mountains bare. Television commercials extolling the virtues and concern of the oil companies flooded our homes while their products did the same to our air and waters.

During the late 1960s ecology was a national fad. Fads are part of the American culture, perhaps the result of a great national energy coupled with a short attention span. But like all fads, the crest of the ecology movement passed, leaving the public with a sense that the environmental crisis was an outmoded topic. This topic is likely, however, to be raised again and again, with increasing urgency, until we realize that the fashions of the well-fed may have rather little to do with the real world.

EXAGGERATION, COMPROMISE, AND UNDERSTANDING

To some extent the demise of this early effort to arouse public concern for our environment was a predictable response to some rather shocking statements by environmentalists—some valid, others less so. One result was that sober scientists in various fields began to offer rebuttals to the often shrill warnings of the "eco-freaks." They told us that the problem had been exaggerated, or that there actually was no real problem. We were also told that by the time any problem did appear, it could be resolved by improved technology.

Even among the environmentalists there were disputes over both the seriousness of the problem and the source of the danger. Some maintained that, with the possible exception of Southeast Asia, the environment was in no real danger of losing its potential to support increased populations. This was in disagreement with those who felt that the major threat to human existence was man's burgeoning numbers. Others said it was misdirected technology that had gotten us into trouble, and a redirection of priorities could get us out. With such arguments going on among those who were supposed to know—those who had seen the data—it was no wonder that the nonscientist was confused. The confusion, unfortunately, led to still further disenchantment in some quarters. Perhaps no one was right! There were even claims that the environmentalist movement was not only misdirected, but sinister. One piece of literature pointed out, among other things, that the first Earth Day took place on Lenin's birthday.

Probably there are few, without economic interests directly at stake, who would seek to undermine ecological efforts. After all, what advantage can there be in despoiling one's own habitat? The important questions today do not concern the benefits of a "good" environment. We all want that. Instead, we hear arguments over what comprises a good environment. Which is the more important component of man's habitat, a clean river or an air-conditioned house? Unfortunately, in many cases we can't have it both ways.

We also hear arguments over what methods to employ in regaining or maintaining a good environment. Should we force industries to produce clean effluent at the cost of higher prices and fewer goods for us all? Should we limit the size of families by law? Should we publicly finance those who are running for public office, so that our elected officials are not beholden to the large corporations that now finance their campaigns?

Again, these are questions that science cannot answer. However, the answers will be forthcoming one way or another, even if it is through silence. What science can do is provide us with the information that is available, so that we can make the best decisions possible in an increasingly complex and precarious world. You won't find all the answers in this, or any other, book. But perhaps you will begin to grasp some of the more critical problems and to understand the ramifications of your decisions. We can no longer afford to respond off the tops of our heads; we must all begin to *understand* our world, and ourselves, at the deepest level possible for us.

Figure 2.8 A free river (above left), the Snake River tumbles over the Upper Mesa Falls on Henry's Fork in the Targhee National Forest, Idaho. We must begin to carefully evaluate the importance of maintaining such areas. A river harnessed (below left). Such rivers provide a stable water supply and in some cases flood control to surrounding areas, but at a cost. Often the cost can be measured in ecological as well as aesthetic terms. (Courtesy: U.S. Department of Agriculture and U.S. Bureau of Reclamation)

3
The Chemistry of Life

Chemistry? Why chemistry?! Often beginning students of biology, particularly those who don't intend to make a profession of the sciences, are aghast at confronting chemistry in what they thought was going to be a biology course. Is chemistry really necessary? The answer is absolutely yes and no!

Biology can be "done" without chemistry. Darwin had only a rudimentary knowledge of the subject, and many modern naturalists do not use it in any way in their work. However, some knowledge of the subject is necessary in order for you to know why you breathe, urinate, and also, perhaps, why your mood changes before a thunderstorm. Some knowledge of chemistry as it pertains to biology may also help you to make critical personal and political decisions in the years ahead. For instance, the pollution-control devices now required on most automobiles trap hydrocarbon molecules, but allow nitrous oxides to pass into our air. Automobile manufacturers argue that if the hydrocarbons were also allowed to pass into the air, they would neutralize many of the nitrous oxide ions by bonding with them to produce relatively harmless compounds. But what are these compounds, and what are the effects of hydrocarbons? Even if you have never had any great yearning to be a scientist, some idea of the properties of these molecules will, if nothing else, make you a more perceptive consumer and a more discerning voter.

WHAT BIOLOGISTS DO FOR A LIVING

To the layman the word *biologist* often conjures up the image of a balding little man whose right shoe squeaks as he pads through aisles of dusty books all written in Latin. Others may visualize butterfly chasers leaping gleefully through the brush with their nets poised, or birdwatchers peering through field glasses in the cold, wet dawn in hope of catching one glimpse of the rare double-breasted seersucker. Perhaps such descriptions do fit some working biologists. But biology is much more than this. Many biologists search for the mysteries of life in clean, well-lit laboratories, amid tinkling glassware. About fifteen years ago it was said that a biologist is one who thinks that molecules are too small to matter, a physicist is one who thinks that molecules are too large to matter, and anyone who disagrees with both of them is a chemist. Perhaps it is true that the biologists whose scope is limited to molecules must periodically be convinced of the existence of the platypus, but no one else has been able to explain to the birdwatcher how a tiny hummingbird has the energy to cross the Gulf of Mexico. The dividing lines between these disciplines will undoubtedly become less distinct as time passes and more and more parts are fitted into the "great puzzle." So far, however, neither the field biologists nor the molecular biologists have been able to tell us, after all, what life is really about. On one hand, observations of the behavior of groundhogs do not provide us with a complete description of the mysterious processes we call life. On the other hand, we don't know nearly enough about where the molecular data fit into the big picture. As molecular biologists search for the integral units of life, they are often left sifting electrons through their fingers, "life" having escaped them somewhere along the way. Rest assured that we're not even going to approach such formidable questions as the "nature" of life. We will, however, consider a few of the basic concepts of molecular biology. As you continue to read, you may gain some appreciation for the problems confronting those dealing with larger questions.

Now, here's the plan. We will first learn what an atom is supposed to look like. Then we will explore the nature of a few specific molecules and compounds. In ensuing discussions we will consider the ways these particles work. You should be forewarned that in following chapters we will frequently refer to the basic principles mentioned here.

Before we go on to the specifics, one other point should be mentioned. We will often consider some highly detailed and incredibly complex features of many kinds of plants and animals in general terms. The reason is that in spite of their great morphological and ecological differences, all living things function biochemically in much the same way. In fact, this similarity is further evidence of the common origin of present life forms.

ATOMS AND ELEMENTS

All matter, whether it comprises an alga, a redwood tree or a distant star, is composed of *elements*. *Atoms* are the smallest indivisible particles of these elements. Atoms can be further divided into component parts, of course, but when this is done, special qualities of that element are lost. Each element, or type of atom, has its own special chemical behavior by which it is identified.

There are 92 elements which occur in nature. Other elements have been synthesized in the laboratory, so that the total number of known elements is now 103 (at least, was at the time of this writing). Each element is named and is designated by letter symbols. For example, hydrogen is designated by the letter H. Nothing is all that simple, however, so we find that sodium is designated by Na, from its Latin name *natrium*. The symbol standing alone usually designates one atom. For example, O refers to one atom of oxygen. The symbol 2O refers to two atoms of oxygen. But the symbol O_2 means that two oxygen atoms are joined together into one *molecule* of oxygen, the form in which oxygen usually exists in nature.

CHNOPS

CHNOPS is a "word" you might like to remember to identify the six elements which make up 99 percent of living matter. However, if you were to drop this word casually at your next gathering, it is not likely that everyone will know you are referring to carbon, hydrogen, nitrogen, oxygen, phosphorus, and sulfur. If they do, you are in the company of organic chemists and should leave immediately.

THE STRUCTURE OF AN ATOM

Now let's consider what an atom is supposed to look like. We have to say "supposed to" because atoms cannot be seen, so all the evidence is circumstantial. It is, however, impressive. People have been imprisoned on less compelling evidence. Atoms, according to theory, are composed of a dense center around which smaller particles spin in orbit. The center, called the *nucleus*, consists of *protons*, which are usually accompanied by *neutrons*. Each proton carries one positive charge, designated by a plus sign. Neutrons, as the name implies, carry no charge. Each electron carries one negative charge, designated by a minus sign. Electrons are much smaller than the protons (about 1/1835 the mass) and are in orbit around the nucleus. An atom of hydrogen, the simplest element, consists of one proton and one electron, with no neutrons.

Helium is the next simplest element and is composed of two protons, two neutrons and two electrons. The number of protons in the nucleus is its *atomic number*. Elements are arranged according to their atomic number. Elements differing by one proton may have vastly differing chemical properties. You may have noticed in the figures that the positively charged protons are balanced by the same number of negatively charged electrons so that the atom

Figure 3.1 Hydrogen atom (top) and helium atom (bottom).

had a neutral charge. Also, it should be mentioned that these figures are not drawn to scale. The reason is that the distance between the nucleus and its electrons is enormous. You can get some idea of the relative distances if you imagine that the period at the end of this sentence is a nucleus. Its nearest electron would probably be across the street somewhere.

HOW ATOMS MAY VARY

Isotopes

The atoms of a given element always have the same number of protons in their nuclei. If this were to change, the atom would be a different element. However, in certain elements the number of neutrons in atoms may vary. For example, most oxygen atoms have eight protons and eight neutrons, but there are also some oxygen atoms which have nine neutrons, and some which have ten. These variant forms of the same element are called *isotopes*. There are eight different neutron variants of oxygen, so oxygen has eight isotopes. Some elements have as many as twenty isotopes.

Generally all the isotopes of an element have about the same chemical properties, but in some cases the tiny difference can be important. Some isotopes of uranium, for example, are much less stable than others.

Ions

An atom may also vary in electric charge. In its ordinary state each atom has the same number of negatively charged electrons as it has positively charged protons; hence, it is electrically balanced. The number of protons, or positive charges, can't change without changing the element. However, an atom can gain or lose electrons, and when this happens, it is no longer electrically balanced. If the atom loses an electron, it is left with a net positive charge, since its protons now outnumber its electrons. If it should happen to gain an extra electron, it would have a net negative charge. Charged particles—atoms which have somehow acquired either a positive or a negative charge—are called *ions*. Ions, as we will see later, function importantly in living systems.

ELECTRON SHELLS

Electrons are in orbit around the atomic nucleus and are moving at almost unbelievable speed. They do not circle about the nucleus at random, but move in definite energy paths, with each path at a specific distance from the nucleus. The paths of electrons are usually depicted as concentric spheres around the nucleus, called *shells*. The electrons in the outer shells, it turns out, have to orbit

Figure 3.2 Electron energy level.

much faster—that is, at much higher energy levels—than those in the inner shells.

It is possible for an electron to move from one shell to another within the atom. However, in order to move outward to a higher-energy orbit the electron must be "excited" by some external energy source, such as heat, light, or electricity. When the electron falls back into a lower orbit, the energy which had provided its boost is released. Figure 3.2 gives you an idea of the energy relationships.

OXIDATION AND REDUCTION

If an electron is given enough of an energy boost it may escape from its atom altogether. The loss of electrons by this process is called *oxidation* because in some cases the lost electron is recaptured by an oxygen molecule. In fact, some substances cannot lose electrons unless oxygen is available to accept them. The capture of such electrons is called *reduction*. This simple process of electron transfer —the oxidation of one material and the reduction of another—has tremendous implications for life. In some cases, for example, an electron travels in company with a proton—that is, as a hydrogen atom. Thus oxidation and reduction involve the transfer of hydrogen from one element to another. When oxygen is reduced by two such hydrogen atoms, a water molecule is formed.

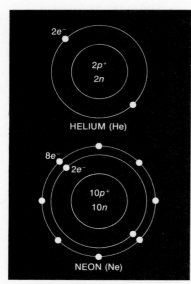

Figure 3.3 Inert atoms helium (top) and neon (bottom).

Figure 3.4 Argon (top)
Figure 3.5 Oxygen (bottom)

Table 3.1—Some Elements of Living Matter

Element	Symbol	Atomic number	Electrons in each shell 1	2	3	4	Percentage of earth's crust
Hydrogen	H	1	1	0	0	0	0.10
Carbon	C	6	2	4	0	0	0.03
Nitrogen	N	7	2	5	0	0	Trace
Oxygen	O	8	2	6	0	0	46.60
Sodium	Na	11	2	8	1	0	2.90
Magnesium	Mg	12	2	8	2	0	2.10
Phosphorus	P	15	2	8	5	0	0.10
Sulfur	S	16	2	8	6	0	0.05
Chlorine	Cl	17	2	8	7	0	0.05
Potassium	K	19	2	8	8	1	2.60
Calcium	Ca	20	2	8	8	2	3.60
Iron	Fe	26	2	8	8	8	5.00

ELECTRONS AND THE PROPERTIES OF THE ATOM

Another important property of atoms is the fact that only a certain number of electrons can occupy any one electron shell. The first (inner) shell will accommodate only two, the second only eight, the third only eighteen, and the fourth only thirty-two. However, no matter how many shells an atom has, the outer shell is the one that is critical. Many of the chemical properties of the atom are determined by this shell. The reason is that the atom will tend to fill this shell with electrons.

When the outer shell of an atom contains its full electron complement the atom is *inert*; it is almost impossible to make it react with any other element. Helium and neon, for example, are inert elements (Figure 3.3). Argon is also an inert element, despite the fact that its outer shell is not full. It works out that if the third or fourth shell of an atom has as many as eight electrons, the atom actually behaves as if this shell were full. These inert elements which are seeking no electrons are not particularly interesting in terms of life processes. With elements, as with people, the most fascinating are the dissatisfied ones.

Why is the number of electrons in the outer shell so important? Let's consider the case of oxygen. A look at Table 3.1 tells us that oxygen has an atomic number of 8; that is, it has eight protons and therefore needs eight electrons. However, since the inner shell will require two of these electrons, this leaves the outer shell with six—two short of its shell requirement. This is why oxygen so readily accepts electrons from other atoms.

The situation is different for sodium. Sodium has an atomic

SODIUM (Na)　CHLORINE (Cl)

SODIUM CHLORIDE (Na⁺Cl⁻)

SODIUM ION (Na⁺)

Figure 3.6 Sodium

Figure 3.7 Sodium plus chlorine becomes sodium chloride (table salt).

number of 11, so it has two electrons in its first shell, eight in the second shell, but only one in the third shell. Since it isn't feasible for a sodium atom to gain another seven electrons, it tends instead to lose its single outer electron. Of course, this ionizes the sodium, giving it a slight positive charge. Sodium ionized in this way is then written Na⁺. For the same reason magnesium, which has twelve electrons, often appears as Mg⁺⁺ (do you see why?). The fact that chlorine has an atomic number of 17 may give you a clue to the reason salt (NaCl) is formed so easily. Work it out.

CHEMICAL BONDING

When atoms join together to form molecules, they do so in specific ways. Moreover, since the atoms of each element have their own peculiar bonding properties, only certain kinds of molecules can form. The qualities of the constituent elements, however, may suggest very little about the nature of the molecule they form. For instance, sodium is a deadly poison, and so is chlorine, but together they form table salt.

IONIC BONDS

Let's take a closer look at how table salt is formed. We know that sodium has two electrons in its first shell, eight in its second, and only one in its third. Therefore sodium has a tendency to give up

Figure 3.8 Ionized sodium chloride. When common table salt, NaCl is placed into water, the constituent atoms separate so that the positively charged sodium and the negatively charged chlorine exist as ions, each acting as independent molecules.

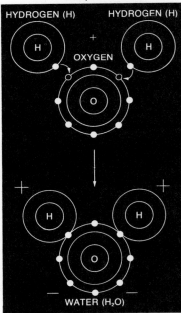

Figure 3.9 Covalent bonds (top) form as two hydrogen atoms move in to share their electron with an oxygen atom. The "unbalanced" structure (bottom) of a water molecule results in a stronger negative charge on one side of the molecule than the other.

its outer electron (at the expense of becoming ionized) in order to meet its outer-shell requirements. We also know that chlorine has two electrons in the first shell, eight in its second, and seven in its third. Hence it can fill its outer shell requirement rather easily by picking up a single electron. Now, when a strong electron donor such as sodium encounters a strong electron acceptor such as chlorine, there is a quick switch of an electron from one atom to the other. This satisfies the outer-shell requirements of both atoms, but it leaves the sodium, which has lost an electron, with a positive charge (Na^+). And of course the chlorine, with its additional electron, now has a negative charge (Cl^-).

It is probably no surprise to you to learn that opposites attract. The two ions, Na^+ and Cl^-, are held together by their opposite charges, and the result is sodium chloride, NaCl. *Ionic bonding,* then, involves the transfer of an electron from one atom to another and the joining together of the resulting oppositely charged ions as a molecule. In some situations ions with opposite charges can exist near each other without forming molecules. For example, when salt is poured into water the Na and Cl atoms dissociate and remain as separate ions in the water (Figure 3.8). This is because water molecules have unevenly distributed charges so they cluster around the oppositely charged ions and tend to segregate them. We'll learn more about this when we get to hydrogen bonding.

COVALENT BONDS

In another type of bonding atoms do not exchange electrons, but instead share them. Let's consider an example. Oxygen, as we have seen, needs two electrons to satisfy its outer shell. One way it can acquire them is by sharing electrons with two hydrogen atoms through *covalent bonding.* This bond, which forms a water molecule (H_2O), satisfies the shell requirements of both the oxygen atom and the two hydrogen atoms (Figure 3.9).

When a covalent bond is formed between two different kinds of atoms, the shared electron is usually attracted more strongly to one element than to the other. In the case of water, for instance, the shared electrons are pulled more strongly toward the oxygen nucleus than they are toward the two hydrogen nuclei. As you see in Figure 3.9, the result is an asymmetrical distribution of charge on the water molecule. In some kinds of covalently bonded molecules this tendency for electrons to be more attracted to one kind of atom is pronounced, and in others no asymmetric charge results at all. Thus there is actually no clear line of distinction between ionic and covalent bonding. To summarize, in ionic bonding an electron from one atom is completely captured by the other atom; in covalent bonding the electron may be almost, but not quite, captured, or it may be shared about equally between the two atoms.

A word should be said here about notation, how chemical structures are written. Bonding is often shown by a line from one chemical symbol to the other. For example, when two hydrogen atoms join together the resulting hydrogen molecule may be written as H—H. When an oxygen atom joins with two hydrogens the bonds may be shown as H—O—H. When two oxygen atoms join together, each of them is seeking two electrons, so their two shared electrons form a *double bond*, O=O. Look back at Table 3.1. Why can molecular nitrogen be written N≡N?

THE MAGIC OF CARBON

We have not discussed carbon in any of our examples of bonding. A glance at Table 3.1 shows why. Carbon has six protons and six electrons. We know that it has two electrons in its inner shell, so it must have only four in its outer shell. In order to satisfy its shell requirements, then, it must either give up these four electrons or gain four more. Either of these changes, however, would throw the charge too far out of balance. So the only way carbon can resolve this dilemma is by sharing its four outer electrons with other atoms through covalent bonding. For this reason, when carbon appears in a molecule, it is usually found sharing four electrons with other atoms.

Moreover, because carbon can meet its shell requirements in two ways—by either a gain or a loss of four electrons—it can combine with either electronegative or electropositive elements. It can share the electrons provided by oxygen, nitrogen, sulfur, or phosphorus atoms, or it can provide the electrons sought by hydrogen ions (which have lost their electrons).

Because of its peculiar bonding properties, carbon serves as the backbone for a wide variety of molecules. For example, if a carbon atom simply adds four hydrogens, the result is methane (Figure 3.10). Chemists would ordinarily represent this as CH_4, or

Figure 3.10 Carbon, because of its peculiar valence properties, may accept four hydrogen atoms to form methane, a gas.

$$H-\underset{\displaystyle \overset{|}{H}}{\overset{\displaystyle \overset{H}{|}}{C}}-H$$

or

If two or more carbon atoms attach to each other, we may get a variety of larger molecules:

ETHANE   PROPANE   BUTANE

The molecules produced by any combination of hydrogen and carbon atoms are called *hydrocarbons*. And as you might guess, it is possible to build some very long hydrocarbon chains. Note that in these kinds of molecules the electron-shell requirements of every atom are satisfied; each carbon atom has four other atoms with which to share electrons and each hydrogen atom is satisfied through a single covalent bond with a carbon atom.

Carbon chains may also branch out to form more complex arrangements:

And their ends may join so that they form rings:

Sometimes the notation for such structures is simplified to show just the carbon backbone.

There are several very important *functional groups* which can attach to hydrocarbons. These include the *hydroxyl groups*, —OH, which can form alcohols:

The amino groups, —NH$_2$, which can form amines:

The *carboxyl groups,* —COOH, which can form acids:

These functional groups, once they have formed, tend to behave as if they were a single atom. They move together as a group when a larger molecule of which they were a part becomes dissociated. When they exist as unattached charged particles they are sometimes referred to as *free radicals.*

Because of carbon's peculiar bonding abilities, it permits the formation of complex molecules, which then interact in still more complex ways to implement life's processes. Thus the chemistry of carbon is, in a sense, the chemistry of life.

HYDROGEN BONDS

There are a number of other, weaker types of bonds which may hold atoms together. The most important of these is the *hydrogen bond.* This bond forms only between a few small electronegative atoms such as oxygen and nitrogen. Strangely enough hydrogen bonds are important to life just because of their weakness. It takes very little energy to form them, and they are easily broken.

A hydrogen bond is formed when two electronegative atoms share a single hydrogen atom—that is, when the hydrogen atom forms a bridge between them. As an example, water molecules may be held together by hydrogen bonds. The bonding takes place because the oxygen atoms draw their electrons in close about them and, as a result, are somewhat electronegative. This means that the hydrogen atoms (the protons from which the electrons were drawn) are left with a slight positive charge. They are therefore attracted to the nearest available electronegative particle, which is likely to be the oxygen atom of the next water molecule. Thus water molecules are joined in a loose, constantly changing latticework.

Hydrogen bonds last for an incredibly brief time—about 10^{-11} seconds. But in their short existence they bestow upon water the unusual qualities of being very fluid and, at the same time, relatively stable. These qualities are critical to life in many ways. Have you ever noticed how long it takes for water to boil as the liquid seems to resist being changed to a gas? As the bonds quickly break and reform they impart a remarkable stability to the water. Hydrogen bonding also makes water resistant to change from its liquid form to its solid form, ice. This stability, or resistance to change, enables us to breathe in a wide range of temperatures by retarding the evaporation of moisture from our lungs, an important factor for terrestrial animals. The unequal charge distribution of water molecules, which permits hydrogen bonding in the first place, also gives water vital qualities as a solvent. The solvent properties of water tend to cause complex molecules to break down, allowing their components to interact in new ways. Thus all biochemical reactions in our bodies take place in water. Also, because of the fluid nature of water, caused by the continually breaking hydrogen molecules, it is able to pass through our tiniest capillaries and to seep deep into the earth's crust to reach the roots of the largest trees.

CHEMICAL REACTIONS

Chemicals, as we know, can interact—so let's take a brief look at how such reactions take place. Basically, chemical reactions involve simply the breaking of bonds and the formation of new bonds. Thus our task here will not be too difficult.

We have already mentioned chemical reactions such as the type represented by

$$A + B \rightarrow AB$$

An example of this sort of reaction is the formation of hydrochloric acid from hydrogen and chloride ions:

$$H^+ + Cl^- \rightarrow HCl$$

Here one ion each of hydrogen and chlorine form one molecule of hydrochloric acid. In many cases, however, we have to designate the number of constituents in the product. For example, the ionized hydrogen and chlorine would ordinarily be in molecular form themselves, as H_2 and Cl_2, so we would get two molecules of hydrochloric acid:

$$H_2 + Cl_2 \rightarrow 2HCl$$

In the formation of water we are also breaking the bonds of hydrogen molecules (H_2) and oxygen molecules (O_2) and forming new bonds in the water molecules. Thus

$$2H_2 + O_2 \rightarrow 2H_2O$$

That is, two molecules of molecular hydrogen and one molecule of molecular oxygen form two molecules of water. Notice that the same number of hydrogen and oxygen *atoms* appear on both sides of the arrow, so that the equation is balanced.

Many reactions of this kind are reversible. For example, we can also break water down into hydrogen and oxygen:

$$2H_2O \rightarrow 2H_2 + O_2$$

We sometimes write reversible reactions with a two-headed arrow:

$$2H_2O \leftrightarrow 2H_2 + O_2$$

In another type of chemical reaction the constituents may simply switch partners. Instead of two chemicals interacting to form a single substance, they may simply exchange components and come out as two completely different molecules. If we combine hydrochloric acid (HCl) with sodium hydroxide ($NaOH$), for instance, we get water (H_2O) and table salt ($NaCl$):

$$HCl + NaOH \rightarrow H_2O + NaCl$$

Figure 3.11 How an enzyme is believed to work.

Reactants

Enzyme

Reactants joined by their attachment to specific configurational sites along the enzyme

Product

Enzyme unchanged

The hydrogen atom from the HCl and the hydroxyl group (–OH) from the NaOH have joined as water and the remaining atoms have formed table salt.

ENERGY OF ACTIVATION

Some chemical reactions take place spontaneously; that is, the atoms themselves contain whatever energy is necessary to make them react when they come in contact with each other. Reactions of this kind have been known to cause explosions as amateur chemists randomly mix chemicals. Other reactions need some extra energy boost to set them off. For example, a rock at the top of a hill may need a shove before it begins to release its energy by tumbling down. And a match can burn, but it can exist quite nicely without ever bursting into flame. To initiate the oxidation process you have to add the heat produced by friction. Once the reaction has started, it will continue, at least long enough for one to light whatever one is smoking. The energy boost required to initiate a given chemical reaction is called its *energy of activation.*

In the case of the match, why should heat be able to provide the critical energy of activation? What does heat do? Heat causes an increase in the kinetic energy of the molecules; that is, it increases the speed of their random movements. The more active the molecules are, the greater the likelihood that they will bump into each other, and it is in such encounters that molecular bonds are altered.

Another way of increasing the likelihood of molecular contact (or chemical reaction) is to increase the concentrations of the two reactants. Or the mixture of reactants can be compressed to force the molecules closer together. Thus an increase in pressure has the same net effect as increasing the concentrations.

All these methods are fine for initiating interactions in laboratory chemicals, but how are chemical reactions activated in a living organism? The application of heat or pressure or drastic increases in the reactant concentrations in living things would be disruptive to say the least. As you may have guessed, there is another way chemical reactions may be initiated—one more compatible with the delicate mechanisms of life.

ENZYMES

Suppose you tried to impress your little brother by mixing hydrogen and oxygen in hope of forming water. He probably won't be very impressed because nothing would happen. However, if you added a tiny piece of platinum to the mixture, the hydrogen and oxygen would react violently enough to blow his hat off. And, interestingly enough, afterwards you could recover your platinum. The platinum served as a *catalyst,* and a characteristic of most catalysts is that they trigger or speed up reactions between other elements without actually entering into the reaction themselves. They do not change the type of reaction that can take place, or the direction of the reaction. They simply make it possible for the reaction to take place by lowering the energy of activation. It is through this process of *catalysis*—chemical reactions initiated by catalysts—that the "cold" chemistry of life can take place.

The special types of catalysts found within living things are called *enzymes* (identified by one memorable student as "Naval officers"). All enzymes are proteins. Proteins, as we shall see later, are large, highly complex molecules, and it is this complexity that provides for the specialized action of different enzymes. Figure 3.11 illustrates the process by which an enzyme is believed to work. Note that the reactants are not an exact fit with the enzyme molecule, so that for an instant certain bonds in the reactants are placed under a bit of stress. The resultant weakening of these bonds renders the reactants susceptible to attack by other agents, such as the functional groups of another molecule or free H^+ or OH^- ions. Once the new molecule has formed, it detaches and diffuses away from the enzyme molecule. This whole process takes place so quickly that a single enzyme may go through several million such cycles a minute in a dazzling display of efficiency. The enzyme in our illustration is shown joining two substrate molecules. This is what happens, for example, in the process of building proteins, but enzymes can also act to break down molecules into their component parts, as in the process of digestion.

All chemical reactions are accompanied by a change from one energy level to another. That is, the product of a chemical reaction will have either more or less energy than the starting material. If the product has more energy, the process is called *endergonic*; if it has less energy, the process is called *exergonic*.

But just as a boulder cannot roll up a hill without a push, neither can endergonic chemical reactions take place without an expenditure of energy. Thus in an endergonic reaction:

$$A + B \xrightarrow{\text{energy}} AB$$

However, once you have gotten your boulder up the hill, it has considerable potential energy, and if it should roll back down, that energy would be released. Similarly, when the product of a chemical reaction has less energy than the starting material, that extra energy must be released during the reaction:

$$AB \xrightarrow[\text{energy}]{} A + B$$

Of course, if the boulder happened to hit you on its way down, it might cause certain changes in your body. In this case the boulder's energy would be doing work. We'll see what kind of work can result from exergonic reactions when we discuss photosynthesis and respiration. Before we consider energy relationships in living things, however, let's take a look at a few key kinds of substances which enter into those reactions. Their names are bandied about rather freely, especially by people on diets, but have you wondered what a carbohydrate really is? Or a fat, or protein, or nucleic acid? Now, at last, you can know.

THE MOLECULES OF LIFE

CARBOHYDRATES

Carbohydrates are molecules which contain carbon, hydrogen, and oxygen in a proportion of one carbon to two hydrogens to one oxygen atom (CH_2O). They are important because they are the principle energy source for most living things and because they serve as the basic material from which many other kinds of molecules are built. Carbohydrates may be very simple, like glyceraldehyde, or they may be more complex, like glucose and fructose. These longer carbohydrates may exist as either chains or rings.

The simpler carbohydrate molecules are called *monosaccharides,* or simple sugars. Two simple sugars may combine to form a *disaccharide,* or many of them may join together to form various

Figure 3.12 Some important simple sugars. The five and six-carbon sugars may exist as either chains or rings. In solution, chains tend to form rings.

GLYCER-ALDEHYDE   GLUCOSE   FRUCTOSE

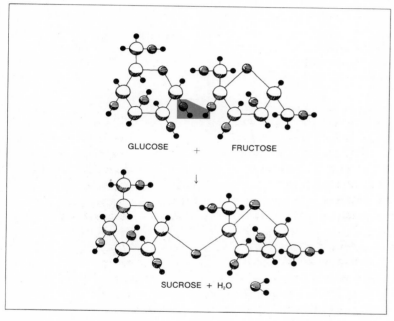

GLUCOSE   +   FRUCTOSE

↓

SUCROSE + H₂O

Figure 3.13 The formation of a double sugar, sucrose, from two simple sugars, glucose, and fructose. The double sugar is formed when the components of water are removed from the simple sugars. When the elements of water are chemically added to sucrose it breaks down again into glucose and fructose. The length of the bond links are not to scale.

polysaccharides. For example, when the elements of water are removed from glucose and fructose, these sugars may combine to form sucrose, or table sugar (Figure 3.13). The process of removing water is called *condensation* or *dehydration*. When water is chemically added to condensation products, a process called *hydrolysis*, the product is broken back down into its component substances. You can see why hydrolysis is an important process in digestion.

Glucose molecules may combine through condensation to form long branching chains of *glycogen* (Figure 3.14). Glycogen, which

Figure 3.14 Glucose units in glycogen, a polysaccharide. The glycogen molecule may contain several hundred glucose units in long branching chains. Glucose is stored in this form and then released through the addition of the components of water.

Section of glycogen chain

Glucose molecule

Figure 3.15 Segments of starch and cellulose molecules. Notice that the units of glucose are bonded differently in the two molecules.

STARCH

CELLULOSE

is a starch, is the principal form in which glucose is stored in the body, chiefly in the liver. When your body is in need of glucose, the glycogen is disrupted by hydrolysis and the component glucose molecules are released into your bloodstream.

Glucose units may also be joined by a somewhat different sort of linkage to form cellulose, a relatively insoluble polysaccharide which is common in plants. Our digestive systems can readily break

Figure 3.16 Molting of the dog-day cicada. As the insect grows and becomes too large for its exoskeleton, it molts, leaving the old skeleton behind. Later, it will redeposit a new skeleton around its vulnerable body. The exoskeleton of insects is composed largely of glucose units linked together in such a way as to render it indigestible by most organisms. (Courtesy: Lee Jenkins from National Audubon Society)

starches down into their component parts, but the particular bonding in cellulose makes it fairly resistant not only to hydrolysis, but also to most digestive enzymes. This is why people don't eat much wood and are rarely seen grazing. Cows and termites have some help: they serve as hosts to certain one-celled organisms which live in their digestive tracts and which can break down cellulose into sugars.

The hard covering, or exoskeleton, of insects is also composed of units of sugar bonded in a way that makes them indigestible. I was once following a young man down a street in Austin, Texas during one of the area's regular cricket invasions when he suddenly scooped a handful of the hapless insects from a wall and stuffed them into his mouth. I suppose I could have explained to him the nature of sugar bondings, but I walked into a trash can instead.

LIPIDS

Like carbohydrates, *lipids* are composed of carbon, hydrogen and oxygen. Lipids, however, have far less oxygen. Also, lipids are insoluble in water.

Fats are only one type of lipid, although they are perhaps the best known. There are a number of kinds of fats, and these are of interest for a variety of reasons. Most important, however, fats and their fatty acid components are the chief forms of energy storage in the body. They contain twice as much energy for their

GLYCEROL + FATTY ACIDS → FAT + WATER

weight as do carbohydrates or proteins. You would think from this that fat people ought to be bursting with strength and energy. You'll see as we go along why they aren't.

Each molecule of fat consists of two types of compounds, the alcohol *glycerol* and *fatty acids.* Glycerol is a chain of three carbon atoms, each with a hydroxyl (OH) group. In the formation of fats, the hydroxyl groups of the glycerol join with the hydrogens of the carboxyl (COOH) units of fatty acids through the process of condensation (the removal of water). This process is outlined for you in Figure 3.17.

It should be pointed out that the fatty acid chains are highly variable. They are usually sixteen to eighteen carbon units long, but they may be much shorter or longer. The three fatty acid components also don't have to be all the same length and fats may vary in the way the carbon atoms are bonded. For example, in some fats the carbon chains are fully saturated with hydrogen atoms. In other fats, however, certain carbon atoms may be joined by a double bond instead of filling their shell requirements by adding hydrogen atoms (see Figure 3.18). The former type of fat is called *saturated.* Guess what the latter type is called.

There are other important lipids besides fats. For example, there are the *waxes,* with their long saturated fatty acid chains. And then there are the *phospholipids.* In phospholipids the third hydroxyl group of glycerol is attached to a molecule that contains phosphorus. Phospholipids are important in the structure of living membranes, as we shall see.

Another important class of lipids is the *steroids.* The steroids are fatlike substances which have a structure quite unlike that of the other lipids we have considered. Whereas the other lipids are formed from the bonding of fatty acids and alcohols, steroids are formed from four interlocking carbon rings with a variety of side groups attached (Figure 3.19). Some people develop a deep and

SATURATED FAT

UNSATURATED FAT

Figure 3.18 Unsaturated and saturated fats. The fatty acids of an unsaturated fat contain less than their potential maximum number of hydrogen atoms.

Figure 3.19 Cholesterol, a steroid which is synthesized in the liver. All steroids are composed of four basic interlocking rings, but they differ in the side groups attached to those rings. Foods such as milk, butter and eggs are high in cholesterol. Too much of this steroid in the diet may result in its being deposited in the aorta (the artery leaving the heart) a possible cause of arteriosclerosis and heart attacks.

abiding affection for steroids when they learn that sex hormones are included in this group.

PROTEINS

The general structure of proteins is simple enough to describe, but this simplicity is deceptive. Proteins are actually immense molecules with incredibly complex structures. Let us not be deterred, however, as we plunge ahead into new realms of knowledge.

We will, of necessity, be dealing largely in generalities. Also, our generalities will be based on many assumptions, since we know the actual structures of relatively few proteins. Fortunately, however, proteins are composed of smaller building blocks which we do know something about.

Amino Acids

Proteins are made up of chains of smaller nitrogen-containing molecules called *amino acids*. As you may have guessed from the name, amino acids contain an amine group (NH_2) and an acid group (COOH). Both of these functional groups are attached to the same carbon. The carbon chain itself comes in a variety of forms, and for the sake of simplicity, we'll just call it *R*, for radical. For a molecule to qualify as an amino acid, then, it need only have the configuration

Remember that the COOH may be written

$$\underset{OH}{\overset{O}{\underset{\|}{C}}} \quad \text{or}$$

There are many known amino acids, but only twenty appear in proteins. Some of these are shown in Figure 3.21. As you can see, the carbon group *R* determines the name of the compound.

Peptides

Amino acids join together to form long *peptide chains* (polypeptides), and one or more of these then make up the protein. Polypeptide chains may consist of as many as 300 amino acids. The amino acids need not be of the same type, but they are always linked together in the same ways; the OH part of the carboxyl group joins with the H of an amine group to form water, as shown in Figure 3.20. The resulting linkage is called a *peptide bond*.

It is the sequence of amino acids in the chain that determines the biological character of the protein molecule. A change in the position of a single amino acid in a sequence of thousands can alter or destroy the activity of the protein. You can imagine the incredible number of proteins possible through the variable sequencing of amino acids. You can also imagine the high degree

Figure 3.20 The formation of a peptide bond.

CARBOXYL GROUP   AMINE GROUP

POLYPEPTIDE

Figure 3.21 A few of the amino acids found in proteins.

GLYCINE   SERINE   GLUTAMINE

ALANINE   THREONINE   PHENYLALANINE

VALINE   ASPARTIC ACID   CYSTEINE

of specificity that is possible in such a system. It turns out, in fact, that although all proteins utilize only twenty amino acids as building blocks, almost every species has proteins which are peculiar to it alone.

Protein Structure

A protein molecule may be described in terms of its amino acid sequence, which constitutes its *primary structure.* As protein chains form, however, they also take on a *secondary structure.* As a result of the hydrogen bonds that develop between every fourth amino acid, the protein molecule spirals into a helical shape, like a circular staircase. Its helical arrangement imparts its secondary structure.

Box 3.2 Unraveling the Structure of Insulin   In 1954, after ten years of painstaking research, Frederick Sanger and his colleagues at Cambridge University in England were able to tell us the full sequence of the amino acids in insulin. This work, for which Sanger was awarded a Nobel Prize, consisted in breaking the insulin molecule into smaller pieces of protein and then analyzing the pieces. First he subjected the entire molecule to strong acid to break it down to its component amino acids. He was then able to identify each of these amino acids by a process called chromatography. This involves filtering the product through a column of an adsorbent material for which different molecules have different affinities. Each amino acid thus adheres to the material at a different rate and so is stopped at a specific point as it moves down the tube. By comparing his results with the known adsorbing point of each acid, Sanger determined which amino acids were present in insulin. He was also able to learn something about their relative amounts.

To investigate the sequence of amino acids in the molecules Sanger used more specific hydrolyzing agents to break certain bonds in the long chain. Pepsin, for example, hydrolyzes bonds only between tyrosin or phenylalanine and other amino acids. To determine which end of the segment was free and which was attached, he added a chemical (dinitrofluorobenzene) which attaches to any terminal amino group and turns the compound a bright yellow.

After analyzing many such short segments, Sanger was eventually able to work out their structures, and where the terminal amino acids of these groups were identical, he could overlap them to reconstruct longer chains. For example, an overlap of the segments

$$A—C—X—N—O—F$$
$$N—O—F—G—O—I$$

yielded the longer segment

$$A—C—X—N—O—F—G—O—I$$

Later he was able to determine how the two component molecules of insulin were joined by bonds between sulfur atoms of cysteine molecules. Thus, after years of tracking down small bits of information, patiently, one step at a time, Sanger was eventually able to put all the pieces together and tell us for the first time the nature of a relatively simple protein.

Proteins which have structural functions in the body—those which protect the body or keep it from falling apart—usually form rope-like fibers or sheets. Hair, nails, cartilage, bones, and tendons all contain such fibrous proteins. These proteins usually have only primary and secondary arrangements.

But not all proteins are fibrous. Some fold back on themselves in a *tertiary structure* to form dense clumps, or globules (see Figure 3.22). Enzymes, for example, are globular proteins with complex tertiary structures. A protein maintains its tertiary structure by means of hydrogen bonds, hydrophobic (water-repelling) forces, and disulfide linkages (bonds between cysteine amino acids along the chain). Bonds of this kind are weak and can be broken easily by such things as heat or acid. Once the tertiary bonds are broken, the molecule can become fibrous (or "denatured"). This is what happens when you cook the protein-rich white of an egg. Such changes are irreversible, as you know if you have ever tried to unfry an egg.

In addition to their tertiary structures, two or more globular proteins may twine around each other to form a *quaternary structure*. Hemoglobin, the blood's oxygen-carrying molecule, has such a configuration.

Now that you have some idea of the basic chemistry of life, let's turn our attention to how such molecules function in living systems.

Figure 3.22 The tertiary folding of a hypothetical protein. R groups extend out in every direction to complete the spherical structure so that globular proteins are actually quite dense.

4
Energy

(Courtesy: Brown Brothers)

If we are ever to understand the nature of the "real world," we must learn more about the behavior of energy. Energy exists in a variety of forms, at different levels, and it ebbs and flows in countless directions throughout the universe. Moreover, energy can be trapped or stored, and then released to do work at some later time. Molecules, physical things, can be rearranged so that they hold more energy. Then, when they are converted back to their former structure, that energy is released. A person holding a jar of nitroglycerine would probably be obsessed by a single hope: that the molecules will not rearrange themselves to a lower energy level.

Potential energy is stored energy; it is doing no work. When potential energy is released to do work, it then becomes *kinetic energy*. The jar of nitroglycerine and the rock at the top of the hill have potential energy which can be released to do work.

Both potential and kinetic energy can take many different forms. A car battery has potential electrical energy. When that energy is released to turn over a starter, it becomes mechanical energy. As the parts of the starter move, friction causes some of the initial energy from the battery to be dissipated as heat energy. Thus we see that not only can energy exist in different forms, but it can also be converted from one form to another.

THE LAWS OF THERMODYNAMICS

There are a few simple laws which govern the behavior of all energy, regardless of its form. According to the *first law of thermodynamics*, energy can be neither created nor destroyed, but can only be changed from one form to another; hence the total amount of energy in a system remains the same, regardless of the physical or chemical changes it may undergo. This means that the energy which is released as our boulder topples down the hill must be directly proportional to the amount of energy we had to exert to get it up the hill. It also means that the energy which is "lost" as the heat of friction is not actually lost; it is merely unavailable

Figure 4.1 A rock which is about to fall has potential energy.

ONLY THE SUN GIVES

WITHOUT TAKING

for use in that system. However, in any energy transformation, according to the *second law of thermodynamics,* energy tends to dissipate. Thus after each energy transformation the available energy in a system will be slightly lower than before. As a result, according to the second law of thermodynamics, all processes act so as to "increase entropy." *Entropy* is a measure of disorder, so this means that a system becomes increasingly disorganized as its available energy decreases.

With a little imagination, you can see that these energy laws have some very interesting implications for living systems. Any organism must expend energy simply to remain organized. And since life can only continue in highly organized states and our environment is essentially disruptive, living systems must have both a source of energy and some efficient means of converting it from one form to another.

In the biological world the ultimate source of all energy is the sun. This energy is captured, changed, shifted about, and filtered through living systems, where it is used to reduce entropy. The sun's energy begins its useful progression in green plants, where it is stored in rather high-energy molecules. Each stage in life's energy cycles extracts its share of energy from those molecules, until finally all that is left are low-energy molecules of carbon dioxide and water. Let us now consider a few of the chemical reactions which result in your having the energy to do whatever it is you would rather be doing.

ATP: THE ENERGY CURRENCY

The standard energy currency of cells is a fairly simple molecule called *adenosine triphosphate* or ATP. ATP is referred to as "energy currency" because it must be "spent" for work to be done. It is spent by breaking one of its high-energy bonds, thus converting the molecule to adenosine diphosphate, or ADP, plus inorganic phosphate, or P_i. In other words,

$$ATP \xrightarrow{\text{energy}} ADP + P_i$$

Now, as you know, in order to have money to spend, work must be done (by someone). So it is with the chemistry of life. In order to produce ATP, energy must be expended somewhere along the line. That energy is used to put a molecule of inorganic phosphate back onto an ADP molecule, a place where it really doesn't "want" to go. It has to be forced back onto the ADP molecule by the expenditure of energy. Thus,

$$ADP + P_i \xrightarrow{\text{energy}} ATP$$

Figure 4.2 The hip saying of the 1960s is true in that the sun probably receives nothing from the earth. The sun, however, produces its radiation as hydrogen is converted to helium. But these elements had to come from somewhere. Such matters probably border more on philosophy than on biology. The point is, you should note the reciprocal and interdependent relationships of living things as our energy story unfolds. (Courtesy: Photophile)

Figure 4.3 ATP is actually adenylic acid (comprised of adenine, ribose and a phosphate) attached to two more phosphates. The last two phosphates are attached by high energy bonds. When these bonds are broken, large amounts of energy are released; energy which is available to do work.

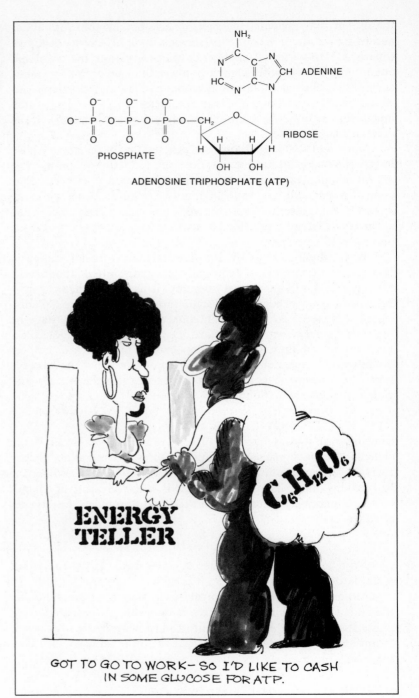

But if ATP provides the energy necessary for life's functions, and energy is necessary for the reconstruction of ATP, we are left with a problem. Where does the energy necessary to put the phosphate on the ADP come from? Unlikely as it seems, that energy comes from the sun. What we will see now is how green plants use sunlight to form ATP, which they then use to make glucose, and how animals, by eating plants, use the plants' stored glucose to form ATP of their own.

PHOTOSYNTHESIS

Photosynthesis is a remarkable process which takes place in green plants and some other organisms. In fact, green plants get their color from a pigment called *chlorophyll*, which drives the photosynthetic mechanism by capturing the energy in light. There are two distinct steps in the photosynthetic process, one which requires light (the light reaction) and one which doesn't (the dark reaction). The net result of this vital drama is that carbon dioxide and water are converted to glucose and oxygen:

$$6CO_2 + 12H_2O + \text{sunlight} \xrightarrow[\text{enzymes}]{\text{chlorophyll}} \underset{\text{glucose}}{C_6H_{12}O_6} + 6O_2 + 6H_2O$$

Many of the details of this process remain cloaked in mystery to this day, especially some of the dark reactions. However, more is known than you would probably care to have described to you, so we will consider photosynthesis only in rather general terms.

THE LIGHT REACTION

Photosynthesis begins when light strikes a chlorophyll molecule and initiates the light reaction. The chlorophyll is activated by the light energy, so that two of its electrons move to a higher energy state and are then captured by an electron acceptor. The chlorophyll replaces its lost electrons by literally tearing electrons from a water molecule, and this disruption causes the water molecule to fall apart, releasing the oxygen as a gas. This entire process takes place within the system designated as photochemical system II (Figure 4.4).

From the electron acceptor in photochemical system II, the electrons are passed through an electron-transport chain. The chain consists of a series of electron-acceptor molecules, each at a lower energy level than the last, so that the electrons progressively lose their energy as they are passed from one molecule to the next. Some of this energy is used to make ATP from ADP and P_i.

Finally the electrons are passed along to another pigment

Figure 4.4 Light energy captured by the chlorophyll-pigment system of photochemical system II boosts electrons uphill to an electron acceptor. The electrons are then passed along a chain of molecules to a lower energy level. As they are passed downhill some of their energy is released to form ATP. At the end of the chain they are captured by the chlorophyll pigment system I, where they are boosted by using the energy of the sunlight to a second electron acceptor. From here, they are passed to $NADP_{ox}$ to form $NADP_{red}$. The end products of the light reaction are ATP and $NADP_{red}$, products which are important in the dark reaction.

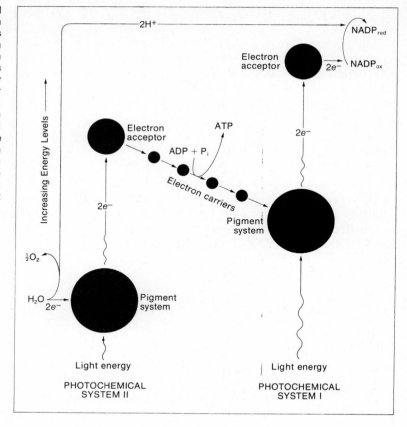

system at the bottom of photochemical system I (designated as I because it is believed to have evolved first). The fact that the first system occurs second is a bit confusing, but it's probably the least of your worries. Then light energy again boosts the electrons to another electron acceptor, this one at an even higher energy level than the last. From this acceptor the electrons are passed along to a nucleotide of adenine called $NADP_{ox}$ (since it is in oxidized form), thus reducing it to $NADP_{red}$. You will notice in Figure 4.4 that the electrons accepted by $NADP_{ox}$ are still at a rather high energy level.

The light reaction is sometimes called the Z reaction because someone thought its diagram formed a Z (actually, it forms an N unless you turn the book sideways).

THE DARK REACTION
The dark reaction is the process by which plants actually manufacture food from water and carbon dioxide. It is called the dark

Box 4.1 Chlorophyll Chlorophyll is a pigment molecule found in the chloroplasts of all green plants. The green color you see in such plants is actually caused by light rays from the center portion of the spectrum, which are reflected back to your eye by the molecular structure of the chlorophyll. The colors you do not see—the reds and purples at either end of the spectrum—are the ones that are absorbed. This scheme is not coincidental. The red and violet light rays are the ones which contain the highest concentration of light energy, energy which the chlorophyll uses in the vital process of converting light energy from the sun into usable energy for all living things.

Chlorophyll is usually a mixture of two different compounds known as chlorophyll A and chlorophyll B:

$$C_{55}H_{72}O_5N_4Mg \qquad \text{and} \qquad C_{55}H_{70}O_6N_4Mg$$
Chlorophyll A $\qquad\qquad\qquad$ Chlorophyll B

Most green plants have more chlorophyll A, although a few, as well as some algae, ordinarily utilize chlorophyll B or some related compound instead.

The chlorophyll molecules are located in the chloroplasts of green plants. Each cell may contain as many as eighty of these small bodies, which hold the chlorophyll molecules stacked in piles within their shelf-like grana. In this way the largest possible reactant surface is made available for the important chemical reactions in which these pigment molecules participate.

CHLOROPHYLL A $\qquad\qquad\qquad$ CHLOROPLAST

reaction because light is not necessary to drive this part of the mechanism. The energy in this case comes from the ATP and high-energy electrons in the $NADP_{red}$ which were produced in the light reaction. The major biochemical pathway in this reaction, also called the *Calvin cycle*, is traced in Figure 4.5.

The dark reaction consists of fourteen or fifteen complex steps which we need not go into here. But, basically what happens is that

Figure 4.5 Summary of the Calvin cycle. Ribulose diphosphate (RuDP), a five-carbon compound, is formed by the addition of phosphate from ATP molecules formed in the light reaction. The addition of carbon dioxide to the RuDP forms a highly unstable six-carbon sugar which quickly splits into two three-carbon molecules (PGA). The action of ATP and $NADP_{red}$ from the light reaction transforms PGA to phosphoglyceraldehyde (PGAL), the end product. Most of the PGAL is converted back to RuDP, but some is converted to glucose, starches, fats and proteins.

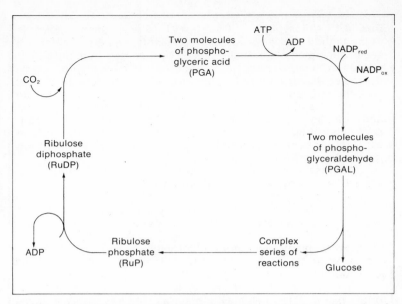

a carbon dioxide molecule is added to a five-carbon sugar molecule which is already present in the cell. This five-carbon sugar is *ribulose diphosphate*, or RuDP:

$$
\begin{array}{c}
CH_2\!-\!O\!-\!P \\
| \\
C\!=\!O \\
| \\
CHOH \\
| \\
CHOH \\
| \\
CH_2\!-\!O\!-\!P
\end{array}
$$

The additional carbon results in a very unstable six-carbon sugar, which quickly splits into two three-carbon sugars. Then there is a complex series of events in which the bonds between carbon atoms are broken and reformed, eventually producing two molecules of phosphoglyceraldehyde, or PGAL. The PGAL is the end product of photosynthesis. PGAL may then enter into a number of metabolic pathways. If the plant is in need of food, it can be used as is. The plant may also convert it into food stores in the form of glucose or starch. Or it may convert the PGAL to proteins or fats. Most of the PGAL, however, is changed back to RuDP, by the expenditure of energy from ATP, so that the cycle begins again.

Following is a discussion of how the food made by plants can be utilized to provide energy for other living things.

CELLULAR RESPIRATION

The glucose manufactured in photosynthesis is now ready to be "cashed in" for ATP molecules to drive the reactions of living cells. The plant may use its own glucose, of course, but that glucose is also available to animals which eat the plants, and then to other animals which eat those animals. Plant glucose is one of the things that the grazing steer is after—as is the man who eats the steak. The process of trading glucose for ATP is referred to as *cellular respiration*, since it takes place within the cell. Cellular respiration takes place in three stages: *glycolysis*, the *Krebs cycle*, and the *electron-transport chain*.

Briefly, what happens during respiration is that the glucose molecule is split, the hydrogen atoms (both protons and electrons) are transferred from the carbon atoms to oxygen atoms, and the energy released as a result of this transfer is used to convert ADP to ATP. Let's look at these steps in more detail.

GLYCOLYSIS

The process of glycolysis takes place in the cell fluid, or *cytoplasm*. At this stage the six-carbon glucose molecule is split up and changed into two molecules of a three-carbon substance called *pyruvic acid*.

In the formation of pyruvic acid the glucose is oxidized by the transfer of hydrogen to two nucleotides of adenine called NAD. The NAD_{ox} molecules thus become NAD_{red}. Also, during this process two ATPs are produced, as shown below. If no oxygen is present, these two ATPs are all the energy an organism can get from a glucose molecule. Not much. However, yeasts and other *anaerobic* organisms, those which do not utilize oxygen in their respiration, derive all their ATPs from glycolysis.

Two molecules of
pyruvic acid

Pyruvic acid is not the end of the line in glycolysis. In yeasts, the respiratory products are carbon dioxide and ethanol, or spirit

Figure 4.6 Pyruvic acid is not the end of the line in glycolysis. In yeasts, glucose is converted to ethanol (the alcohol of spirits). Yeast cells bloom on the skin of the grapes. When the juice of these grapes is left in airtight containers, the glucose of the fruit is changed to ethanol. Notice that CO_2 is released in the formation of ethanol so you should somehow allow for the escape of gas to prevent your "experiment" from blowing up.

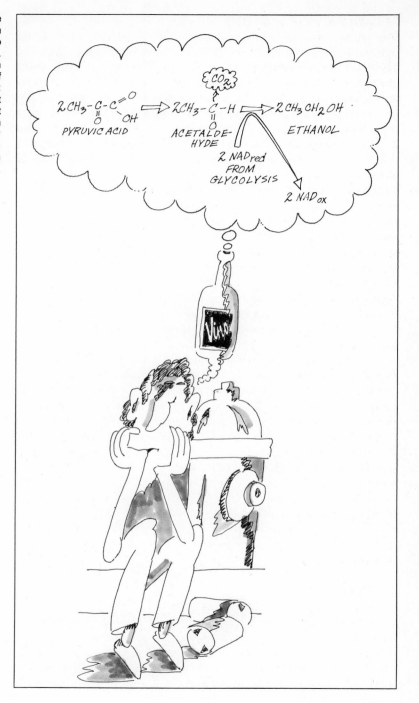

alcohol. The NAD_{red} gives up its two added hydrogens to form the alcohol and is then free to return as NAD_{ox} to pick up two more hydrogen atoms from a new glucose molecule. We have known about the practical aspects of this process for some time, since early man, confronted with all the great puzzles of the universe, busied himself at once with learning how to make alcoholic beverages.

In animals the product of oxygenless respiration is not alcohol, but lactic acid. It is not surprising that anaerobic animals, such as internal parasites, are fairly sluggish. They just don't have the ATPs to be frisky. When *aerobic* animals—those which require oxygen for respiration—are unable to get enough oxygen during prolonged exercise, some of their energy may be derived from glycolysis (since aerobic and anaerobic processes can occur simultaneously in these animals). In this case lactic acid accumulates in the muscles, and they weaken with fatigue. If enough lactic acid accumulates, the muscles will fail to contract, or contraction will be sufficiently impaired to affect coordination. This is why a runner who is not in condition is usually gasping for air at the end of a race and may stagger across the finish line. A conditioned runner is likely to have developed his cardiovascular system along with his muscles, so that it will provide him with sufficient oxygen to prevent the buildup of lactic acid.

In mammals lactic acid is removed from the muscles by the blood system and carried to the liver, where it is reconstituted to glucose and stored as glycogen. This transformation back to glucose is uphill, of course, and thus requires an expenditure of energy in the liver. However, this is no great hardship, since the energy level of lactic acid is still fairly high. In other words, the boost to glucose is a short one.

In aerobic animals, the pyruvic acid formed in glycolysis has a continuing and critical role. It goes through a series of reactions in which almost all the energy it holds is extracted from it. That energy goes to make new ATP's, many more than can be formed in glycolysis. Let's take a closer look as the energy-loaded electrons of pyruvic acid are passed along until they end up as mere components of humble water.

THE KREBS CYCLE

We know that in the absence of oxygen a glucose molecule provides an aerobic organism with only two molecules of ATP to carry out its functions. But if oxygen is present, it's a whole new ball game. The pyruvic acid will go through a series of reactions which generates thirty-six more ATPs. In aerobic respiration the two pyruvic acid molecules formed from the glucose are shunted into the *Krebs*

Figure 4.7 The person who builds up his cardiovascular system through exercise such as jogging is better able to bring oxygen to his muscles as his circulatory system develops along with his muscles.

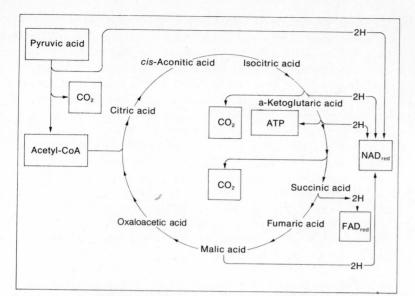

Figure 4.8 The Krebs cycle. To trace the pathway of activity, begin with pyruvic acid in the upper left of the diagram. Notice that oxaloacetic acid actually cycles, continually joining with acetyl-CoA. Further, the CO_2 and H leaving the cycle are actually the remains of acetyl-CoA.

cycle, named for its British discoverer Sir Hans Krebs. Eventually the high-energy pyruvic acid is broken down completely to the simple end products carbon dioxide and water, having released most of its energy along the way. Both the Krebs cycle and the next stage of respiration, the electron-transport chain, take place in tiny structures within the cell called *mitochondria.* (You'll have a chance to look at these in the next chapter.)

If oxygen is present the three-carbon pyruvic acid molecules produced by glycolysis oxidize—with NAD molecules again accepting two hydrogens—and lose one carbon in the form of carbon dioxide. The remaining two-carbon fragment, acetic acid, then unites with a protein substance called *coenzyme A*, or CoA, to form *acetyl-CoA*:

$$
\begin{array}{c}
CH_3 \\
| \\
C \\
O \diagdown \; CoA
\end{array}
$$

It is actually the acetyl-CoA which enters the Krebs cycle. It should be stressed that the body doesn't get all its energy from glucose. Fats and some amino acids also convert to acetyl-CoA, although they do so by a less direct route, and they too enter the Krebs cycle.

In the Krebs cycle the acetyl group separates from its CoA enzyme and combines with a four-carbon molecule called *oxalo-*

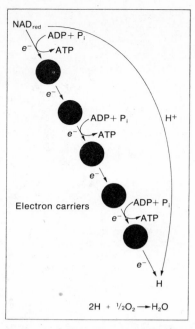

$2H + \frac{1}{2}O_2 \rightarrow H_2O$

Figure 4.9 Diagram of an electron-transport chain. As the electrons flow downhill from one carrier to the next, (which is at a lower energy level) the energy which the electrons possessed when they were held by the NAD is released in a stepwise manner. That energy goes to form ATP from ADP. Finally the spent electrons combine with protons and are accepted by an oxygen molecule (now you can see why you breathe) and water is formed. Some organisms, such as certain insects and desert rodents, can derive all the water they need from this "metabolic water."

acetic acid. The result is the six-carbon compound citric acid, the common acid of lemons and limes. The citric acid is then oxidized which results in the splitting off of two carbons in the form of carbon dioxide, and the remainder is again oxaloacetic acid, which is recycled to pick up another acetyl group from the next acetyl-CoA molecule. The cycle continues as long as acetyl-CoA is available. This process is illustrated in Figure 4.8. As you can see, the stored energy in the acetyl group has been used to convert three molecules of NAD_{ox} to NAD_{red}, one molecule of FAD_{ox} (another electron acceptor) to FAD_{red}, and one molecule of ADP to ATP. The glucose molecule has by this time been completely oxidized. Its products, however, will go on to provide the cell with more energy as they are passed along to the electron-transport chain.

Note that the electron acceptors NAD and FAD are all nucleotides, and each of these substances readily accepts and releases two electrons at once. Do you see a certain parsimonious efficiency in a system that makes multiple use of nucleotides in a continuous oxidation and reduction process? What do you suppose is the advantage of a cycling system in living things?

THE ELECTRON-TRANSPORT CHAIN

The last stage of aerobic respiration is the *electron-transport chain.* In this process the electrons which had been accepted by the NAD and FAD molecules are now passed along to a series of electron carriers similar to those referred to in photosynthesis. Here again each acceptor is at a lower energy level than the preceding one, a system which permits the energy of the electrons to be freed in a stepwise process, The energy which is released is used to form more ATP (Figure 4.9).

At the end of the chain the electrons join with protons, and the resulting hydrogen atoms combine with oxygen to produce water. In the series shown in Figure 4.9 the electrons are passed along in pairs from NAD, and three molecules of ATP are formed. However, if the electrons are instead passed along from FAD, which has a slightly lower energy level, only two ATPs are formed.

Let's stop at this point to retrace the respiratory reactions. During glycolysis the six-carbon sugar glucose is split into 2 three-carbon molecules. These molecules are then rearranged as pyruvic acid, and in the process a small quantity of ATP is made available for energy. In the absence of oxygen the pyruvic acid is converted in yeasts to carbon dioxide and alcohol and in aerobic animals to lactic acid. In the presence of oxygen in animals the pyruvic acid is instead converted to acetic acid, which combines with coenzyme A.

Figure 4.10. The basic cycle of life on earth. Chloroplasts capture the sun's energy and convert it to glucose. The mitochondria utilize the products of glucose to form ATP. The CO_2 and water formed by the mitochondria are utilized by the chloroplast.

Figure 4.11 The kangaroo rat is a tiny desert-dwelling rodent which is able to derive its moisture from the food it eats and from metabolic water. Can you think of two reasons its eyes are so large? (One has to do with saving water, the other with saving its hide.) (Courtesy: Bureau of Sports Fisheries and Wildlife, Photo by E. R. Kalmbach.)

The resulting acetyl-CoA then enters the Krebs cycle, where carbon dioxide is produced, along with some ATP, and electrons are transferred to FAD and NAD molecules. The FAD_{red} and NAD_{red} then release these electrons to the electron-transport chain, and as the electrons travel down the chain to lower energy levels, ATP is generated at each step. The remaining electrons are finally deposited in the "electron dump," oxygen, to form water. The energy left over after these reactions appears as metabolic heat. That's why the person next to you is so warm.

When we trace the energy of the sun through living systems, we can envision a cycle like that shown in Figure 4.10. Energy from the sun is utilized by the chloroplasts in plant cells to make oxygen and glucose, and some of the sun's energy is stored in the glucose. A number of things can happen to this glucose, but eventually some of it reaches the cells of aerobic animals. The mitochondria in these cells transform certain constituents of the glucose to carbon dioxide and water, and in so doing they release the stored energy of the glucose to make ATP, a chemical which provides usable energy to the body.

All this takes place in cells. So let's have a closer look at these incredible structures.

5
Cells

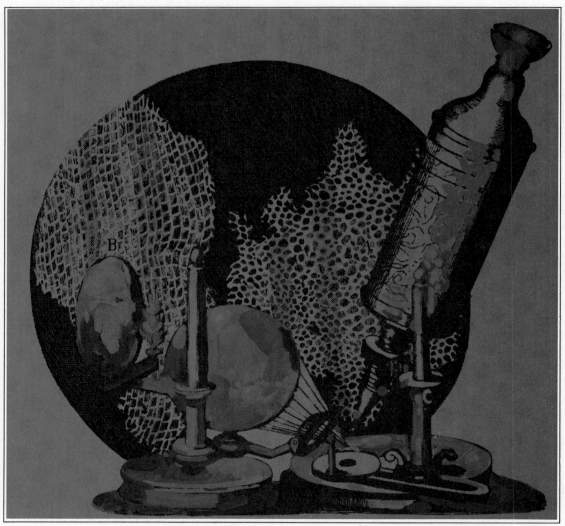

(Courtesy: The Granger Collection, New York)

In the mid-1600s Robert Hooke was appointed Curator of Experiments for the prestigious Royal Society of London. At the group's weekly meetings Hooke was to provide some sort of demonstration for them, something of scientific interest for their entertainment and information. This was a difficult task, since the society comprised the elite of the scientific community—a rather jaded lot, to be sure. As it happened, however, Hooke was interested in the recently developed microscope and managed to employ this fascinating new device most effectively in several of his demonstrations. One night, in an attempt to find something new to examine under his microscope, he cut a very thin slice of cork with his own sharpened penknife. When he peered through the microscope, he was somewhat surprised at what he saw. Hooke wrote at the time that the cork seemed to be composed of tiny "boxes" or "cells"—and a new field of science was born.

Many years were to pass before the scientific world would grasp the extreme importance of Hooke's "little boxes." In fact it was not until the nineteenth century that a comprehensive cell theory was advanced. The idea that all living things—from oak trees to squids, from tigers to men—are made up of cells is owing primarily to the work of two Germans, the botanist Matthias Jakob Schleiden and the zoologist Theodor Schwann.

Today *cytology*, the study of the cell, has been enhanced by an array of sophisticated techniques and by the invention of the electron microscope. A cell, we now know, is the smallest unit which can exist as an independent organism. It has, at some point in its development, everything necessary for life. A cell which is part of an organism cannot ordinarily survive independently, but many kinds of isolated cells have lived and reproduced under carefully regulated laboratory conditions. Also, some organisms, such as the protozoa, consist of only a single cell. Here we are concerned primarily with the cells of multicellular animals, but many of the principles we will discuss apply to protozoa such as the amoeba in Figure 5.1.

Figure 5.1 A protozoan or one-celled animal (in this case an amoeba) has just ingested another protozoan, a paramecium (the elongated body). The circular structure is the nucleus of the amoeba. The amoeba moves by flowing into any of its extended "arms." (Courtesy: Carolina Biological Supply House)

ONE 3-INCH "CELL"
Surface area of membrane
54 square inches

TWENTY-SEVEN 1-INCH "CELLS"
Surface area of membrane
162 square inches

Figure 5.2 Both "cells" have the same volume but the bottom one has a far greater surface.

CELL SPECIALIZATION AND CONTROL

Cells come in a variety of sizes and shapes, each with its own particular limits and abilities. Not only do the cells of various organisms differ, but the cells of a single organism may be quite different from each other. The tiny boxlike cells beneath the skin of your hand are not much like the nerve cells which extend all the way from the base of your spine to your foot.

Cells which differ in appearance may also behave quite differently, as would be expected. Some are highly irritable—that is, they respond quickly to environmental changes. Some are contractile, others secrete fluids, and still others have long tails and can swim. Such variety indicates one great advantage of cellularity which might have made the condition worth developing through evolution. That advantage is *specialization*. An organism living in a complex environment would be better equipped to meet its many varied demands if it consisted of a number of kinds of cells, each with its own special abilities. Think of the advantage of having some cells which can react to light, others which can distinguish particular types of pain, and then others which can contract and move you toward the light and away from the pain. As different kinds of cells developed and then became increasingly coordinated with each other, it is easy to see how the organism they comprised could come to fit more precisely into its part of a competitive world.

Another important advantage to an organism in having many cells, instead of simply being one large cell, is greater *control* over interactions between the organism and the environment. The environment, as you know, is likely to be disruptive; the successful organism is one which is best able to minimize its entropic effects. This greater control is made possible by the increased membrane area provided by many cells. The membranes which bound all cells are more than sacks which hold the fluids in. They are living, responding structures which regulate what moves into the cell from the environment and also what moves out from it. Hence the more membrane area an organism possesses, the greater its control over its internal environment. As you can see from Figure 5.2, a 200-pound cell would have less regulatory membrane surface than a 200-pound organism composed of many cells. This is one reason why there is no such thing as a 200-pound cell.

PROCARYOTIC AND EUCARYOTIC CELLS

Before we become involved with the description and behavior of cells, you should first understand what kinds of cells we will be talking about. We will not be discussing the bacteria and blue-green algae. These are one-celled organisms which differ in several basic

Box 5.1   The Electron Microscope   The limits of the light microscope, the one with which you are probably the most familiar, lie in its power of resolution. *Resolution* refers to the smallest distance by which two points can be separated and still be distinguished as separate points. If they are separated by a distance smaller than the *resolving power* of the microscope, they will be seen as a single point. The resolving power of the electron microscope is far greater than that of the light microscope. As a result, objects magnified to the same degree can be seen more clearly, and with the greater magnification afforded by the electron microscope, objects can now be seen that could not be seen at all with the light microscope.

With the light microscope the image is seen as contrasting light and dark areas as light passes through the specimen from below. The problem of resolution is therefore compounded by the inherent limitation of light diffraction, the bending of light rays as they pass through an object. The length of the light wave itself (that is, the color of the light) is a critical factor here. As the wavelength increases, so does the angle of diffraction, and its resolution is thus reduced. Under ideal conditions the best light microscopes have a resolution of about 0.25 microns with ordinary white light.

The electron microscope, which became available in the 1930s, works on an entirely different principle. Excited electrons are drawn from a heated filament and are then directed, or focused, by magnetic fields. As the electrons pass through an object, they are deflected or absorbed to various degrees by differences in the density of the matter. They then produce an image on a coated screen or special film below the object. Electrons are easily deflected, so the specimen must be cut exceedingly thin and the operation must be carried out in a vacuum.

The wavelength of an electron beam is only about 0.05 Ångstroms (an Ångstrom is a hundred-millionth of a centimeter) so, theoretically, resolutions of this power are possible. A major disadvantage is that heavy-metal stains must be employed, and these are usually highly poisonous to living things. Because of this and the necessity for extremely thin sections, there is no way living material can be examined. Thus there is no way to know how much of what we see only with an electron microscope has resulted from the method of preparation.

(Courtesy: Coates and Welter)

Figure 5.3 Molecules escaping from a perfume bottle will soon diffuse throughout the entire room.

ways from all other cells. They are called *procaryotic cells* to differentiate them from the *eucaryotic cells*, which are our primary concern. The differences between the two types of cells are summarized in Table 5.1. The terms are defined later in the chapter.

Table 5.1—Differences Between Procaryotic and Eucaryotic Cells

Structure	Procaryotic cells	Eucaryotic cells
Nuclear membrane	Absent	Present
Inclusion bodies (mitochondria, Golgi body, etc.)	Absent	Present
Flagellae	Lack 9-2 structure	Have 9-2 structure
Chlorophyll	Not in chloroplasts	In chloroplasts
Cell wall	Contains amino acids and muramic sugars	Does not contain amino acids and muramic sugars (when wall is present)
Chromosomes	Composed entirely of nucleic acids	Composed of nucleic acids and proteins

HOW MOLECULES MOVE

In the regulation of complex systems of cells, one problem is moving things around in those systems. So now let's see how molecules can move into, out of, and through cells.

DIFFUSION

The molecules of any substance—gas, liquid, or solid—are constantly in motion and continually bumping into each other and rebounding, thus changing their direction of movement. The principle becomes apparent if you open a vial of perfume (or if you're the baser sort, try hydrogen sulfide) in a corner of a room. Soon the smell will be detectable in all parts of the room. If the solution is allowed to evaporate completely, all parts of the room will eventually smell the same. Until that happens, however, there will be a *concentration gradient*, with a greater concentration of perfume molecules nearer the vial.

The reason for the shift of molecules away from their center of concentration is that their movement is random and they bump into each other with greater frequency where they are more densely packed. When they collide, of course, they rebound and change direction. Their chances of collision are lessened when they are moving outward toward areas of lesser concentration. As a result, this is what they tend to do. The *net* movement of these molecules is outward, away from the area of concentration. Their individual movements will be random, and by chance collisions with other molecules they will diffuse randomly from areas of higher concentration to areas of lower concentration.

Molecules may also be transported by their own random motions through the fluid environments in which living cells exist. While certain kinds of molecules may encounter barriers in the form of membranes or other structures, other types of molecules can pass with relative impunity through such obstacles. In some cases the difference depends simply on the size of the molecules in relation to the size of the pores in the membrane; smaller molecules pass through more easily. A few substances, such as carbon dioxide, oxygen and water, are able to diffuse rather freely through almost any cell membrane, while proteins usually cannot.

OSMOSIS

Battlefield nurses in World War II were taught to bathe exposed viscera wounds only with a salt solution, not with plain water. If you can muster the courage to place a drop of your own blood into water and watch the results under a microscope, you will see why. The blood cells will quickly swell and rupture, releasing their contents and leaving only colorless, limp "ghosts" of themselves. Plain water can have the same effect on other living cells.

Osmosis is the net movement of water through a membrane separating two solutions—a membrane which is more permeable to water than to whatever is in solution. Remember that the membrane separates both water molecules and molecules of the solute. Suppose we have two sugar solutions separated by a membrane, and that one solution has a higher *concentration* of sugar than the other (see Figure 5.4). Now we can discuss the movement of the water molecules and the sugar molecules separately. If the membrane is more permeable to water than it is to sugar, then the sugar molecules can go through the membrane only with difficulty, or not at all. Water molecules, however, pass through easily and will follow the law of diffusion: they will move from the side of their higher concentration to the side of their lower concentration. The lower concentration of water, of course, is on the side with the higher sugar concentration. Therefore water from one side will pass through the membrane to the other side where its concentration is lower.

The result is that water will *rise* on the side which started with the highest sugar concentration until the concentrations are equalized, or until the weight of the greater volume forces it back through the membrane at the same rate it enters. (Some recent data indicate that the movement of water through the membrane may be due to other factors besides simple diffusion, but you will be pleased to learn that these concepts are beyond the scope of this book.)

Figure 5.4 In osmosis, the water will move through the membrane covering the mouth of the funnel. It will raise the water level within the funnel if the concentration of the solute is higher within the funnel than outside it.

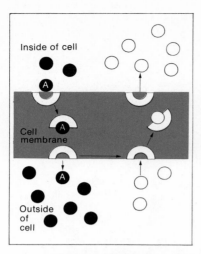

Inside of cell

Cell
membrane

Outside
of
cell

Figure 5.5 Cell membranes may move particles in and out of cells against osmotic gradients. Energy is expended but the exact mechanism is unknown.

Cytoplasm, or cell fluid, is an aqueous (water-based) solution of a large number of types of molecules. When a cell is immersed in pure water or in a solution with fewer solute molecules than the cytoplasm, then there is a tendency for water to move into the cell, causing it to swell and possibly rupture. This is what happened to the blood cells we mentioned above. When the fluid outside the cell has a higher concentration of solute molecules (and therefore a lower concentration of water), the effect is the opposite. If a cell is placed into a more concentrated solution, the water from the cell moves outward, causing the cell to dehydrate. However, a salt concentration of about 0.85 percent would have no effect on the volume of a human cell. In this case the number of particles in solution would be about the same both inside and outside the cell. Note that the important factor here is the concentration of the solution, not the type. We have discussed the concentrations of both solutes and water. In order to understand how osmosis affects cells you must clearly differentiate these two concentrations in your mind. You may, or may not, like to know that different types of cells vary widely in their permeability to water. A human blood cell is highly permeable to water. In contrast, a one-celled amoeba is able to live at the bottom of fresh-water ponds partly because it forcibly expels excess water which enters it, but also partly because its membrane is rather resistant to water.

ACTIVE TRANSPORT

In some cases a living cell can move molecules across its membrane from a place of lower concentration to a place of higher concentration. Such a chore is obviously "uphill," in that it goes against the normal tendency of the molecules to move in the opposite direction. As an example of such movement, consider cells which contain far less sodium than does their surrounding medium. Membranes are permeable to sodium to some degree, and so some sodium manages to diffuse in. As fast as it moves in, however, the cells pump it back out. The mechanism of the "pump" is not completely understood, but it is generally assumed that some sort of carrier molecule is involved—a membrane molecule which picks up sodium from inside the cell, diffuses with it across the membrane, and releases it on the other side (Figure 5.5). A variety of molecules are subject to active transport, and some, such as potassium, may be carried the other way—into the cell.

CELL COMPONENTS

We will now consider some of the components of cells. The treatment is of necessity a bit artificial because of the complexities of cell composition. Just keep in mind that neither the descriptions nor illustrations are likely to depict the situation for all cells, nor for any cell all the time.

Cells are living things, so they change. In fact, the inside of a living cell is a seething, roiling mass of viscous, grainy fluid with tiny bodies of every description—bodies which are constantly growing, extending, moving, multiplying, appearing and disappearing. We know that cells must maintain and repair themselves, and that they must divide and grow. We also know that the cell's sensitive cytoplasm responds to specific environmental stimuli, and as those stimuli change, so does the behavior of the cell. Thus the vigorous panorama we see inside living cells should not be entirely unexpected. It should serve to remind us, however, that the goings-on within cells are incredibly complex, and that we cannot really hope to understand everything that is happening there. At least not yet.

Figure 5.6 A micrograph (top) and illustration of micrograph (bottom) of the plant cell wall. (Courtesy: Science Software Systems, Inc. Los Angeles)

CELL WALLS

Only plant cells have cell walls; animal cells do not. Whereas cell membranes are living things, cell walls are dead, rather inflexible, and highly permeable products of cellulose and related compounds which exist in a mat of fibers.

The cell wall of plants is commercially important in a number of ways. It is the tough cell wall which makes plant tissue useful as a building material. It is for cellulose that paper companies have leveled vast areas of American forests in response to our carefully cultivated demand for "disposable" commodities. Cellulose is also a major component of celluloid, rayon, cotton, and hemp. (Hemp, which is the Cannabis plant, was supplied in the U.S. by the cultivation of *Cannabis sativa* or marijuana. This is the same plant which is presently giving Kansas law-enforcement officers fits.) The lignin found in cell walls was once considered a totally useless by-product of paper manufacturing and presented a disposal problem. Researchers were hired to find a way to alter it so that it could be sold to us somehow, and the clever fellows managed to find a use for it in the manufacture of synthetic rubber, vanillin, and adhesives.

THE CELL MEMBRANE

All cells are bounded by a delicate structure called the cell membrane or plasma membrane. The membrane is so thin that its presence was postulated on circumstantial evidence alone until the advent of the electron microscope. In 1940 J. F. Danielli, a biologist at King's College in London, reported that the cell membrane consists of a rather wide area lying between two thinner, denser

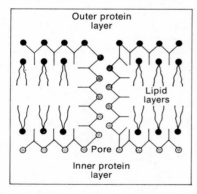

Figure 5.7 A micrograph (top) and illustration (middle) of the cell membrane. A hypothetical construction of the cell membrane (bottom). This construction takes into account most of the available data. (Courtesy: Science Software Systems, Inc., Los Angeles)

areas. The outer layers are believed to be protein and the light inner layer phospholipids and steroids. Together these three layers comprise the *unit membrane*. The molecules of the inner layer are apparently oriented parallel to the cell's radius, while those of the two outer layers lie perpendicular to the cell's radius (Figure 5.7).

Danielli's description was based on an elaborate series of permeability studies. In recent years more direct methods, such as electron microscopy and x-ray diffraction techniques, have been employed. The newer evidence at first seemed to confirm Danielli's model, but later reports provided disconcerting discrepancies. It now seems that membranes are generally composed of lipids and proteins, but the arrangement of these molecules may vary widely. For example, membrane thicknesses may be quite different from cell to cell, and the ratios of different lipids may vary from one type of cell to another. The most likely explanation of these differences is that there are different structural arrangements of the various lipids and proteins among cells. For example, perhaps in some cells the sandwich is reversed (in effect, making the lipid the "bread"), or maybe in some cases the proteins are globular instead of extended. A great many alternative descriptions have been suggested, but at this point the basic model of the unit membrane still appears to be correct in its essentials.

You will recall that cell membranes are selectively permeable; that is, they permit certain molecules to cross through and prohibit the passage of others. Thus this membrane regulates the constituents of the cell's interior, helping to protect it from an essentially hostile environment. Not all particles entering or leaving a cell must pass through the membranous material, however. Electron microscopy has revealed that the membrane surface is marked with many tiny pores through which some materials may pass. The specific roles of such pores are not well understood at present.

MITOCHONDRIA

Mitochondria are tiny structures, about the size of bacteria, which occur in almost all types of eucaryotic cells. They appear in a variety of shapes—round, elongate, and even threadlike. In some cells they seem to squirm around and move through the cytoplasm and in other cells they are fairly stationary. In either case, however, they tend to aggregate in places where work is going on—places where the greatest amount of energy is required. Not only are they found in the most active parts of cells, but they also occur in greater numbers in more active types of cells. There are more mitochondria in liver cells, for instance, than there are in less active cells such as cartilage. The concentration of mitochondria at places where work is being done is not coincidental; they are the "power-houses" of the cell. You may remember that it is in the mitochondria of aerobic animals where cellular respiration takes place, so it is here that ATP, CO_2, and H_2O are generated. (Glycolysis primarily takes place in the cytoplasm.)

A mitochondrion is composed of two membranes, with the inner membrane folded into shelf-like *cristae* (Figure 5.9). The mitochondrion is filled with a fluid *matrix* in which other reactions necessary to the synthesis of ATP occur. When ATP is produced, it is released into the cytoplasm at the surface of the mitochondrion. The mitochondria of some protozoa and insects have tube-like structures rather than cristae, but the advantage is the same: they increase the surface area.

Some interesting hypotheses have been advanced to account for the evolutionary development of mitochondria in cells, but the question has not yet been resolved. One idea is that they pinched off from the cell membrane. As the story goes, the first cells were probably no larger than the mitochondria themselves are now. But as cells grew, their volume increased more than their surface area (in a sphere, as surface area is squared, volume is cubed). Cells developing inpocketings in their membranes became more efficient since this provided greater surface area. Some of these pinched off and gradually became modified into mitochondria. Another hypothesis is that mitochondria were originally bacterialike organisms which moved into the interior of cells as parasites. One point which may support such an idea is that mitochondria have their own DNA, or hereditary material, and undergo their own replication in coordination with the cell's division.

GOLGI BODIES

In 1898 Camillo Golgi, an Italian cytologist, was experimenting with some cell-staining procedures and discovered that when he used certain stains, such as silver nitrate, peculiar "bodies" became apparent in the cells. The "reticular apparatuses" he described had never been noticed before, but when other workers looked for them, they turned up in a variety of cells. However, because they could not be seen in living cells, there was a great argument over whether they really existed or were just artifacts or debris produced by the staining process itself.

The electron microscope resolved the debate. Indeed, these strange bodies did exist. It was found that they had a characteristic and identifiable structure no matter what kind of cell they were found in. In every case they appeared as a group of tiny flattened vesicles, lying roughly parallel to each other, in the cell's cytoplasm.

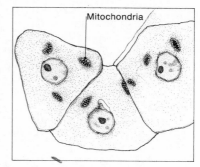

Figure 5.8 A mitochondria. (Courtesy: Carolina Biological Supply House)

Figure 5.9 Cutaway of a mitochondrion. The inner membrane is highly folded with the folds (cristae) projected inward. On the surface of the cristae are tiny granules which are believed to be the site of enzymes involved in electron transport. Enzymes involved in the Krebs cycle are thought to be located in the fluid matrix inside the mitochondrion.

Even after their existence was confirmed, an argument continued over their function. What did they do for a living? In spite of a great amount of research on the function of these strange bodies, we're still not sure. But there is some evidence that they serve as a sort of packaging center for the cell. Enzymes and other proteins, as well as certain carbohydrate molecules, are collected in these bodies, and here they are enclosed in "sacks" so that they are held apart from the rest of the cell. In some cases, after these molecules are packaged they are transferred to the cell membrane, where they are excreted out of the cell.

LYSOSOMES

The mysterious *lysosomes* are roughly spherical structures about the size of mitochondria. Their insignificant appearance, however, belies the rather startling role they play in the history of a cell. It is believed that lysosomes are packets of digestive enzymes which are synthesized by the cell and packaged by the Golgi bodies. If these digestive enzymes were floating free in the cell's cytoplasm, the cell itself would be digested. In fact, Christian de Duve, who first discovered the lysosomes, called them "suicide bags." The dramatic description is not entirely unwarranted, since they actually serve the function of destroying the cell which bears them. In some cases the destruction of certain cells is beneficial to the organism. For example, a superfluous cell could be an old one which is not functioning well, or one in a part of the body which was undergoing reduction as part of a certain developmental process.

Lysosomes may also act to selectively dispose of mitochondria, red blood cells, or bacteria. Fragments of these structures have been found within these puzzling sacks. Malfunctions of lysosomes have been found in connection with a number of human diseases, including cancer.

PLASTIDS

Plastids are found in plant cells, but not in animal cells. They are similar in some respects to mitochondria. For example, they are double-membraned, and they also have their own DNA. Strangely, their DNA may differ somewhat from that of the cell. Plastids come in a greater range of sizes than do mitochondria. However, the most significant difference between these two structures lies in their roles: mitochondria make free energy available from food and special plastids make food from solar energy. Plastids, then, are the bodies in which you would expect to find chlorophyll.

There are two major types of plastids—those which are colored

Figure 5.10 The Golgi bodies. The role of these bodies in the cell's activity still isn't entirely clear. (Courtesy: Carolina Biological Supply House)

Plastid

Vacuole

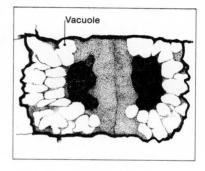

Vacuole

and those which are white or colorless. The best known of the colored plastids are the green chlorophyll-containing *chloroplasts*. Other chloroplasts may contain yellow or carrot-colored pigments, which, logically enough, are called *carotenoids*. The colorless plastids are storage bodies in which fats, starches, or proteins may be kept until needed by the cell.

VACUOLES

Vacuoles are fluid-filled sacks found in the cells of both plants and animals. It is in plant cells, however, that they reach their greatest development; in fact, they may be the most conspicuous body in the cytoplasm. There are a number of types of vacuoles, each with a different function. Some, for example, are highly active in the cell's metabolism, and others are simply storage vessels.

In plants the vacuoles are filled with a "cell sap," a fluid whose volume increases through osmosis. The pressure of the swelling vacuole forces the cell membrane against the cell wall and imparts firmness to the plant tissue. A plant wilts when there is not enough fluid to keep its huge vacuoles filled. Besides water, the vacuole sap may contain sugars, organic acids, proteins, and pigments. The acid component is what gives oranges and lemons their tart taste.

Protozoa may contain *contractile vacuoles*, which enable them

Figure 5.11 Plastids (left) in parechyma cells. (Courtesy: Carolina Biological Supply House)

Figure 5.12 Vacuole (middle) in an amoeba. (Courtesy: Carolina Biological Supply House)

Figure 5.13 Vacuole (right) in a plant cell. (Courtesy: Carolina Biological Supply House)

Endoplasmic reticulum — Nucleus

Figure 5.14 Rough endoplasmic reticulum. (Courtesy: Science Software Systems, Inc. Los Angeles)

to squeeze waste material or excess water out through the cell membrane. Such vacuoles may appear and disappear in response to the organism's needs or they may be relatively permanent structures—a useful feature in our identification of these tiny organisms.

THE ENDOPLASMIC RETICULUM

The *endoplasmic reticulum* (ER) varies widely in appearance in different cells (Figure 5.14). It may exist as elongate vesicles or as long tubelike structures. In any case, however, it always appears as spaces surrounded by membranes. Some of these membranes are studded with tiny granular particles called *ribosomes*. Those with ribosomes are called *rough ER*, and those without, *smooth ER*. Ribosomes may also be found in the cytoplasm, independently of the ER. We'll discuss these fascinating little structures later. For now, though, you will have to settle for knowing that they are the sites of protein synthesis. They appear to be almost round, and in rapidly growing cells they may occur in small clusters of five to ten.

In early electron-microscope studies it was found that the ER somehow had a close association with the cell nucleus. There is now an abundance of evidence of a connection between the ER and the outer portion of the nuclear membrane. Since the nuclear membrane differs in important ways from other cell membranes, this finding has several implications. For example, the nuclear membrane may not be as "special" as it was once thought to be; it may be simply an extension of the general membrane component of the cell. Because of its large surface area, the ER could conceivably have the function of expediting interactions between the cell's nucleus and its cytoplasm by providing an expanded interface between them.

The ER membrane may be confluent not only with the nuclear membrane, but with the cell membrane as well. (See Figure 5.21.) This means that there may be an open channel from the nucleus to the outside of the cell. The cell might thus be able to transport products manufactured in the nucleus directly to the outside. But more important, with such a connection, the nucleus might be able to react more quickly to changes in the cell's environment.

CENTRIOLES

Centrioles are small, dark cylindrical bodies just outside the nucleus in an area of specialized cytoplasm. They are normally found in the cells of animals, algae, and fungi; they are rarer, however, in the cells of higher plants.

Centrioles have a characteristic structure of nine fibrils (or little

Flagellum

fibers) running along their length, just below the surface (Figure 5.15). Usually they are paired, with each centriole lying at right angles to the other. We will see shortly how the centrioles function in cell division.

CILIA AND FLAGELLAE

Cilia and *flagellae* are hairlike projections from certain plant and animal cells. They are not distinguishable from each other except in terms of length; cilia are about 10 to 20 microns long while flagellae run up to thousands of microns long. Both cilia and flagellae function either in moving the cell or in moving substances across the cell surface. For example, a sperm cell swims by beating its whiplike flagellum, and upward sweeping motions of cilia clean your breathing passages (unless you have killed them by smoking).

The cilium or flagellum is actually an outpocketing or extension of the cell membrane and contains a cytoplasmic matrix. The rather consistent arrangement of such structures has generated much speculation about their origin. There are usually eleven groups of fibrils extending the length of the appendage, and these are arranged in a very particular order. Nine of the fibrils lie just under the surface of the structure, just as in the centrioles, but there are two additional fibrils running along the center (Figure

Figure 5.15 Cross-section (left) of a centriole. (Courtesy: Science Software Systems, Inc. Los Angeles)

Figure 5.16 A Euglena, (middle) a protozoan which is propelled by a single flagellum. (Courtesy: Carolina Biological Supply House)

Figure 5.17 Cross-section (right) of a cilium. Note the 9-2 arrangement characteristic of cilia and flagellae. (Courtesy: Science Software Systems, Inc. Los Angeles)

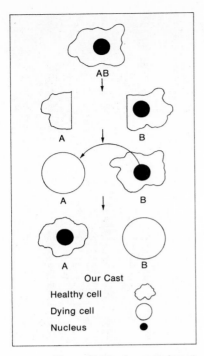

Figure 5.18 An experiment showing the critical role of the nucleus. See the text for a more complete explanation.

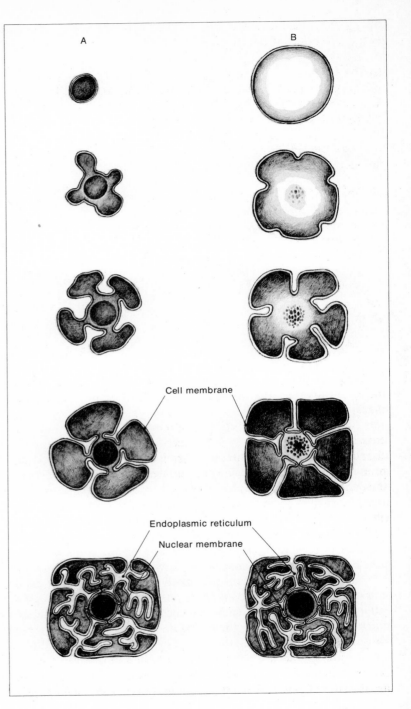

Figure 5.19 Two theories of the relationship of the nucleus to the cell in the evolution of the cell. In A, the nucleus exists first and the ER and the cell membrane are extensions of the nuclear membrane. In B, the primal structure is the cell. The nucleus develops later as the cell membrane folds in to surround the nucleus and form the ER.

5.17). Partly for this reason, it is believed that each cilium develops as the extension of one of a pair of centrioles.

THE NUCLEUS

The physical prominence of the nucleus among the cell's various structures is matched by the role it plays in the cell's history. This large, fairly obvious body is intimately involved with processes such as the cell's internal regulation and reproduction.

The nucleus is surrounded by a membrane; this membrane differs from the cell membrane in that it is doubled (Figure 5.19). The inner and outer parts of the nuclear membrane pinch together in certain places over the nuclear surface to form scattered thin-walled areas called *nuclear pores*. Although they are not actually pores, since there is no pore opening, they provide areas of high permeability which probably facilitate the flow of molecules between the nucleus and cytoplasm.

A number of relatively straightforward experiments have been done to show how critical the nucleus is to the cell's welfare. For example, if an amoeba is cut in half, the half with the nucleus continues to function as before. The half without the nucleus is unable to move about or capture food, and it soon dies. However, if the nucleus is surgically restored to this half within a few days, it recovers and is as good as new (Figure 5.18).

This doesn't mean that the nucleus has absolute control over the cytoplasm. As you might expect, biological stories usually aren't that simple. There are indications that the cytoplasm also has certain influences on the behavior of the nucleus. In addition, other environmental factors may influence the nucleus. For example, a change in the environmental temperature causes the nuclei of hair-producing cells to turn the hare's coat from one color to another as the seasons change.

Inside the nucleus is a fluid matrix in which a number of different types of bodies are suspended. When cells are stained in certain ways, a netlike structure becomes visible within the nucleus. This material is called *chromatin* because of its affinity for the stains (*chromatin* is from the Greek word for "color"). At a certain period in the cell's cycle the chromatin shortens and thickens and forms *chromosomes*. Chromosomes occur in a variety of simple rodlike, curved, or forked shapes. Each species, however, is characterized by a specific number of chromosomes. In normal human body cells there are exactly 46.

Chromosomes are composed of protein and a molecule called DNA, which we will consider shortly. It is these structures which are the sites of *genes*. You are undoubtedly aware that genes are the

Figure 5.20 The varying hare in its winter coat (top). The skin cells of the hare are able to respond to certain environmental factors, through a chain of mediating mechanisms, so that its color corresponds to its surroundings. In the summer it is brown (bottom) and hard to see among the forest leaves. (Courtesy: Michigan Department of Natural Resources)

Figure 5.21 A typical animal cell.

Nuclear membrane | Cell membrane | Golgi body | Nucleolus | Nucleus | Endoplasmic reticulum

Chromatin net | Vacuole | Lysosome | Centrioles | Mitochondrion | Ribosomes

determiners of hereditary characteristics. They determine the characteristics of the cells, and hence of the organism. Basically, genes operate by determining what kinds of proteins the cell will produce. This is an extremely critical role, as we will see. But we're getting ahead of our story.

Another structure found in the nucleus is the *nucleolus*. Each

Chromatin net | Nucleus | Nucleolus | Golgi body | Chloroplast | Cell wall

Cell membrane | Mitochondrion | Ribosomes | Endoplasmic reticulum | Vacuole | Nuclear membrane

Figure 5.22 A typical plant cell. What are some of the differences between this cell and the animal cell?

nucleus usually contains only one nucleolus, but it may contain more. The nucleolus is an oasis of a certain type of molecule called RNA, which will play an important role in another story—the one about how protein is made.

Now, to pull together what we have learned about the cell and its constituents, Figure 5.21 depicts the general structures of an animal cell and Figure 5.22 a typical plant cell. Keep in mind that these are stylized representations and that few real cells would look like these. However, most real cells, at some time in their life cycle, would include the various bodies shown.

CELL REPLICATION

It may not surprise you to hear that when carrot cells reproduce, they form more carrot cells. And oyster cells form more oyster cells. An oyster cell is very good at being an oyster cell, so when it reproduces, its best bet lies in making a very close approximation of itself. If a good functioning oyster cell were to generate new kinds of cells, there would be no guarantee that those new cells would work—and in fact, they probably wouldn't. So there is an advantage in having cells form new cells which are very similar to themselves. Such accurate duplication is assured by the exacting mechanism of cell reproduction.

Let's see, now, how cells go about reproducing themselves so that highly regular offspring result. Later we will learn something about the process at the molecular level.

MITOSIS

The process of cell reproduction is called *mitosis.* In general terms, this is a process in which each cell replicates its chromosomes, along with some cytoplasm, and then divides into two new cells. Each new cell receives one of the replicated chromosomes, with the result that it has a full chromosome complement identical to that of the parent cell. As the cell divides, the cytoplasm is distributed about equally between the "daughter" cells. Some of the inclusion bodies, such as the mitochondria, also reproduce themselves at cell division. Others, such as ribosomes, are synthesized anew by the daughter cells. The result is new cells which are *about* the same size, have *about* the same inclusion bodies in the cytoplasm, and have *identical* chromosomal complements.

For purposes of discussion the process of mitosis is divided into four stages called *prophase, metaphase, anaphase* and *telophase.* When the cell is not dividing it is said to be at *interphase.* This was once called the "resting" phase of a cell, but we now know that this description is inaccurate. At interphase many of the cell's components must double prior to the next mitotic division. The distinction between one phase and another is in a sense arbitrary, since there is a continuous progression from one phase to the next. Those who find comfort in labeling sometimes add further breakdowns, such as late prophase or early anaphase.

At interphase the chromosomal material of the cell may be difficult to distinguish, since it is dispersed throughout the nucleus in fine netlike strands which have a granular appearance. At the beginning of prophase, however, the chromatin becomes more visible as it begins to gather together and thicken into long twisted strands. The strands may now be called chromosomes. During interphase the genetic component of the cell has doubled, so that now each chromosome is made up of paired strands called *chromatids,* which are joined at some point along their length by a centromere (Figure 5.23).

The animal cell at this stage generally becomes rounder as the cytoplasm becomes heavier and more viscous. Other changes are occurring as well. The centromeres become markedly conspicuous, and the chromosomes continue to shorten and thicken as they move toward the nuclear membrane. At about this time a peculiar *spindle* appears in the cytoplasm—a structure consisting of tiny tubules which are seen as thin lines radiating across the cell from opposite poles. It is about then that the nuclear membrane begins to vanish. As it disappears, the chromosomes move toward the center of the cell and line up across the spindle. This is now the metaphase stage.

Anaphase begins with the division of the centromeres and the separation of the chromosomes into their separate chromatids, each now a single chromosome. Subsequently each of these single chromosomes moves along the spindle lines to opposite poles of the cell. It seems at this point as if the lines of the spindle have attached to the centromeres and are pulling the chromosomes apart. The result is that the chromosomes which converge at each end of the spindle are identical. The clumping together of the identical groups of chromosomes at each spindle pole marks the beginning of the telophase stage.

The chromosomes of each group now begin to stretch out and unwind as they revert to their own interphase condition. In animal cells, as the chromosomes become longer and thinner, a nuclear membrane begins to form around each group of chromosomes. The cytoplasm of what was the parent cell is divided as the membrane between the two groups of chromosomes pinches in and finally closes together, dividing what are now two separate cells. In plants, with their rigid cell walls, there is no pinching in. Instead a new cell wall begins to form in the cytoplasm between the divided chromosomes. Each daughter cell may at first be somewhat smaller than the mother cell because it has less cytoplasm, but such differences are quickly reduced by water intake and the generation of new inclusion bodies. The daughter cells may also be of different sizes, since the cytoplasmic division may not be precise, but each daugh-

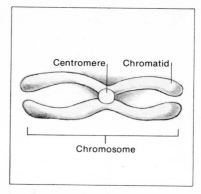

Centromere   Chromatid

Chromosome

Figure 5.23 At the beginning of prophase, the chromatin material becomes twisted into strands which are now called chromosomes. Chromosomes are composed of two chromatids joined at the centromere. The separate chromatids are distinct only at certain times.

Figure 5.24 Mitosis—Onion root tip.
A. Interphase: chromatin material diffuse; prominent nucleolus.
B. Early prophase: chromatin condensing. C. Late prophase: nucleolus disappearing; chromatin becoming distinct. D. Metaphase: chromosomes are distinct and lining up across spindle. E. Early anaphase: centromeres divided; single-stranded chromosomes move to opposite poles. F. Late anaphase: chromosomes clustering. G. Early telophase: new cell wall beginning to form.
H. Late telophase: new nuclei beginning to form. I. Interphase of daughter cells. (Courtesy: Carolina Biological Supply House)

A

B

F

G

C

D

E

H

I

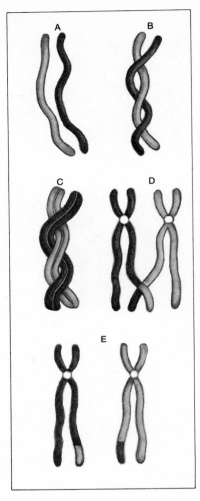

Figure 5.25 In meiosis, homologous chromosomes synapse, or join, together in early prophase (see A and B). During this time, genetic components may be exchanged. Following this, each chromosome appears to divide longitudinally, forming chromatids (C). This is followed by a separation of homologous chromosomes with the crossover of chromatids as they separate (D). The result is a pair of chromosomes that have exchanged genetic information (E).

ter cell has an identical genetic component. The chromosomes in each new cell will then replicate themselves from nuclear material as preparations for the next division occur. See Figure 5.24 for illustrations of mitosis.

Before we take a look at how all this is done at the molecular level, let's consider another process—one which seems similar to mitosis in its procedure, but which has important differences with exceedingly critical consequences.

MEIOSIS

The description of meiosis should answer some puzzling questions for you if you have thought much at all about the developmental process, especially if you are aware that eggs and sperm are simply specialized cells. At fertilization the nucleus of an egg and a sperm are joined, and from the subsequent divisions of the fertilized egg, a new organism is formed. Because of the regularity of the mitosis process, each new cell of the developing individual is genetically identical to its parent cell. This means that all the cells of the developing organism have their full complement of chromosomes. Since eggs and sperm are both cells, when they combine at fertilization, why don't the resulting offspring have twice as many chromosomes as their parents?

We know that the number of chromosomes is identical in every cell of the body, since all the cells of any organism spring from that first fertilized egg. We also know that chromosome number is important, since every species has its specific number. For the union of an egg and a sperm to produce the correct number of chromosomes in the cells of the offspring, each of them must have exactly half the chromosome complement of other cells, so that the *sum* of their chromosomes equals the full complement. As it turns out, each egg and sperm has exactly half the number which is characteristic of the species. When the egg and sperm are joined, the result is a primal cell with the correct number of chromosomes. Then through millions of divisions of that cell (in conjunction with cell specialization, which we'll worry about later), a new individual is formed.

Now, egg and sperm cells start out with the full complement of chromosomes. Somewhere along the way half the chromosomes are lost. The process by which certain cells in the *gonads* (ovaries or testes) lose half their chromosomes on the way to becoming eggs and sperm is called *meiosis*.

An important point should be noted here. As you see in Figure 5.25, chromosomes come in *pairs*. For every long chromosome, there is another identical long chromosome in that nucleus.

This relationship prevails because the egg and sperm contribute identical kinds of chromosomes to the new individual. Thus, for any pair of chromosomes in the body, one can be considered the descendant of a paternal chromosome (from the father's sperm) and the other the descendant of a maternal chromosome (from the mother's egg). The result is a contribution by each parent for every genetically determined characteristic of the offspring (with special exceptions, as we shall see). So, since chromosomes come in pairs— 23 pairs in humans, 9 pairs in cabbages, 47 pairs in goldfish—if we have a single gene for eye color from one parent, we will have another gene for eye color from the other parent (although not necessarily for the same color).

Now let's see how meiosis works. As we follow the development of a cell in the gonad through its meiotic divisions as it becomes an egg or a sperm, it may seem at first that we are retracing mitosis. But the sharp-eyed, critical student will quickly see important differences.

First of all, the meiotic process is divided into two parts, meiosis I and meiosis II (Figure 5.26). Before meiosis begins there is a replication of the genetic material in the interphase, just as in mitosis. The chromatin net then shortens and thickens to form distinct chromosomes, and the spindle begins to appear as the nuclear membrane begins to disappear. It all sounds familiar so far.

In early prophase I each pair of chromosomes comes together. Each chromosome consists of two identical chromatids, so each chromosomal group is made up of four chromatids at this point. As the chromosome pairs line up side by side, an important—and somewhat surprising—event may take place. The chromosomes may exchange parts. This exchange is important in providing variation among offspring. Other ways of providing this variation and the importance of variation in evolution are discussed in later sections.

In metaphase I, as you would expect if you have managed to remember our discussion of mitosis, the chromosomes line up across the developing spindle. In meiosis, however, the chromosome *pairs* separate, with one member of each pair moving to opposite spindle poles, marking anaphase I. In this case the centromeres do not divide, as they do in mitosis. Thus the chromatids of each chromosome now travel together to the spindle poles. At telophase I, the end of the first stage of meiosis, the chromosomes begin to elongate and revert to their netlike or grainy condition, and a nuclear membrane may form as the cell enters interphase II. Each nucleus now contains only half the number of chromosomes of the parent cell, although each chromosome is two-stranded. Not only are there fewer chromosomes in each cell, but they are probably qualitatively

Figure 5.26 Meiosis—Lily anther.
Meiosis I A. Prophase I: chromosomes have synapsed, forming tetrads. B. Metaphase I: tetrads line up across spindle. C. Anaphase I: centromeres do not divide. Homologous chromosomes separate and move to opposite poles. D. Early telophase I: chromosomes clustering. E. Late telophase I: new nuclei forming. New cell wall forming.
Meiosis II F. Early prophase II: diffuse chromatin condensing. G. Metaphase II: distinct chromosomes lining up across spindle. H. Anaphase II: centromeres divided. Single stranded chromosomes move to opposite poles. I. Telophase II: new nuclei and cell walls beginning to form. Four haploid cells result. (Courtesy Carolina Biological Supply House)

A

B

F

G

C

D

E

H

I

Figure 5.27 A comparison of mitosis and meiosis in animal cells, showing *two pairs* of chromosomes. The maternal and paternal chromosomes are identifiable by their different color intensities. Notice the specific behavior of the chromosomes as you compare metaphase and anaphase between mitosis and meiosis. Also notice that a second division is required to complete the meiotic process. As a final observation, compare the chromosome number in telophase of mitosis ·with that in telophase of meiosis II.

MITOSIS

1. INTERPHASE
Chromosomes not visible; division preparations in progress.

2. PROPHASE
Centrioles migrated to opposite sides; spindle forms; chromosomes become visible as they shorten; nuclear membrane, nucleolus fade.

3. METAPHASE
Chromosomes aligned vertically on cell equator. Note attachment of spindle fibers from centromere to centrioles.

4. ANAPHASE
Centromeres divide; single-stranded chromosomes move toward centriole regions.

6. DAUGHTER CELLS
Two cells of identical genetic (DNA) quality; continuity of genetic information preserved by mitotic process

These cells may divide again after growth and DNA replication has occurred.

5. TELOPHASE
Cytoplasm divides; chromosomes fade; nuclear membrane, nucleolus reappear; centrioles replicate (reverse of prophase).

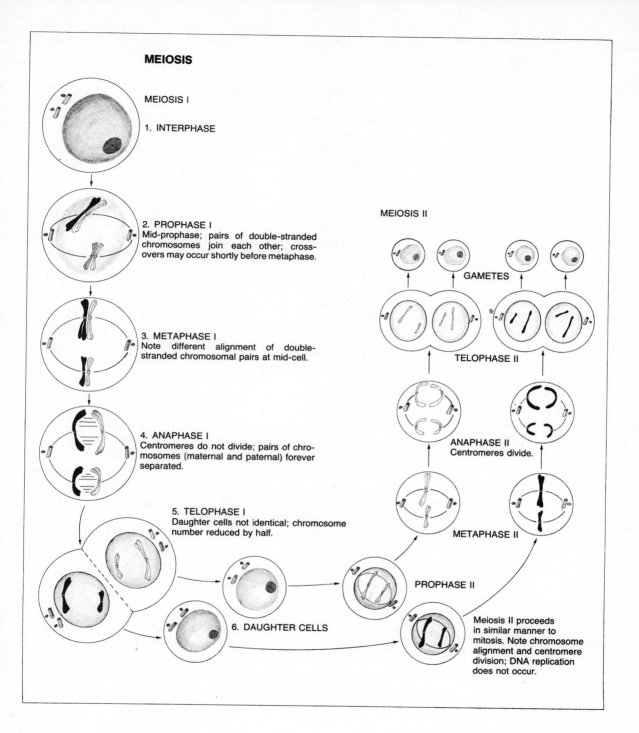

MEIOSIS

MEIOSIS I

1. INTERPHASE

2. PROPHASE I
Mid-prophase; pairs of double-stranded chromosomes join each other; crossovers may occur shortly before metaphase.

3. METAPHASE I
Note different alignment of double-stranded chromosomal pairs at mid-cell.

4. ANAPHASE I
Centromeres do not divide; pairs of chromosomes (maternal and paternal) forever separated.

5. TELOPHASE I
Daughter cells not identical; chromosome number reduced by half.

6. DAUGHTER CELLS

MEIOSIS II

GAMETES

TELOPHASE II

ANAPHASE II
Centromeres divide.

METAPHASE II

PROPHASE II

Meiosis II proceeds in similar manner to mitosis. Note chromosome alignment and centromere division; DNA replication does not occur.

different from any of the original chromosomes as a result of crossovers.

Interphase II differs from the interphase preceding meiosis I in an important way: the chromosomes do not replicate themselves. Aside from this difference, the second meiotic division proceeds in the usual way. In prophase II the chromosomes shorten and thicken, and the nuclear membrane (if it is present) begins to disappear. At metaphase II the chromosomes line up across the equatorial plate as spindle fibers again appear. Now, at anaphase II, in contrast to anaphase I, the centromere breaks apart, and the chromosomes separate into their chromatids, which move toward opposite poles. At telophase II a nuclear membrane forms around each set of chromosomes. The cytoplasm is divided as new cell membranes appear, thus dividing each group into a separate cell. So where there was one cell with a full complement of chromosomes, there are now four cells, each with half the chromosomal complement. You can take a closer look at the difference between meiosis and mitosis in Figure 5.27.

If the meiotic process has been taking place in a testicle, each of four resulting cells will undergo certain changes, separate, grow a tail and become a sperm. If the meiosis was in an ovary, however, three of the resulting cells would have degenerated into tiny, nonfunctional *polar bodies*, with only one cell surviving to become an egg. It is not known what determines which of the four cells this will be.

There are two major take-home lessons from the story of meiosis. One is that meiosis halves the chromosome number of eggs and sperm. Second, meiosis provides a means of shuffling and reorganizing chromosomes, thus increasing variation in the offspring. This reorganization takes place in crossovers; in the apparently random lineup of chromosomes at metaphase I, so that paternal and maternal chromosomes are mixed by the time a cell enters meiosis II; and in the apparently random selection of polar bodies in egg formation. You may come to have an abiding appreciation for this process when you realize that it is because of meiosis that you don't look like your little brother.

THE DOUBLE HELIX

As more and more information on the behavior of chromosomes was produced, a growing question was, What are they? What are they made of? Chemical analysis showed that the chromosomes themselves were composed of protein and an acid called deoxyribonucleic acid, or DNA. By 1950 it was pretty well accepted that the hereditary material itself was the DNA component of the chromosomes. But how was the DNA arranged in the chromosome? And how did it duplicate itself at mitosis? And in particular, how could it direct protein synthesis? Several American and British groups were attempting to answer these questions, and the prevailing notion was that there were three chains in each DNA molecule. Linus Pauling, the great American chemist, and Rosalind Franklin of England both favored this idea but disagreed over whether the nitrogen bases stuck out of the molecule or projected inward.

Then a brilliant and highly confident young American post-doctoral fellow ran across an ebullient and talkative English graduate student in biophysics at Cambridge University. The results of what transpired were described by James Watson in his controversial best-seller, *The Double Helix*.

The Englishman, Francis Crick, had noted that when the DNA portion of chromosomes was photographed by x-ray diffraction techniques, a peculiar image appeared—one which could only be produced by molecules arranged in a helical, or spiral, pattern. Watson's biological intuition told him that only two strands were involved. After all, reproduction usually involved only two parents and there are paired chromosomes and pairs of genes. Then, on the basis of x-ray diffraction information provided by Rosalind Franklin and others, after careful consideration of a wide range of information, and after building several huge DNA replicas in their laboratories, Watson and Crick stated in 1953 that DNA is composed of two chains coiled around each other as a double helix. You can get the idea if you visualize a ladder twisted so that the rungs remain perpendicular to the sides.

According to their model, the two longitudinal sides of the ladder are alternating sugar and phosphate molecules. The "rungs" of the ladder are formed by four nitrogenous bases—the purines adenine (A) and guanine (G) and the pyrimidines thymine (T) and cytosine (C). These bases, with their sugar and phosphate components, are called *nucleotides* (Figure 5.29). There is one base per sugar-phosphate, and two bases join together to form each rung. The bases extend inward from the ladder's sides and are held together by weak hydrogen bonds. It was found that two of the larger bases (purines, with two nitrogen-containing rings) were too large to fit between the sides of the ladder, and two pyrimidines, with their single nitrogen-containing rings, were too short to reach across. The arrangement could work only if a pyrimidine joined with a purine to form the rungs (T to A and C to G).

According to Watson and Crick's model, the nucleotides along one of the strands of the double helix could follow any sequence. For example, one might "read" along a strand and find ATTCG

Figure 5.28 James Watson, who along with Francis Crick first described the DNA double helix. (Courtesy: Dr. James D. Watson, Harvard University)

Figure 5.29 Nucleotides of DNA. Note the sugar (2-deoxy-D-ribose), the phosphate group and the nitrogen-containing base. If the base has one ring, it is called a pyridimine. If it has two rings it is called a purine.

PYRIDIMINES

Phosphate — Thymine (base)

Cytosine (base)

PURINES

Phosphate — Adenine (base)

2-Deoxy-D-ribose (sugar)

Guanine (base)

TAACGCGT in one segment. Also, the strands were found to be very long, so that the necessary complexity for life could be provided by the variation in sequence and the long numbers of nucleotides in any chain. If the DNA from a single cell were extended in a straight line, it would be about five feet long! There are about 10 billion nucleotide pairs in man's 46 chromosomes. Think of the number of possible sequences!

Now, if the sequence in one strand of DNA is known, the sequence of the other strand can be deduced. The purine adenine (A) will always be bound to the pyrimidine thymine (T) of the other strand. And wherever cytosine (C) exists along one chain, it will be attached to guanine (G) on the other strand. So what would be the nucleotide sequence of the complementary segment of the strand described above?

DNA REPLICATION

As you recall from our discussion of mitosis, after the cell has divided each chromosome replicates itself, so that the chromosomal component is doubled before the next division starts. But how does the chromosome, with its intertwining strands, do this?

Actually, Watson and Crick's model of DNA structure suggested the mechanism—conceptually a simple one. When the time comes for the chromosome to replicate itself, its DNA strands separate. The weak hydrogen bonds holding them together are broken, and the molecule opens up. Purines and pyrimidines are now exposed to the nuclear sap, which is loaded with many kinds of molecules. From among these free-floating molecules, adenines, guanines, cytosines, and thymines attach to the exposed bases of the open DNA strand according to their base-pairing formula (T to A, C to G). The new and the old strand of bases then bind together by forming weak hydrogen bonds along their length. Thus, step by step, an identical DNA molecule is produced from each strand of the original, using the raw materials from the nucleus. Each new DNA strand will become a chromatid at the next cell division.

HOW CHROMOSOMES WORK

Chromosomes function not only in reproductive processes by the action of eggs and sperm, but in all the other cells of the body as well. After all, skin cells have nuclei; so do nerve cells and muscle cells and almost all the other cells which comprise organisms. And in those nuclei are chromosomes. In fact, it is those chromosomes which direct a developing cell to become skin or muscle or nerve. Now let's find out how they do it.

At the same time we can answer another question. You may

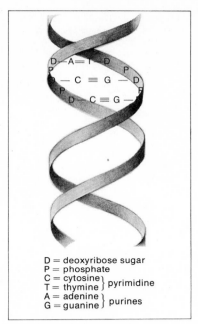

D = deoxyribose sugar
P = phosphate
C = cytosine ⎫
T = thymine ⎬ pyrimidine
A = adenine ⎫
G = guanine ⎬ purines

Figure 5.30 How DNA strands are joined.

Figure 5.31 As the DNA unwinds, complementary strands are formed along the length of each coil, thus forming new DNA.

have been astonished—or at least mildly interested—to learn that practically all the cells in your body have the same chromosomal component. After all, those cells all arose as a result of mitotic divisions of that first fertilized egg. Since we know that mitosis results in an equal distribution of chromosomes, how can those same chromosomes cause a cell to become muscle in one case and nerve in another?

Before we launch into this discussion, you might note that cells become specialized into muscle or nerve or whatever as a direct result of differences in their component enzymes. Enzymes are biocatalysts, you will recall, and their role is to facilitate certain very specific biochemical events. It is through the precise summation of such events that cells become specialized. So what we are really asking here is how do chromosomes make those peculiar proteins called enzymes? How do events in the nucleus alter the developmental pathway of the rest of the cell?

The first thing that happens is that the chromosomes unwind, just as they did in chromatid replication in mitosis. As they unwind, the hydrogen bonds between purines and pyrimidines are broken, exposing these nucleotide bases to the nuclear fluid, and free nucleotides attach to the exposed bases along the DNA strands. As before, free cytosine attaches to exposed guanine, free guanine to exposed cytosine, and free adenine to the exposed thymine. However, where adenine lies exposed along a DNA molecule, it is attached, not to free thymine, but to *uracil*, another nucleotide in the cytoplasm. Here, then, the DNA strands are not building complementary strands of DNA, as they do when they replicate themselves in mitosis. They are building another substance— *ribonucleic acid*, RNA. RNA has one more oxygen atom in the sugar component of its nucleotides, and in place of thymine it has the nucleotide uracil (Table 5.2). Also, most RNA exists as a single strand, and not as a helix.

So, as the DNA unwinds, RNA is manufactured along its length. Sections of this RNA will then separate from the DNA strand and move from the nucleus into the cytoplasm, each one carrying a message encoded in its nucleotide sequence. This RNA is called, appropriately enough, *messenger RNA*, or *m*RNA.

The *m*RNA, then, is a rather long, and, as we will see, somewhat unstable molecule which contains guanine, cytosine, adenine and uracil and the sequencing of its nucleotide bases is very precisely directed by the DNA inside the nucleus. Before we see what this very specific *m*RNA does in the cytoplasm, let's learn something about how its message is carried—the mysterious genetic code. The genetic alphabet consists of only four nucleotide bases, yet the

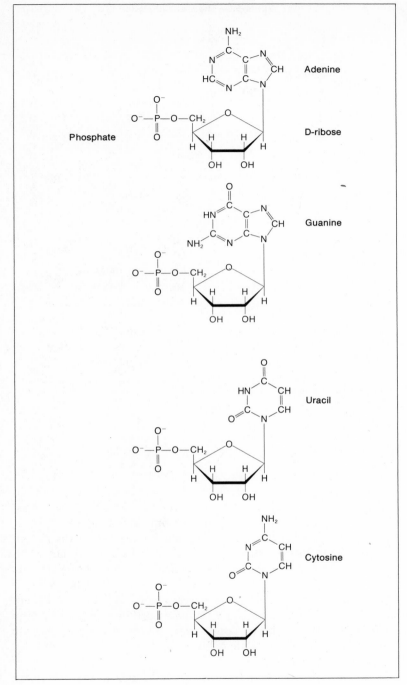

Figure 5.32 Nucleotides of RNA. Compare with Figure 5.29.

Table 5.2—The Genetic Code

Second letter

	U	C	A	G	
U	UUU UUC } PHE / UUA UUG } LEU	UCU UCC UCA UCG } SER	UAU UAC } TYR / UAA stop / UAG stop	UGU UGC } CYS / UGA stop / UGG TRP	U C A G
C	CUU CUC CUA CUG } LEU	CCU CCC CCA CCG } PRO	CAU CAC } HIS / CAA CAG } GLN	CGU CGC CGA CGG } ARG	U C A G
A	AUU AUC AUA } ILE / AUG MET (start)	ACU ACC ACA ACG } THR	AAU AAC } ASN / AAA AAG } LYS	AGU AGC } SER / AGA AGG } ARG	U C A G
G	GUU GUC GUA GUG } VAL	GCU GCC GCA GCG } ALA	GAU GAC } ASP / GAA GAG } GLU	GGU GGC GGA GGG } GLY	U C A G

First letter (left) / Third letter (right)

Phenylalanine	Serine	Tyrosine	Cysteine
Leucine	Proline	Histidine	Tryptophan
Leucine	Threonine	Glutamine	Arginine
Isoleucine	Alanine	Asparagine	Serine
Methionine		Lysine	Arginine
Valine		Aspartic Acid	Glycine
		Glutamine	

messages are sufficiently precise to code for every kind of protein that makes up the highly complex structures of living organisms. How is this complexity achieved with only a four-unit code?

Actually, we already have enough information to figure out the answer to this question. Although proteins are very large and complex, they are made up of amino acids—of which there are only twenty kinds. So the problem reduces to simple mathematics. If there are only four nucleotide bases, obviously one base cannot code for each amino acid, since this could produce only four amino acids. A two-base combination for each amino acid would provide four times as many possible combinations ($4 \times 4 = 16$), but this is still not enough.

So, early in the search for the genetic code it was postulated that a three-base combination was necessary to code for each amino acid. At first this was just a guess; the verifying data came later. It was reasoned, however, that with four nucleotide bases, a combination of any three as a code for a particular amino acid would provide sixty-four possible combinations ($4 \times 4 \times 4 = 64$). But there are only twenty amino acids. So, as you may have guessed, some nucleotide triplets are coded for the same amino acid.

Of the sixty-four nucleotide triplets, called *codons,* it is known that sixty-one code for specific amino acids. But what about the other three, UAA, UAG and UGA? These particular codons, interestingly enough, serve as punctuation marks in the language of life; they signal the end of a given message. The codon AUG apparently doubles as an initiator, indicating the start of a message.

Now that we have some idea of the nature of the genetic code, let's see how it operates to produce the astonishing array of proteins found in living things. When the newly formed *m*RNA leaves the nucleus and enters the cytoplasm, it diffuses through the cytoplasm until it encounters one of the cell's many ribosomes. It joins temporarily with the ribosome to form a ribosome-*m*RNA complex, the site of protein formation. The function of the ribosomes is to decode the message which was inscribed by the DNA. The amino acids from which the new protein will be formed are escorted to the ribosome-*m*RNA site by another RNA nucleotide called *transfer* RNA, or *t*RNA. (There is also a *ribosomal* RNA in the ribosome; we don't know yet exactly what it does.)

There are at least as many kinds of *t*RNA molecules as there are amino acids, and each kind of *t*RNA molecule can attach only to one type of amino acid. The specificity of the *t*RNA molecule is determined by the three nucleotides which are found at one end. For example, if these three bases all happen to be adenine, an AAA sequence, then the *only* amino acid that the *t*RNA molecule can attach to is one called *phenylalanine.* Now, when a *t*RNA with this nucleotide arrangement arrives at the *m*RNA-ribosome site with its phenylalanine, where do you think the phenylalanine will be inserted on the ribosome-*m*RNA complex? Of course! Only at a place on the *m*RNA where there are three complementary uracils (UUU). So both the kind of amino acid and its position in the developing protein are dictated by the DNA.

Figure 5.33 shows the process by which a new protein is synthesized. The ribosome attaches to one end of the *m*RNA strand and moves along its length, stopping to read off each nucleotide triplet to the *t*RNAs, which have diffused to the *m*RNA site with their amino acids. When it pauses at a triplet, a *t*RNA molecule with the proper complementary sequence drifts to that position,

Figure 5.33 Summary of protein synthesis. The ribosome attaches to one end of the mRNA molecule and moves along it, pausing at each codon to "read it off" to the tRNA molecules with their amino acids in tow. The proper tRNA moves to the codon it matches as soon as the ribosome touches the codon, briefly plugs in its triplet to the template codon at that site, leaves its amino acid and then moves away to be used later. The ribosome then moves along to the next site, reads off its codon and the next tRNA zeroes in to that spot. As the ribosome moves along the mRNA, it may be followed by another ribosome making the same polypeptide. The mRNA is rather short-lived. After making its specific polypeptides, it disappears. The tRNA molecules, on the other hand, may survive for a much longer period and be reused many times.

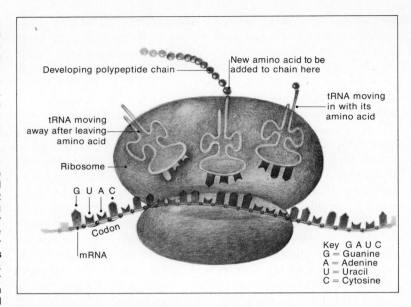

briefly plugs into the *m*RNA codon, and then moves away, leaving its amino acid behind. Meanwhile, the ribosome has reached the next codon site, and another *t*RNA falls into place, leaving its amino acid behind the first one. Thus amino acids are aligned, one by one, in proper sequence for the formation of a specific protein. When all the slots in the sequence have been filled with the appropriate amino acids, and these amino acids have attached to each other to form polypeptides or proteins, the protein moves away into the cytoplasm, and the *m*RNA begins to break down.

The resulting protein may be a structural material or it may be an enzyme, depending on the *m*RNA sequence along which it was constructed. If it is an enzyme, it will participate in the biochemical activities of that cell, and hence will help to determine its nature—what the cell looks like and what it does (perhaps the same things one might worry about a blind date). We know that the chromosomes are identical in every cell, but how can the same DNA component result in cells of entirely different types?

We know that different cells are specialized by different enzymes. We also know that enzyme production is directed by the DNA of the chromosomes through *m*RNA strands formed along its length, which then move into the cytoplasm to direct protein formation. Since proteins, even within a single cell or organism, are of many different kinds, we can assume that many kinds of *m*RNA are formed along a single DNA chain. Thus it is clear that a specific *m*RNA is produced only along a certain sec-

Figure 5.34 Why do we age? The old Harry Truman (right) looked different than the young Harry Truman (left) because of cell changes. (Courtesy: Wide World Photos)

tion of the long DNA molecule. It would be inappropriate for DNA to continually synthesize all of the varieties of *m*RNA, since this would mean all of the cell's proteins would be produced. So only certain segments in each DNA molecule are active in *m*RNA production. In other words, parts of the chromosomes are "shut down" while other parts are busy forming *m*RNA, with the working segment depending on which kinds of proteins are to be synthesized.

This explanation of cell differences, as you will immediately see, simply places the question of control one step further back. We are now faced with the question of how the active section of a chromosome is determined. At this point the data become rather varied and complex, but there is some evidence that the position of the cell in the body may be critical. The idea, in such an explanation, is that the milieu of a cell may have an effect on its chromosomes.

PROBLEMS IN CELL BIOLOGY

The sight of complex flowcharts and remarkable electron-microscope photographs of cell processes often leads us to believe that the great questions have been answered, and all we need to do now is to tie up a few loose ends. Nothing could be further from the truth. In fact, we have barely touched the "hem of the garment" in many areas of cell biology.

As an example of our ignorance, we are unable to control aging at almost any level. What causes cells to lose their tone?

Box 5.2 Viruses   Viruses are remarkable little organisms, about 1/20,000 the size of a bacteria, which reproduce entirely inside cells. Viruses are unique among living things—if, in fact, they are alive. Outside of a living cell they are completely inert. In some cases they look like crystals, and as lifeless as ordinary table salt. As soon as a virus enters a living cell, however, it begins to take over the metabolic machinery of that cell. It reorganizes the cell's processes so that the cell begins to engage in producing more viruses instead of continuing its normal activities. Finally, when the cell ruptures, new viruses are released to take over new cells.

Single or double strands of DNA or RNA are usually found inside the virus. Outside, it is coated with a protein layer which protects its nucleic acid from the cell's normal enzymes. This protein layer also gives the virus a special affinity for certain kinds of host cells. Some viruses, for example, have an affinity for bacteria. When such a bacteriophage, or bacteria-eater, accidentally bumps into a proper host cell, it attaches by its tail through covalent bonding. An enzyme of the virus digests a hole in the cell wall, and the virus then contracts its "head" and releases its nucleic acid contents into the bacterium.

(Courtesy: Carolina Biological Supply House)

Why do fat cells under the eyes give way as we grow older? We just don't know! The only consolation for spiteful agers is that the day will also come for the ebullient youths around them—unless a method of cellular control is developed. In our youth-worshipping culture we are presented with a vast array of moisturizers, hormone creams, lotions, and exotic "aging remedies." The search goes on. Meanwhile, we can really say little more than "lead a good life"—and we don't even know what *that* is.

In spite of what may have seemed great assurance in the foregoing discussions, we actually know very little about such matters as mitochondrial replication, spindle formation, cell-environment interaction, cell differentiation (what makes cells different), cell growth, chromosome structure, or the permeability of membranes. We know only a precious few of the simplest enzymatic pathways.

Cancer is certainly one of our most dreaded and puzzling cellular problems and continues to absorb a great deal of research attention throughout the world.

Some cancer is the result of an uncontrolled mitosis on the part of certain cells. Cell division, as we know, is a tightly regulated procedure in the living organism, from the time of fertilization onward. Once the body has stopped growing, cell division must be carefully geared to replace only the cells which have worn out or died. In some cases, however, cell division begins to lurch out of control, and produce new cells with different qualities. These new kinds of cells may spread rapidly as cancer.

Viruses, certain chemicals, irradiation, and automobile exhaust products have all been linked with cancer, but in some cases the disease seems to appear spontaneously, perhaps through some small coding mistake in DNA replication. We may not be able to prevent cancer until we have a complete understanding of chromosome structure and the process of mitosis. In the meantime, techniques of virus control and antibody synthesis may enable us to control the symptoms.

Little by little we are accumulating bits of information about the nature and processes of cells, but there seem to be few shortcuts. Our hope of ever understanding these matters lies, now, in the slow accumulation of hard-won data by diligent researchers. Perhaps we have finally begun to ask some of the right questions. But with all the areas in which we still know too little even to formulate questions, much of the knowledge we need is likely to come from "pure" research—research which is not prompted by a specific utilitarian goal, but is done simply to "find out." The breaking of the genetic code and the description of DNA were the results of pure research. Thus it may prove scientifically tragic that government funding of pure research has been cut so drastically in our country. The federal funding of the sciences is being channeled into heart and cancer research at present. These are, of course, worthy causes, but the fact is often overlooked that those people involved in solving very specific problems for mankind must draw on the wealth of background information provided by researchers who simply wanted to know.

6

Inheritance

Occasionally you will hear someone just back from a trip to Mexico or Spain boasting of his bullfighting experiences. You might be disillusioned to learn that he hadn't fought any bulls at all—that his "bullfight" was with a heifer. At certain times of the year guests are invited to the ranches where the valiant "brave bulls" are bred to watch the testing of the cows—to determine which stock will be used for breeding. As part of the festivities guests may then be invited to try their hand at caping a young cow. The breeders are interested in the cows because it is axiomatic in such circles that a fighting bull gets its strength from the sire, but its courage from the cow. The origin of such beliefs is lost in tradition, but it is evident that man utilized the principles of genetics long before he became interested in their basis.

Figure 6.1 Fighting bulls are bred for their courage. (Courtesy: Mexican National Tourist Council)

You will recall that Darwin knew something about artificial selection from his familiarity with livestock breeding. The offspring of heavier animals, he knew, tend toward heaviness; heifers which were daughters of high-yield milk producers tended to produce more milk. The principle seemed clear enough to animal breeders: some traits are inherited. However, in Darwin's time it was believed that the traits of both parents blended in the offspring, just as two colors of ink do when they are mixed; hence these characteristics could never show up in "pure" form in subsequent generations. In other words, a black horse and a white horse would produce a gray colt, and when these gray horses were mated, they would continue to produce gray offspring, but the original black or white could never reappear.

Although such thinking was firmly entrenched at Darwin's time, it ignored a body of contradictory evidence. For example, as far back as the first century Plutarch had noted that some children have qualities obviously inherited from their mothers and fathers, but that others often display the characteristics of their grandparents which had been missing in their parents. Also, there was no satisfactory explanation for the common observation that when black and honey-colored cocker spaniels are crossed the

Figure 6.2 Gregor Mendel was a highly trained scientist of his time. His work was published only seven years after Dárwin's *Origin of Species*, and it is perhaps for this reason that his enormously important studies on the principles of genetics received little attention as the scientific world focused on the fascinating idea of natural selection. (Courtesy: The Bettmann Archive)

pups are not usually a blend, but instead are either black or honey-colored.

MENDELIAN GENETICS

It is interesting that the answers to many of the questions about heredity which plagued Darwin were actually published during his lifetime by a monk named Gregor Johann Mendel (1822–1884), who was a member of an Augustinian order in Brünn, Moravia (now Czechoslovakia). Unfortunately, Mendel's work was given so little notice at the time that Darwin either never learned of it or its implications escaped him.

The Abbot Gregor Mendel was a rather remarkable person in his own right, although he is usually depicted as a kindly old man of the cloth who, while puttering around in his monastery garden, somehow stumbled on important laws governing genetic transmission. Our penchant for building myths to embellish the memory of already worthy people has led us astray again. Early in his life Mendel began training himself, and he became a rather competent naturalist. To support himself during those years he worked as a substitute high school science teacher. The professors at the school, noting his unusual abilities, suggested that he take the rigorous qualifying examination and become a regular member of the faculty. Mendel took the test and did surprisingly well, but he failed to qualify. The conditions for teaching high school were rigorous indeed.

The monastic order to which Mendel belonged, confident of his abilities, sent him in 1851 to the University of Vienna for two years of concentrated study in science and mathematics. When Mendel returned to the monastery, he began his plant-hybridization studies in earnest. He was especially noted for his high level of intelligence and his vigor, and he applied these qualities to the study of plant breeding. He developed new varieties of fruits and vegetables, kept abreast of the latest developments in his field, and became active in community affairs.

In 1865, while Darwin was still puzzling over the enigma of heredity, Mendel presented a single paper at a meeting of the Brünn Natural Science Society. The paper was published in the society's proceedings the following year. Apparently, however, no one understood what Mendel was even talking about, and the work for which he was to go down in history was met with a resounding silence.

Actually, Mendel said three things:

1. When parents differ in one characteristic, their offspring will be *hybrids* for that particular characteristic; however, the

offspring may resemble one parent or the other instead of showing a blend of both traits. This means that the trait of one parent must be dominant over the trait of the other parent. This phenomenon he called the principle of *dominance*.

2. When a hybrid reproduces, its gametes (eggs or sperm) will be of two types. Half will carry the dominant trait provided by one parent and half will carry the recessive trait provided by the other parent. This he called the principle of *segregation*.

3. When parents differ in two or more characteristics, the occurrence of any characteristic in the next generation will be independent of the occurrence of any other characteristic, so that any combination of the parental components may appear in the offspring. This is called the principle of *independent assortment*.

Now, at this point you are probably as mystified as was Mendel's audience, but let us continue. Perhaps the best way to clarify things is to see exactly how Mendel arrived at his conclusions.

THE PRINCIPLE OF DOMINANCE

To begin with, Mendel based his information on a very carefully planned series of experiments and, more important, on a statistical analysis of the results. The use of mathematics to describe biological phenomena was a new concept. Clearly, Mendel's two years at the University of Vienna had not been wasted.

The careful planning that went into Mendel's work is reflected in his selection of the common garden pea as his experimental subject. There were several advantages in this choice. Pea plants were readily available and fairly easy to grow, and Mendel had already developed some thirty-four pure strains. These strains differed from each other in very pronounced ways, so that there could be no problem in identifying the results of a given experiment. Mendel chose seven different pairs of characteristics to work with:

1. Seed form—round or wrinkled
2. Color of seed contents—yellow or green
3. Color of seed coat—white or gray
4. Color of unripe seed pods—green or yellow
5. Shape of ripe seed pods—inflated or constricted between seeds
6. Length of stem—short (9 to 18 inches) or long (6 to 7 feet)
7. Position of flowers—axial (along the stem) or terminal (at the end of the stem)

Mendel's approach, a novel one at that time, was to cross two true-breeding strains which differed in only one characteristic, such as seed color. Peas ordinarily self-fertilize, so for this cross the pollen was transferred by hand. He called this original parent generation P_1 and designated their first-generation offspring the F_1 (first filial) generation. When the F_1 plants were allowed to self-pollinate, so that they crossed with each other at random, the off-spring resulting from this cross were called the F_2 generation, and so on.

Now, when Mendel crossed his original P_1 plants, he found that the characteristics of the two parents didn't blend, as prevailing theory said they should. When plants with yellow seeds were crossed with plants that had green seeds, their F_1 offspring did not have yellow-green seeds. Instead, all of them had yellow seeds. Mendel termed the trait which appeared in the F_1 generation the *dominant* trait and the one which had failed to appear the *recessive* trait. But he was now left with a vexing question. What had happened to the recessive trait? It had been faithfully passed along through countless generations of pea plants so it couldn't have just disappeared.

THE PRINCIPLE OF SEGREGATION

In the next stage of his experiments Mendel allowed the F_1 plants to self-pollinate, and lo and behold, the missing recessive trait reappeared in their F_2 offspring! Also, the *ratio* of F_2 plants with recessive traits to those with dominant traits was fairly constant, regardless of the particular characteristic involved. The exact results for each of the seven pairs of characteristics Mendel used in his tests are shown in Table 6.1.

Finally, in the third year of the experiments, Mendel allowed the F_2 plants to self-pollinate. He found that all those with recessive

Table 6.1—Mendel's Pea Plant Experiment

Dominant form	No. in F_2 generation	Recessive form	No. in F_2 generation	Total examined	Ratio
Round seeds	5,474	Wrinkled seeds	1,850	7,324	2.96:1
Yellow seeds	6,022	Green seeds	2,001	8,023	3.01:1
Gray seed coats	705	White seed coats	224	929	3.15:1
Green pods	428	Yellow pods	152	580	2.82:1
Inflated pods	882	Constricted pods	299	1,181	2.95:1
Long stems	787	Short stems	277	1,064	2.84:1
Axial flowers	651	Terminal flowers	207	858	3.14:1

traits produced only recessive F_3 offspring. However, the F_2 plants which showed a dominant trait produced both types of F_3 offspring. One-third of them produced *only* dominant offspring. The other two-thirds produced both dominant and recessive offspring, but they produced three times as many offspring with the dominant trait as they did with the recessive trait. In other words, all the F_2 recessives and one-third of the F_2 dominants bred true. The other two-thirds of these dominants produced mixed offspring—but in the same 3:1 ratio of dominant to recessive as in the plants their F_1 parents had produced.

Let's consider a specific example of a cross between two pure lines—plants which breed true when they self-pollinate and are identical in all characteristics except one. In this case suppose one strain always produces round seeds and the other always produces wrinkled seeds. We have already seen that the round form is dominant, so we'll designate that form as R and the recessive wrinkled form as r. Since both plants are true-breeding, we can assume that in each plant the genetic components (from their own parents) are identical for their respective traits. That is, in one plant the components are RR and in the other they are rr. Such plants are said to be *homozygous* for that particular characteristic, meaning that all the gametes they produce will be the same.

Figure 6.3 shows what happens when these two plants are crossed. Remember that they are identical in all characteristics except one, so only the differing characteristic is diagrammed. Since half the gene component in the F_1 generation comes from each parent, the genetic composition, or *genotype*, of any F_1 plant would have to be Rr. Hence these plants would be *heterozygous* for the characteristic in question. Now, when these plants produce gametes (eggs and pollen), each gamete will carry either an R component or an r component, and the combination which occurs when two gametes come together at fertilization will determine the genetic characteristic of the resulting F_2 plant.

The results we should expect in the F_2 generation are diagrammed in a form called a *Punnett square*. As you can see in Figure 6.3, the physical characteristics, or *phenotypes*, of the F_2 seeds occur in a 3:1 ratio. However, a moment (or several moments) of reflection will show that the F_3 ratio Mendel obtained in the third year of his experiments could have been gotten only if one-fourth of the F_2 plants were homozygous dominant for the characteristic, one-half were heterozygous, and one-fourth were homozygous recessive. So the genotype of this F_2 generation is described as 1:2:1.

These results led Mendel to conclude that every heritable

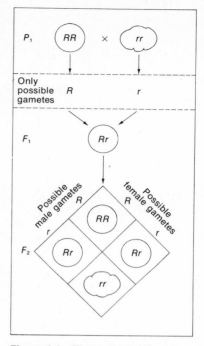

Figure 6.3 The expected F_2 ratio when plants differing in only one characteristic (seed shape) are crossed. The F_1 generation has been inbred.

characteristic is determined by two components, one from each parent. (Before you start feeling smug about the fact that you knew this, remember that it was Mendel who told you.) Thus if two homozygous parents were represented by *RR* and *rr* (as for seed form), their offspring would have to be heterozygous *Rr*, since one component would have come from each parent. And when this *Rr* individual produced gametes, half would contain an *R* component and the other half would contain the *r* component. A cross between two *Rr* individuals could then be expected to yield offspring with the genotypes *RR*, *rR*, *Rr*, and *rr*, or a genotype ratio of 1:2:1. What would be the genotypes and the phenotypes of the offspring if an *Rr* crossed instead with a homozygous recessive?

Mendel's experiments with monohybrid crosses enabled him to show that the genetic element for each characteristic, now called a *gene*, consists of two components (or *alleles*), one contributed by each parent. However, only one of these two components may be expressed in physical form. Hence an individual carrying both components may show no sign of the other parental characteristic at all. The principles of dominance and segregation implied by these results ruled out (in Mendel's mind, at least) the earlier notion that heritable characteristics were always blended in the offspring.

THE PRINCIPLE OF INDEPENDENT ASSORTMENT

In his next set of experiments Mendel crossed *dihybrids*, true-breeding plants which differ in two characteristics. In one such experiment plants with *round* (*R*) *yellow* (*Y*) seeds were crossed with a strain that had *wrinkled* (*r*) *green* (*y*) seeds. We saw in Table 6.1 that round and yellow are both dominant, so we shouldn't be astounded to learn that all the F_1 plants had round yellow seeds. Since one P_1 parent was *RR YY* and the other was *rr yy*, all the F_1 plants would have to be *Rr Yy*, or heterozygous for both characteristics. When these plants were then allowed to self-pollinate, we would expect a random and independent assortment of the various characteristics to produce the F_2 generation shown in Figure 6.4.

It was apparent to Mendel from the results of this dihybrid cross that the segregation of genes for one characteristic was not affected by the segregation of genes for the other characteristic. Thus he was able to deduce that genetic combinations follow the principle of independent assortment. Now, it should be pointed out here that if you were to run a single experiment to test Mendel's results, you might end up with all wrinkled green seeds in the F_2 generation—just as when you toss a coin ten times, it *might* come up tails each time. You could expect the results shown in Figure 6.4 only if you ran enough tests for the law of averages to apply.

Figure 6.5 (top) Four-o'clocks of red and white may combine to produce pink flowers. (Courtesy: Stecher-Traung Schmidt Corp., George Fox)

Figure 6.4 (left) A dihybrid cross showing independent assortment. Only one possible recombination is shown. Can you determine the rest? There are nine possible genetic combinations. See page 129.

Before we leave the subject of Mendelian genetics, you should know that there are the inevitable exceptions. Not all characteristics are expressed in dominant-recessive form. In some cases a characteristic may appear in the offspring in a form intermediate between the characteristics of each parent. For example, four-o'clocks may have either red or white flowers. A cross between the two produces pink-flowered plants. A situation in which one characteristic is not dominant over the other is called *intermediate inheritance*. As before, one component, or allele, of the gene is contributed by each parent, but in this case neither allele gives way to the other. Instead they both exert an influence on the physical characteristic determined by that gene. This means, of course, that the pink-flowered plant is heterozygous for flower color. Could a cross between two such plants produce red or white flowered plants again? What kinds of gametes would these plants produce?

There are also other ways in which intermediate effects arise,

Figure 6.6 Sizes of woodpecker bills in millimeters.

Percentage of each size

Woodpecker bill sizes, millimeters

and such variations within species may have a number of parameters. For example, people do not come in only two sizes, large or small. They come in a wide range of heights, weights, and shapes, just as they differ widely in bone size, nose shape, and any number of other physical characteristics. Such variation can be accounted for if we assume that certain traits are coded for by more than one gene.

Suppose some characteristic such as the length of bill in a woodpecker were coded for by four separate genes, with "longness" indicated by a capital letter and "shortness" indicated by a lower-case letter. Then a bird with the genotype *AA BB CC DD* would be expected to have the longest bill, provided environmental factors such as diet were conducive. A bird with the genotype *Aa BB Cc DD* would have a somewhat shorter bill, and one with the genotype *Aa bb cC dd* would have a still shorter bill. Now, since there is a bill size which is "best" for a woodpecker, we would expect to find more birds with bills in the intermediate range than at either size extreme. Thus if the bill length of a population of woodpeckers were plotted on a graph, we might expect something approaching a "normal" curve (Figure 6.6).

CLASSICAL GENETICS

It is unfortunate that those biologists who did read Mendel's reports were skeptical of his findings. It is also unfortunate that no one apparently saw enough significance in them to bother repeating his laborious experiments. However, six years after Mendel had died in obscurity, his work was rediscovered by three separate researchers who had independently arrived at essentially the same conclusions. At this point the field of genetics burst into an era of dizzying expansion. Although Mendel didn't live to see this happen, his experiments provided much of the basis for what is now called post-Mendelian, or classical genetics.

Thus far it had been assumed that the chromosomes of any organism were identical, and that meiosis was simply a way of dividing them up so that the gametes would receive the same number. It was also believed that during meiosis the maternal and paternal chromosomes separated and each set traveled together at anaphase, so that any gamete would contain chromosomes from one parent or the other, but not both. In the early 1900s, however, it was found that chromosomes are not identical, but come in a variety of sizes and shapes, and that they also come in pairs—two of each type. Chromosomes which comprise the pair are called *homologs*. In addition, it was found that meiosis divides chromosomes not only quantitatively but also qualitatively, so that each gamete receives both the same *number* and *kind* of chromosomes.

A young graduate student at Columbia University, Walter S. Sutton, began a study to try to relate Mendel's findings to what was known about meiosis. His results were surprising. For example, he found that when chromosome pairs lined up at meiosis, there was no way to determine which way a paternal or maternal chromosome would go at anaphase. Some maternal chromosomes moved to one pole and some to the other, but there was no way to predict their direction. Now, at this time it was not known that chromosomes were the carriers of genes. The independent assortment of chromosomes, however, confirmed a hunch Sutton had that hereditary units are associated with chromosomes. Also, since there were far more hereditary characteristics than there were chromosomes, he reasoned that each chromosome must be responsible for many genetic traits.

In 1910, after Sutton's report, an innocuous little insect entered the world of genetics. Thomas Hunt Morgan, also at Columbia, began a program of breeding experiments with the fruit fly (or vinegar fly), *Drosophila melanogaster* (Figure 6.7). These little flies are ideal subjects for genetic studies. They are easy to maintain in the laboratory; they mature in twelve days and reproduce prolifically; and they have a variety of features which can be readily identified. They also have giant chromosomes in their salivary glands, on which specific areas, or loci, are readily distinguishable. In addition, they have only four pairs of chromosomes. They have little economic importance, and they don't seem to bother man much where the species come in contact. In fact, it has been said that God must have invented *Drosophila melanogaster* just for Thomas Hunt Morgan.

SEX DETERMINATION

One of the first things Morgan did was to attempt to cause genetic changes, or mutations, in his flies. He subjected them to cold, heat, x rays, chemicals, and radioactive matter. At first he found nothing, but then he noticed among a group of normal red-eyed flies that there was a single male with strange *white* eyes. Morgan was convinced that this was a mutation (a subject we will take up later). He carefully nurtured his little white-eyed specimen and crossed it with several of its red-eyed sisters. In these crosses he found that the F_1 generation were all red-eyed, just as you would expect if white eyes were recessive. Simple enough so far—but then the story became a bit more complicated.

When the F_1 flies were allowed to interbreed, the result was 3,470 red-eyed and 782 white-eyed flies. This is not the simple 3:1

Figure 6.7 The fruit fly, *Drosophila melanogaster*. This tiny insect is a vital part of research in the genetics field. Compare its wing shape with the mutant type in Figure 6.14.

Answer to Figure 6.4.
You should get:

GENOTYPE	PHENOTYPE
1 RRYY	9 Round Yellow
2 RRYy	3 Round green
1 RRyy	3 wrinkled Yellow
2 RrYY	1 wrinkled green
4 RrYy	
2 Rryy	
1 rrYY	
2 rrYy	
1 rryy	

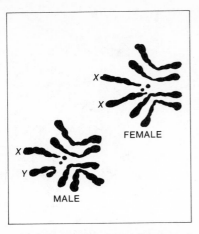

Figure 6.8 The chromosomes of the male and female drosophila. Note the difference in the shape of the sex chromosomes.

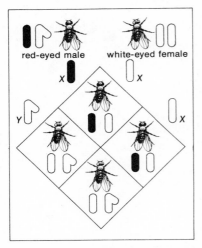

Figure 6.9 This breeding experiment shows that eye color is sex-linked.

ratio you might have expected according to Mendelian principles. Also, there was another troubling factor. All the 782 white-eyed flies were males. The red-eyed flies, however, were of both sexes—2,459 females and 1,011 males. At this point you may have come up with the obvious answer: only males are white-eyed. Alas, when the original white-eyed male was crossed with its F_1 red-eyed daughters, the result was 129 red-eyed females, 132 red-eyed males, 86 white-eyed males—and 88 white-eyed females! If you have a penchant for puzzles, can spare a few minutes, and think you are smart, try to figure out how this could have happened. In case you need a clue, keep in mind that sex is determined by chromosomes. Meanwhile, the rest of us will trudge along.

Morgan surmised that, although chromosomes may segregate independently, the genetic determiners (or genes) may not. He then surmised that the sex-determining factor and the eye-color factor are somehow linked together. Hence, as one goes at anaphase, so must go the other.

By this time Morgan knew something about the chromosomes which determine the gender of an offspring. He knew that of the four pairs of chromosomes in *Drosophila*, only one pair was responsible for gender. He even knew which ones were the *sex chromosomes* and which were the *autosomes*, the chromosomes responsible for other characteristics. In females the sex chromosomes are a rod-shaped pair of chromosomes, for some reason called *X chromosomes*. In males the sex chromosomes consist of only one *X* chromosome and a J-shaped chromosome called a *Y chromosome* (Figure 6.8). The *X* chromosomes are usually larger and contain far more genes than the *Y* chromosome. This fact will be critical in understanding certain of the effects of sex linkage in heredity.

In the production of sperm in males the *X* and *Y* chromosomes line up together at meiosis, so that half the gametes will contain an *X* and half will contain a *Y*. Since females have only *X* chromosomes, the eggs will always contain an *X*. If a *Y*-bearing sperm reaches the egg first and fertilizes it, the offspring will be *XY*, and thus male. If an *X*-bearing sperm fertilizes the egg, the offspring will be *XX*, or female. This is the case in many animals, including mammals, but in other organisms the situation may be different. For example, in birds the sex chromosomes are identical in males (*ZZ*) and different in females (*ZW*).

Now, regarding the question of eye color in *Drosophila*, Morgan showed that the genetic allele for eye color is located only on the *X* chromosome, and white eye color is recessive. Therefore females will have white eyes only if both eye-color alleles are for white eyes. In males, however, when a white-eye factor turns up

MALE					FEMALE				
1	2	3	4	5	1	2	3	4	5
6	7	8	9	10	6	7	8	9	10
11	12	13	14	15	11	12	13	14	15
16	17	18	19	20	16	17	18	19	20
21	22	X Y			21	22	X X		

Figure 6.11 (top) Tell-tale structures in the cells of women.

Figure 6.10 (left) The human chromosomes. Note that pairs are identifiable. Also note the size difference in the X and Y chromosomes.

on the *X* chromosome, there is no dominant counterpart on the *Y* chromosome, so the white eye color will be expressed (Figure 6.9).

SEX LINKAGE IN HUMANS

There have been several interesting applications of the findings regarding sex inheritance in the past few years. For example, the cells in women have tiny specks in their nuclei called *Barr bodies*. Also, white blood cells in women show a small "drumstick" attached to the nucleus (Figure 6.11). Cells from normal men lack both the Barr body and the drumstick. You may remember a stir at a recent meeting of the Olympic games when a European woman who showed remarkable strength in the field events was found to be genetically male, to lack the Barr bodies and drumsticks in her cell nuclei. For certain athletic competitions, sex tests are now mandatory.

Color-blindness in humans is caused by a recessive allele on the *X* chromosome. Thus a woman who is heterozygous for this condition will show no symptoms of color-blindness. Among her children, however, she can expect one son in two to be color-blind and one daughter in two to be carriers of color-blindness like herself. In her sons the recessive gene on the *X* chromosome is not overridden because the tiny *Y* chromosome bears no allele for color discrimination. Figure 6.12 shows the expected occurrences of color-blindness in this woman's grandchildren, the F_2 generation.

There are a number of other types of phenomena associated

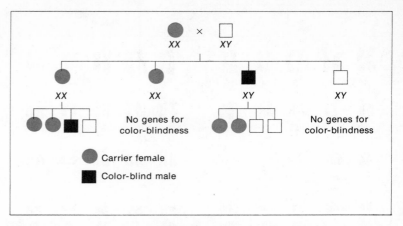

Figure 6.12 How colorblindness is inherited. The original mating is between a carrier female and a normal male. Spouses (not shown) are presumed to be free of the color-blindness gene.

with sex-linked genes which have great impact on people's lives. For example, muscular dystrophy is caused by a recessive gene on an X chromosome. The effects of this abnormality are so pronounced, however, that males carrying the gene usually die in their teens before reproducing. As a result, females rarely have the disease, since the probability of receiving the allele from both parents is very low. However, they can carry a single gene for the condition, so that when they reproduce they may pass it along to their offspring. Another such disorder is hemophilia, which is also a recessive characteristic carried on an X chromosome (Box 6.1).

In other instances, there may be an imbalance of X and Y chromosomes. For example, certain females may have cells with no Barr bodies or drumsticks. In such cases there is only one X chromosome instead of two. The physical characteristics that accompany this condition are female genitalia, but underdeveloped breasts and tiny ovaries—and for some reason, short stature. The presence of an extra X chromosome in males (XXY) also results in certain anomalies. Such males usually have sparse body hair, some breast development, and cells which show both Barr bodies and drumsticks.

There has been some discussion of the observation of an XYY condition in certain criminals. Of course, there is no information on its frequency in the general population of noncriminals. Nevertheless, several legal appeals have been entered on behalf of men convicted of murder and other violent crimes based on the argument that these men had been "genetically driven" to their crimes and could therefore not be judged accountable for them.

GENE LINKAGE AND CROSSOVER
Recall from our discussion of Mendel's work that the assortment of

Figure 6.13 Down's Syndrome (top) and Turner's Syndrome (bottom). A syndrome is a group of symptoms. Down's Syndrome is also called Mongolism (someone thought the peculiar eyefold looked oriental). The disorder includes mental deficiency, a short, stocky body with a thick neck and, often, abnormalities of the heart. Down's Syndrome may occur as a result of an extra chromosome (number 21) and most frequently occurs in children born to older women. The syndrome is detectable by analysis of the amniotic fluid when the embryo is 16 weeks old.

Turner's Syndrome is produced by an error in assortment at meiosis when a female ends up with only one X chromosome. The result is short stature, undeveloped ovaries, widely spaced nipples, and a webbed neck. (Courtesy: Science Software Systems, Inc., Los Angeles)

Box 6.1 The Disease of Royalty Because it was the practice of ruling monarchs to consolidate their empires through marriage alliances, hemophilia was transmitted throughout the royal families of Europe. Hemophilia is a sex-linked recessive condition in which the blood does not clot properly, so that any small injury can result in severe bleeding, and if the bleeding cannot be stopped, in death. Hence it has sometimes been called the "bleeder's disease."

Hemophilia has been traced back as far as Queen Victoria, who was born in 1819. One of her sons, Leopold, Duke of Albany, died of the disease at the age of thirty-one. Apparently at least two of Victoria's daughters were carriers, since several of their descendents were hemophilic. Hemophilia played an important historical role in Russia during the reign of Nikolas II, the last Czar. The Czarevich, Alexis, was hemophilic, and his mother, the Czarina, was convinced that the only one who could save her son's life was the monk Rasputin—known as the "mad monk." Through this hold over the reigning family, Rasputin became the real power behind the disintegrating throne.

chromosomes is independent. Thus the assortment of genes on one chromosome of a pair would have to be independent of those on the homologous chromosome. If two genes are on the *same* chromosome, however, it is likely that they will be passed together to the gamete. The occurrence of two or more hereditary units on the same chromosome is called *gene linkage*.

In *Drosophila*, for example, the recessive alleles for a black body (*b*) and a curved wing (*c*) are carried on the same chromosome. The dominant alleles are for a gray body (*B*) and a normal wing (*C*). So if we diagram a cross between a homozygous normal fly, *BB CC*, and a homozygous recessive fly, *bb cc*, we will get the results shown in Figure 6.15. The only gametes the gray, normal-winged fly can produce are *BC* and the only gametes the black, curved-wing fly can produce are *bc*. Hence their F_1 offspring would all be heterozygous for both characteristics—*Bb Cc*—and would, of course, show gray bodies and normal wings.

Now, if this *Bb Cc* fly were crossed with a homozygous recessive, *bb cc*, you would expect half of their offspring to be gray with normal wings and half to be black with curved wings (this is worked out for you in the Punnet square). In all likelihood, however, this cross would produce about 37 percent gray, normal-winged flies; about 37 percent black, curved-wing flies; 13 percent *black*, normal-winged flies; and 13 percent gray, *curved*-wing flies. We would have to have two conditions to account for this: (1) the genes for both body color and wing shape would have to be on the same chromosome, with their alleles on homologous chromosomes, and (2) the chromosomes would have to be able somehow to exchange genes.

You can probably believe condition 1 easily enough on the basis of our previous discussions, so let's consider how condition 2 might exist. Recall that in meiosis the homologous chromosomes line up during metaphase I, and as anaphase begins, the homologs of each pair move toward opposite poles. While they are lined up side by side in prophase, however, the homologs may stick together at various places, and then exchange parts of their chromatids. The opposite chromatids of the homolog cross each other and then break off at their point of contact. Each broken part then fuses with the opposite portion of its homolog (Figure 6.16).

You might wonder how the chromatids manage to break at *exactly* the same place on each one, so that one chromosome does not end up with more genes than the other chromosome. The geneticists investigating the phenomenon of crossover are asking the same question—but with no answer yet. We do know, however, the basic mechanism by which crossover occurs, and also that some crossover takes place at virtually every meiosis.

Figure 6.14 Curved-wing drosophila. The abnormal wing shape is the result of a genetic mutation.

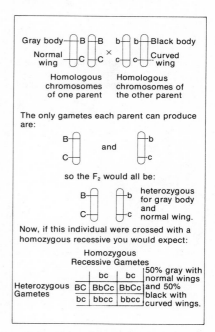

Figure 6.15 Because some genes are linked on the same chromosome, you would expect these results from the cross shown above. Actually the linkage may be disrupted by crossovers.

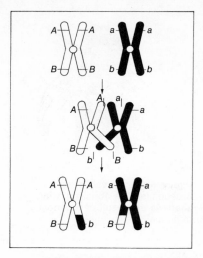

Figure 6.16 The rearranging effects of crossovers between homologous chromosomes. Thus, genes can be "reshuffled" and linkages can be disrupted.

CHROMOSOME MAPPING

If genes are arranged along the chromosomes and specific alleles are precisely exchanged through crossovers, then the genes for certain characteristics must lie at specific points along each chromosome. But how do we know which genes lie where? Finding the answers involves "mapping" the chromosome—a technique which is based on information from observations of crossovers.

Thomas Morgan and A. H. Sturtevant, one of the brilliant young graduate students attracted to Morgan's laboratory in those days, hypothesized that if genes were arranged linearly along the chromosome, then those lying closer together would undergo crossovers less often than those lying farther apart. The idea assumed randomness, or at least a large degree of variability, in the point at which chromosomes would break. Genes lying closer together would thus have a greater *probability* of being passed along as a unit. Stated another way, the percentage of crossover is proportional to the distance between two genes on a chromosome—so *percent crossover* is the number of crossovers between two genes per 100 prophase opportunities.

As an example, suppose two characteristics, which we can call *A* and *B*, show 26 percent crossover. We can assign 26 crossover units to the distance between the two genes. Then if some characteristic *C* turns out in breeding experiments to have 9 percent crossover with *B* and 17 percent crossover with *A*, it would be located between *A* and *B* at a point 9 units from *B* and 17 units from *A*. After the information from many such experiments has been

compiled, a chromosome map can be constructed which indicates the position along the chromosome of the genes which code for certain characteristics. Figure 6.17 shows chromosome maps developed by this technique for chromosomes in the salivary gland of *Drosphila melanogaster*.

MUTATIONS

Mutations, or changes in the hereditary material, may be caused by x rays, chemicals, ultraviolet radiation and a number of other agents. In fact, as you are sitting peacefully reading this fascinating account, your body is being bombarded with tiny subatomic particles from outer space—radiation. If one of these particles should strike you right in the gonad, and alter a nucleotide in a developing gamete, the structure of that gamete could be changed. Recall that a small change in a nucleotide sequence could result in an entirely new codon. Hence if that gamete happens to enter into the reproductive process, whatever change has taken place will be passed along to the next generation, unless the change is lethal or otherwise prevents reproduction, or unless it is reversed through a second mutation, an unlikely event.

GENIC MUTATIONS

There are different kinds of mutations. One is associated with alterations in the genes and is called *genic mutation*. Genic mutations occur during the duplication of DNA. They may involve the omission of nucleotide units, the repetition of nucleotide units, or mistakes in bonding between these units. Or they may involve the substitution of one nucleotide for another—either the substitution of one purine or pyrimidine for another or the substitution of a purine for a pyrimidine or a pyrimidine for a purine.

Not all such changes result in a change in the characteristic. For example, the sequence GAC codes for the insertion of aspartic acid into a protein, and a change from GAC to GAU causes no alteration in the code. Also, other mutations sometimes occur which compensate for the effect of the first change and thus conceal it.

It should be apparent that the addition or deletion of a single nucleotide could change the entire amino acid product, since the codons are read in groups of three nucleotides. For example, ACG-GCC-GGA codes for threonine, alanine, and glycine. If the first G should be repeated, the sequence would instead read ACG-GGC-CGG-A . . . and the amino acids which would be inserted into the protein would be threonine, glycine, and arginine. Thus this one added nucleotide could change every amino acid from that point on in the sequence.

CHROMOSOME MUTATIONS

A second class of mutations involves some physical change in the structure of chromosomes. These, logically enough, are called *chromosomal mutations*. Such changes may occur when the chromosomes are moving about or undergoing changes—for example, during meiosis. At such times a part of a chromosome may be broken off and lost (a *deletion*), or it may break off and reattach

Figure 6.17 Map of the chromosomes of the salivary gland of the *Drosophila melanogaster*. (Adapted from Biological Science by William T. Keeton. Illustrated by Paula Desanto Bensadoun. © 1967 by W. W. Norton & Co., Inc.)

Inheritance 137

Figure 6.18 Diagrams of chromosome mutations.

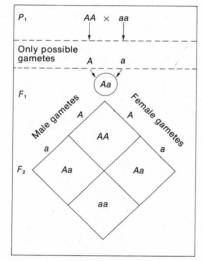

Figure 6.19 From looking at the F$_2$ generation, it would seem that the recessive gene was disappearing from the population. However, this is not the case. Compare the number of dominant and recessive genes in each generation.

at its other end (an *inversion*), or a segment may be duplicated, resulting in a doubling in the number of certain genes (a *duplication*). The most serious of these mishaps is a deletion, since the absence of a large number of genes is likely to result in death.

Of course, not all mutations are bad. In fact, mutations are the source of much of the variation we see between individuals in a population. Among these variations, certain characteristics will enable an individual to fit into the environment better than other individuals. Thus these better-equipped individuals will be more likely to survive and reproduce as the environment exerts its effect on the population. This phenomenon of natural selection is one we will discuss in more detail later.

In some cases mutations may have little effect on the phenotype of the individual. For example, a slight difference in hair color as the result of a mutation might scarcely be discernible. Some mutations may also be highly beneficial in some environments, but not in others. A change which produces blue eyes makes them more sensitive to light, but this sensitivity ceases to be an advantage in environments subject to glaring sun. In general, however, random changes are not likely to be helpful to a population which has become finely attuned to its habitat through countless generations. Suppose you raised the hood on a sportscar, and knowing nothing about motors, you made an adjustment. It may be that the change you made was just the one necessary to make the car run even better, but the odds are a lot greater that you screwed up something. The odds with random mutations are about the same.

POPULATION GENETICS

Brown eyes are dominant over blue eyes in humans. So why doesn't everyone have brown eyes by now?

The answer is not a simple one. In fact, at the beginning of the century many geneticists raised this point as an argument against the validity of Mendel's principles. A fuller understanding of the implications of Mendel's findings, however, reveals that the principles he described can, in fact, readily account for such seeming paradoxes.

To approach such questions we must begin by considering genes in populations. We have already been using the term *population*, and you may have taken it to mean a group of individuals. You were not wrong, but at this point we can give it a more precise meaning. In biology, a *population* designates a group of interbreeding or potentially interbreeding individuals. With this definition in mind, let's now consider the ratios of different alleles for a specific characteristic, such as eye color, in a population.

First, imagine a population of only two individuals in which one sex is homozygous for dominant trait A and the other sex is homozygous for recessive trait a, as shown in Figure 6.19. Now, we know that all their F_1 offspring will be heterozygous for that characteristic. The Punnett square then shows us that in the F_2 generation three out of four individuals will show the dominant trait and only one will show the recessive trait. So it might appear that we are on the way to eliminating the recessive gene from the population. However, if you now plot the F_3, F_4, F_5, and so on, generations, you will find that the proportion of dominant and recessive genes in the population has not changed at all. In fact, look again at the F_1 and F_2 generations. There is a 1:1 ratio of the two alleles even at these stages.

A population of two is unrealistic, but it serves to point out how the phenomenon occurs in larger populations. We can show by means of Punnett squares that the frequency of alleles for any characteristic will remain unchanged in a population through any number of generations—unless this frequency is altered by some outside influence.

This interesting fact was pointed out independently by G. H. Hardy, a British mathematician, and W. Weinberg, a German physician. The principle they formulated is called the *Hardy-Weinberg law*. According to this law: *In the absence of forces that change gene ratios in populations, when random mating is permitted the frequencies of each allele (as found in the second generation) will tend to remain constant through the following generations.* We will consider forces which change gene ratios in populations later, but for now let's consider the basis for the law.

First of all, note the ratios of the combinations AA, Aa, and aa in the F_2 generation, which is our first opportunity to see all the possible combinations. We see in the F_2 that one-fourth of the population is AA, one-half is Aa, and one-fourth is aa. So, to determine what fraction of the F_3 generation will be offspring, say, of AA and Aa crosses, we simply multiply ¼ × ½ and get ⅛. In the third generation, then, assuming random mating, we can expect one-eighth of the population to have the genotypes resulting from this cross. What part of this generation will be the offspring of AA and AA crosses? Of Aa and aa crosses?

Now, there are four possible types of crosses which could have produced AA individuals in the F_3 generation. You can work out for yourself what they are or conserve your energy and read the answer on page 140. For any of these four mating possibilities the boxes can be filled in accordingly. Each of these combinations will produce one-eighth of the F_3 generation, so what part of the

Figure 6.20 The English mathematician G. H. Hardy. Hardy and the German physician W. Weinberg, working independently, discovered in 1908 what has come to be called the Hardy-Weinberg principle. The idea is that, in the absence of selective forces, the gene frequency in a population will remain the same indefinitely. (Courtesy: The Clarendon Press, Oxford)

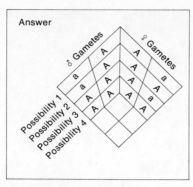

Answer

♂ Gametes

♀ Gametes

Possibility 1
Possibility 2
Possibility 3
Possibility 4

Answer to the question in text, page 139.

population will be heterozygous, or *Aa*? Simple enough: ⅛ + ⅛ + ⅛ + ⅛ = ½. And what was the fraction of *Aa* individuals in the F_2 population? One-half. If you work out the results for all possible combinations for the F_3 generation, or for any combination in any generation past the F_3 generation, you will see that the ratio of genetic components in a population remains stable. This is why we continue to have blue eyes in our population, even though they are a recessive trait.

Now, if this is the case, how can gene ratios in populations change? After all, evolution is nothing more than that—changes in gene ratios. If there were some great advantage in having brown eyes which gave brown-eyed people a better opportunity to survive and reproduce, then the ratio of brown eyes to blue eyes could shift. It could also change if only people with eyes of a certain color were considered sexually attractive, or if mutations were more likely to occur in one direction than another—say, from *a* to *A* rather than from *A* to *a*. If the population were small, shifts in gene frequency could occur rapidly, since in a small group only a few accidental deaths could result in a great change in the overall ratios. Also, gene frequencies could change as a result of the flow between populations through immigration and emigration. Any of these things can occur in human populations, so the Hardy-Weinberg model is to some extent an artificial one. However, it helps to explain the relative constancy we see in populations around us.

Now for those burning with a fierce love of mathematics, let us consider the mathematical statement of the Hardy-Weinberg law. The law is stated as

$$(p + q)^2 = 1$$

which can be written in expanded form as

$$(p + q) \times (p + q) = 1$$

or

$$p^2 + 2pq + q^2 = 1$$

where

p = frequency of the dominant allele (*A* in our example)
q = frequency of the recessive allele (*a* in our example)
p^2 = frequency of the homozygous dominant genotype (*AA*)
$2pq$ = frequency of the heterozygous genotype (*Aa*)
q^2 = frequency of the homozygous recessive genotype (*aa*)
1 = a population in genetic equilibrium (one in which the above conditions are met)

Figure 6.21 An albino with his dark-skinned brothers at a feast in the Coral Islands about 100 miles east of New Guinea. Albinism is a recessive trait and therefore is more likely to appear in small isolated groups. Do you see why this should be the case? (Courtesy: Wide World Photos)

The Hardy-Weinberg law has very specific and important implications for us. For example, if we know the prevalence of the trait of albinism, the absence of pigment, in the population, we can predict, within limits, the probability that a couple will have an albino baby. Here's how this would work. Normal skin and eye pigment in humans is dominant over the albino condition a. The genotype aa occurs in about 1 in every 20,000 people. According to the Hardy-Weinberg equation, this frequency would be given by q^2, so the frequency of genotype aa is

$$q^2 = \frac{1}{20,000}$$

and the frequency of a single allele for this trait is thus

$$q = \sqrt{\frac{1}{20,000}} = \frac{1}{141}$$

The frequency of the dominant allele, A, would then be

$$p = 1 - q$$

$$\text{or} \quad p = 1 - \frac{1}{141} = \frac{140}{141}$$

The heterozygous condition, *Aa*, would therefore occur in the population with a frequency of

$$2pq = 2 \times \frac{140}{141} \times \frac{1}{140}$$

$$= \frac{1}{70} \quad \text{or 1.4 percent}$$

Since 1.4 percent of 20,000 is 280, this means that only 280 people in every 20,000 will be carrying a recessive allele for albinism. Hence, in the absence of a family history of this characteristic in either parent, the chance that any couple will have an albino child is very slim.

In contrast, among American blacks a disease called sickle-cell anemia, produced by the recessive condition *ss*, appears in about one person in 400. Using the Hardy-Weinberg equation, we can see that the heterozygous carrier condition (*Ss*) can be expected in almost one person in every ten in this population. With this kind of information governments and health organizations can establish priorities in dealing with health problems. Whereas the bureaucratic or political mind might view one in 400 as too low a ratio to warrant urgent attention, one in ten is a ratio which certainly cannot be ignored. However, decisions about what steps to take in such a situation are exceedingly difficult. For example, should all the available funds be spent on efforts to overcome only the symptoms of the disease, or should extremely rigorous controls be exercised over marriage—the only way we now know to reduce the occurrence of the disease? Or is there some possibility of controlling the problem at a level somewhere between these extremes? In any case, the number of individuals whom we now know must be carriers for this condition serve as an impetus for some sort of immediate action.

GENETIC ENGINEERING

In a rare disease called phenylketonuria (PKU), the body lacks an enzyme needed to break down a substance called phenylalanine, which is a product of protein metabolism. The phenylalanine thus accumulates in the blood and causes mental retardation. We know that this condition arises from a single gene alteration which inhibits the production of a single enzyme—the one which breaks down phenylalanine. We also know which gene is responsible for the enzyme and the change in this gene which produces phenylketonuria. It is not beyond the realm of possibility, then, that a

Box 6.2 Sickle-cell Anemia   Sickle-cell anemia is a recessive condition characterized by fragile red blood cells which collapse into a sickle shape. These cells are unable to carry oxygen and may clog the blood vessels, so that persons who have the disease suffer painful symptoms and usually die at an early age. The cause of sickle-cell anemia has been traced to a mutation in a single gene, an alteration of one nucleotide in a triplet. The result is that in one of the four complex polypeptides which comprise hemoglobin, a single amino acid is changed—and so is the capability of the entire molecule.

In persons homozygous for this condition two of the four polypeptides are affected. These people do not normally survive long. In heterozygous persons only one polypeptide is affected, and the red blood cells are able to function under normal oxygen requirements. Under conditions in which the body demands additional oxygen, as during strenuous exercise or at high altitudes, these blood cells are likely to collapse. Nevertheless, the heterozygous condition also carries a compensatory advantage in certain environments. Persons who are heterozygous for sickle-cell anemia are more resistant to malaria than those with normal hemoglobin. Thus sickle-cell anemia is maintained in malarial areas from Africa to India.

Normal
cell

Sickle
cell

(Courtesy: National Institute of Health)

Box 6.3 The Attempt to Recreate Leonardo da Vinci Leonardo da Vinci was the illegitimate son of Piero, a notary from the town of Vinci, and a peasant girl named Caterina. The third wife of Leonardo's father later bore a son named Bartolommeo who idolized Leonardo although he was forty-five years younger. After the death of the legendary Leonardo, Bartolommeo attempted an amazing experiment. He studied every detail of his father's relationship with Caterina. Then Bartolommeo, himself a notary by family tradition, returned to Vinci and found another peasant wench who seemed similar to Caterina, according to all Bartolommeo knew. He married her and she bore him a son whom they called Piero. Strangely, the child actually looked like Leonardo and was brought up with all the encourgement to follow in the great man's footsteps. Stranger yet, the boy became an accomplished artist and was becoming a talented sculptor. However, he died young, thus ending the experiment.

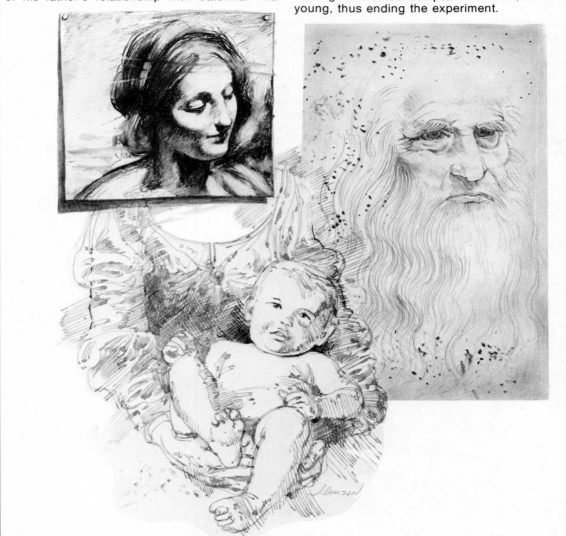

means will be found to "correct" the aberrant gene by adjusting the nucleotide sequence in the chromosomes. Such programmed alterations in chromosomes fall into a branch of research called *genetic engineering*, a field whose possibilities, though remote, are vast.

Cloning is also being considered as a means of genetic engineering. A clone is a group of cells which have been produced by mitotic divisions of a single parent and are therefore genetically identical. A group of protozoa which are descended from divisions and subsequent divisions of a single cell are considered a clone. In cloning experiments single carrot cells excised from adult carrots have been grown in laboratories and, through divisions, have produced normal adult carrots.

These results have naturally led to the speculation that cells cloned from Uncle Albert might mean that Uncle Albert would always be with us (that might be good news or bad news). As a person grew older, a cell might be excised and grown in a suitable medium, eventually producing another individual with the same kinds of cells as the old one. Would this in fact be the same person? The cloned individual would be genetically identical with the old one, but what about all the experiences which made Uncle Albert the character he was? And would the new Uncle Albert remember he owed you money?

The possibilities of cloning assume more practical dimensions if we pass over the issues of personality and immortality and consider the idea that certain qualities other than physical characteristics may be the result of purely genetic combinations. The first studies of genius, in fact, were prompted by the observation that qualities of genius seemed to appear in certain families with a much higher frequency than could be accounted for by pure chance. What if we could increase the frequency of such qualities in our population by reproducing certain genetic combinations through cloning? Suppose, for example, we had been able to clone cells from Leonardo da Vinci. (If you don't immediately see the implications here, you should learn more about the man.)

The concept of genetic engineering, of course, has many ramifications. How do we determine which characteristics will be of benefit to future generations? And if certain genetic characteristics should be reduced or screened out of the population, who should decide whether your particular genetic components should or should not be reproduced? Do you see any other problems associated with genetic engineering?

7
Reproduction

The story of reproduction is an interesting one—at least, it should be, judging from the attention we give to its mechanisms. In this chapter we will consider the process of reproduction in a variety of organisms. However, we will focus particularly on reproduction in our own species, partly because it's more interesting to most of us, but also because it is important for us to have some knowledge of the subject if we hope to limit our population.

When we think of reproduction we tend, as humans, to think of sexual reproduction, the fusion of male and female gametes. And, as we look around us, it seems that most other living things are sexual reproducers—the trees in our forests, the fish in our rivers; the dogs in our yards, the cockroaches on our floors. However, not all organisms reproduce sexually. It stands to reason that reproduction that involves two individuals must be a more complicated process than one that is done by an individual acting alone. It may also be more interesting. But let us begin with some of the principles of *asexual reproduction.*

ASEXUAL REPRODUCTION

VEGETATIVE FISSION

Have you ever wondered how a flatworm reproduces? We'll tell you anyway. The flatworm can reproduce either sexually or by a process called *vegetative fission* (Figure 7.2). The flatworm simply pinches itself in two about midway along its length, and cells then move into the severed area and regenerate the missing parts of each half. The result is two new flatworms.

It is apparent that the cells which rebuild the missing structures in such cases must undergo some sort of fundamental reorganization. After all, these cells had already specialized and were functioning elsewhere in a different role. In some cases, such as regeneration in flatworms or wound healing in higher animals, cell development is reversible; cells have the ability to despecialize, or revert to a more primitive state, and then initiate a new development pathway.

Figure 7.1 Pronghorn antelope. This buck and doe will stay together for a brief mating period, after which they will separate. In some species the bond may be shorter, just long enough for copulation. In other species the bond may last a lifetime. (Courtesy: Leonard Lee Rue III from National Audubon Society)

Figure 7.2 Flatworm reproducing through vegetative fission.

Figure 7.3 Asexual budding in hydra is only one way this animal can reproduce. The "trunks" continually produce swellings which develop into miniature versions of the parent. Finally they break away and float off on their own.

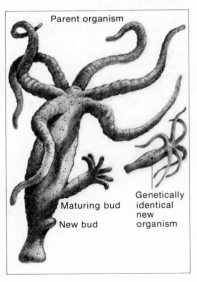

If a flatworm is cut in half, each half will regenerate the missing parts, but when it is cut into smaller pieces, there is a point beyond which a regenerated individual will no longer develop normally. Still smaller pieces are unable to regenerate at all, and the pieces simply die.

Reproduction through regeneration is possible in some species which do not normally reproduce by such means. For example, starfish reproduce sexually, but not long ago vengeful oystermen chopped up starfish which they found in their dredges because the starfish were eating the oysters (an example of man's usual reaction to competing predators). In many cases, however, a new starfish grew from each part the oystermen threw overboard. As a result, the starfish population flourished, and the oyster catches dwindled in response to the increased level of starfish predation.

BUDDING

Another method of asexual reproduction is *budding*, in which a new organism is produced as an outgrowth of the parent organism. The new organism is well developed by the time it splits off from the parent as a new individual. Numerous plants reproduce in a similar way, as do many of the lower animals. One example is the tiny hydra, a freshwater relative of the sea anemone (Figure 7.3).

SPORULATION AND VEGETATIVE FISSION

In *sporulation* the new organism also develops from a bud of the parent organism. However, in this case the new organism does not remain attached to the parent until it is well developed. Instead, it is shed as a rudimentary cell—a *spore*—which then grows into a new individual apart from the parent. Some spores are capable of surviving in a dormant stage for a considerable time until conditions are favorable for them to begin to mature.

Since sporulation requires so small an expenditure of energy by the parent organism, you might imagine that spores would be produced in great numbers. And, in fact, this is usually the case. Spore-producers manufacture prodigious numbers of the tiny particles.

In *vegetative fission* the new individual leaves the "parent" at some developmental state between that in sporulation and budding. The "offspring" then is off on its own at a relatively undifferentiated stage of its development and new cellular organization follows.

ALTERNATION OF GENERATIONS

There are a great many species of animals and plants which have both sexual and asexual phases in their reproductive cycle. Often the phases alternate in what is called an *alternation of generations*. In some plant cycles spores grow into organisms that release, not more spores, but eggs and sperm. When a sperm and an egg unite, the fertilized egg will grow into an adult individual that looks entirely different from the one that had grown from the spores. This organism, when it matures, will then produce spores. The genetic basis for this difference in the adult organisms is illustrated in Figure 7.5. The spores are produced by meiosis, hence they have only half the complement of chromosomes characteristic of the other phase, and grow into individuals of generation A which then release eggs and/or sperm. Of course, both eggs and sperm also have only "half" the number of chromosomes. When these join, generation B is produced which has the full chromosomal complement, and which, when mature, releases spores to initiate generation A again. Gamete-producing generations are called *gametophytic*; spore-producing, *sporophytic*.

Nearly all higher plants, most algae, and some animals go through an alternation of generations. Depending on the species, however, one generation may be much more prominent than the other. For example, mosses have a long gametophyte generation (Figure 7.7), whereas ferns have a much longer sporophyte generation (Figure 7.8). As you may have noticed on your treks into the woods, at certain times ferns have small dark spots on the underside of their fronds. These contain the *sporangia*, the spore-producing structures of the plant.

Figure 7.4 In certain simpler plants the nuclei may divide to form genetically identical spores. When the cell wall ruptures, the spores are released. Then, when a spore reaches a suitable place, it may establish itself and begin formation of a new plant.

Figure 7.5 Diagram of the alternation of generations.

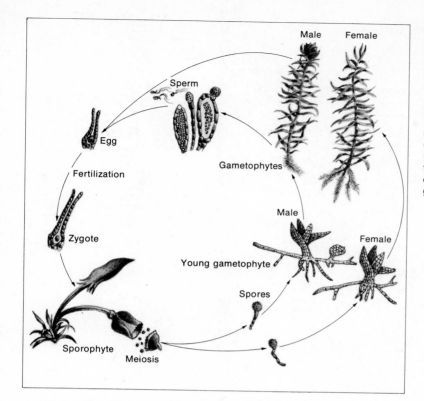

Figure 7.7 The life cycle of a typical moss. The dominant phase is the haploid gametophyte, which can be either male or female and which makes up the familiar velvety moss mat. Gametophytes can reproduce vegetatively or can produce gametes. The fertilized egg becomes the diploid sporophyte, a small fruiting body which grows out of the body of the female gametophyte. Meiosis occurs within the sporophyte. Dispersed spores develop into either male or female gametophytes.

THE DEVELOPMENT OF ALTERNATING GENERATIONS

In species which show an alternation of generations it has been theorized that the sexual or gametophytic phase has been developed in response to drastic cyclical periods in the life cycle. ("Drastic" might be defined as harsh, variable, or unstable.) Under such conditions the variability produced in sexual reproduction might be highly advantageous. The asexual phase, then, might be expected under more optimal, stable, or sheltered conditions. It has been demonstrated, in many species, that sexual reproduction does, indeed, increase in periods of stress. In earlier days steel girders were placed around surviving trees in burned-over areas. The idea was that when the tree grew and came to exert pressure against the girder, the stress would result in greater seed production.

Many species of animals, after experiencing conditions which reduced their numbers, undergo a rapid surge of reproductive activity (which might conceivably be a response to either stress or low numbers). In humans, the baby booms which usually follow great wars have been attributed to stress (but the possibility of high randiness in returning servicemen cannot be discounted). It

Figure 7.6 Ferns (left) may be found in abundance on the sodden floors of rain forests. They are the sporophytic generation of their species. (Courtesy: U.S. Department of the Interior, National Park Service Photo)

Figure 7.8 The life cycle of a typical fern is essentially similar to that of the moss, but it is the sporophyte that is dominant. The sporophyte at first grows out of the gametophyte, but then develops roots and leaves and becomes independent. The haploid gametophyte remains a relatively insignificant, fragile and ephemeral structure. A single gametophyte usually produces both eggs and sperm.

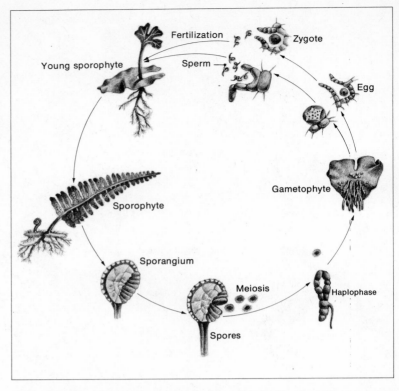

is argued that no sound case can be made at this time for a general increase in sexual reproduction as a response to stress. However, this is not to say that the development of the sexual phase in the alternation of generations may not have been a response to such conditions. The idea is interesting, anyway.

THE ADVANTAGES OF SEXUAL REPRODUCTION

Before going into a discussion of mechanisms of sexual reproduction we might ask, since asexual reproduction is so simple and efficient, why do some animals rely on sexual methods alone? After all, sexually-reproducing forms must normally produce enormous numbers of gametes among which only a few will ever enter into fertilization. So there is great waste in sexual reproduction. Also, there is a risk that some individuals will never be able to interact with another individual so that reproduction may take place. A male bobcat sometimes can't find a female bobcat. Actually, the probability of breeding failure for members of some species is very high. So, if you can reproduce all by yourself, you eliminate the risks associated with sexual reproduction. But sexual reproduction is a common biological phenomenon. Why should this be?

We can begin to understand the adaptiveness of sexual repro-duction when we recall that meiotic processes yield gametes which carry different kinds of genes and that there is normally no way to predict which of these gametes will eventually enter into fertiliza-tion. Also, in most species, mating behavior is a rather indiscrimi-nate process; "mating" animals which respond in specific ways to members of the opposite sex of their species are likely to choose any of a number of individuals from among the population. In "nonmating" animals, reproduction may be an even more random process. For example, some fish simply release their eggs and sperm into the water at spawning time, and the eggs and sperm mix. In some cases, the water may be filled with the eggs and sperm of many individuals. There is, of course, no way to predict or control which sperm will fertilize which egg. Both meiosis and randomness in mating, therefore, tend to increase *variation* in a population.

Variation may be an unimportant quality in a homogeneous and unchanging environment; however, in a heterogeneous or unstable environment the degree of variation in a population may be a critical factor. For example, under one set of conditions a certain genotype may be favored while another meets with little success. When conditions change, however, the second genotype may be favored while the first suffers a loss in numbers. Thus with a genetically variable population the interplay of environment and genotype results in an increased likelihood of survival for some members of the group.

Sexual reproduction has other benefits as well. As mutations continue to appear in a population, some will result in an advantage to their bearers. When an individual bearing a new "good gene" mates, by some happy chance, with an individual bearing another adaptive mutation, the combination may produce offspring with a marked competitive advantage. And in time that particular combi-nation may become predominate in the population. In contrast, in asexual reproduction each daughter organism will be exactly like the parent, with no possibility of different gene combinations. Random mutations might arise in a single individual, but there is no mechanism by which these genes might eventually combine with other new genes in the population, to give rise to still other new combinations.

FERTILIZATION IN SEED-BEARING PLANTS
The seed-bearing plants are a highly diverse group in which the sporophyte generation dominates. In fact, the gametophyte genera-tion is so diminished that it exists entirely within the tissues of the

Figure 7.9 The mating behavior of ruffs is a rather frantic affair with individually marked males crouching and whirling madly on the ancestral breeding ground, each seeking to intimidate his rivals and be accepted by a female. This is his only way of leaving his genes in the next generation. In other species mating may be accomplished much more judiciously, or it may be completely random. The bottom bird is not exhibiting mating behavior.

Figure 7.10 Reproduction in a hypothetical seed plant.

Meiosis in spore

Pollen: male gametophyte

Male sporangium

Sperm nucleus

Egg cell (by meiosis)

Divided sperm nuclei

Pollen tube

Seed

Tube nucleus

Female sporangium

Female gametophyte

sporophyte. Many plants, therefore, propagate by sexual reproduction—that is, through the union of a male gamete (a sperm) with a female gamete (an egg) to form a *zygote*, the primal cell from which the new plant will develop. This union is called *fertilization*. In some plants, such as the willow, the male and female structures are on two different trees. In other plants, such as the conifers (cone-bearing trees), they occur on different parts of the same tree. In flowering plants, however, both structures are usually in the same flower. These structures are the male and female sporangia (since this is the *sporophyte* generation). The male sporangia are the *anthers*, found at the tips of the filaments in the center of the flower (Figure 7.11). They produce spores, which then divide once or twice to become grains of *pollen*, the male gametophyte generation. The female sporangium is the *ovule*, located just above the base of the flower. Certain cells in the ovule undergo meiosis, and these divide a few times to form eight-celled female gametophytes.

Pollen is transferred in a variety of ways to the receptive female part of the plant, the *stigma*. At this point the pollen grain grows a long hollow tube which burrows down toward the female gametophyte. The cells of the pollen grain, the sperm, then move down through the tube, and one of them fertilizes the egg. The fertilized egg, now called a zygote, is the first cell of the new sporophyte generation.

Actually, two sperm enter the female gametophyte. One fertil-

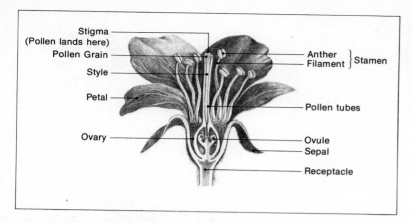

Figure 7.11 Flowers usually contain both male and female reproductive parts. A few species of plants do have separate flowers for males and females. Pollen, the male gametophyte, is formed within the anthers. The ovules comprise the female sporangia. The stigma, ovary, and style make up the pistil.

izes the egg which has developed from one of the eight cells of the female gametophyte. Meanwhile, two of the other female gameto-phyte cells have fused, and a second sperm joins with them to form a cell with three times the chromosome complement of the gametes. This cell will form the *endosperm* tissue around the zygote, which will nourish it as it develops inside the seed. This endosperm tissue is the nutrient substance you see inside a kernel of corn or a grain of wheat. The seed, protected by a hard seed coat, is then released by the plant to await the environmental conditions that will enable the zygote to begin its growth. This waiting period varies tremen-dously with different kinds of seeds. The seeds of the mangrove may start to germinate while the fruit is still on the tree, whereas seeds of the Indian lotus have been known to lie dormant for as long as 400 years.

PLANT DEVELOPMENT

Plant development differs from animal development in two impor-tant ways. First, because of the rigid walls of plant cells, new cells cannot develop in the interior tissue of plants, as they can in ani-mal tissue. Instead, plants have to grow by adding new cells to the outside of the established cells. Second, whereas most animals stop growing when they reach maturity, plants continue to grow throughout their lives.

When the time comes for the seed to *germinate*, or begin growth, the endosperm nucleus goes through a period of rapid division and then the zygote itself begins to divide. As shown in Figure 7.13, the first division of the zygote is uneven, so that one daughter cell is larger than the other. One cell continues dividing to form a chain or stalk of cells; the other cell divides in all direc-tions, producing a globular embryo at the end of the stalk. With

Figure 7.12 A micrograph of the developing *Capsella*. Compare to Figure 7.13C. (Courtesy: Carolina Biological Supply House)

Embryonic epidermis

Vascular tissue — Cortex

Cotyledons

Basal cell

Figure 7.13 The development of *Capsella*, called the shepherd's purse.

progressive divisions the cells in the embryo begin to specialize. Some will eventually produce the *epidermis*, or protective skin; others will produce the *vascular* tissue, which transports nutrients; and others still, will produce the *cortex*, or structural tissue. The structural tissue is that tissue which supports the vascular cells. As each type of structure becomes differentiated, the globule flattens and elongates to form the embryo leaves. These leaves are called *cotyledons*.

There are two parts of the plant which remain permanently embryonic. It is in these areas that future growth will take place. The two growth areas, called *apical meristems*, are at the root tips and the stem tips. The cells of these areas will continue mitosis throughout the life of the plant. As the roots and stems lengthen, the older meristematic cells are left behind to mature and specialize. Figure 7.14 shows the structure of a typical root apical meristem. The apical meristem of the stem differs slightly from that of the root; the stem tip lacks a protective cap and may have leaves. Also, the stem apex is divided by *nodes* which will be the sites of new leaf formation.

Just as apical meristems account for growth in length, *lateral meristems* contribute to growth in girth. Lateral meristems are also called *cambia* (singular, *cambium*). Growth in diameter presents special problems. For example, the vascular system must be enlarged by adding cells to the *xylem*, which carries water up from the roots, and to the *phloem*, which carries food down from the leaves. Since these areas lie deep within the plant, how can new cells be added without rupturing the plant's surface? This is possible because there are two kinds of cambium. The *vascular cambium* produces new xylem and phloem and the *cork cambium* produces new cork, a protective tissue which functions as an epidermis or skin. The vascular cambium lies between the xylem and phloem, and as it produces new cells, those toward the inside of the stem form xylem and those toward the outside form phloem (Figure 7.15).

Plants contain many other types of cells with specialized functions. The fundamental tissue, for example, includes a variety of cell types, such as the thick-walled *sclerenchyma* and *collenchyma*, which function as structural supports (Figure 7.16). Other types of cells, called *parenchyma*, are thin-walled and have large vacuoles. Some of these cells contain chloroplasts.

The leaves of a flowering plant are the sites of photosynthesis, but during its early development, before the leaves are fully exposed, the plant must get its food from some other source. In plants such as peas and beans, food is absorbed from reserves in the cotyledons or "seed leaves". The first true leaves will arise from a region at the base of the cotyledons known as the epicotyl (Figure 7.17). As the plant grows, the role of food production will be taken over by the leaves. Leaves occur in a variety of sizes and shapes, as anyone knows, but in Figure 7.18 we have a diagram of a stylized leaf.

ANIMAL FERTILIZATION

In animals which reproduce sexually, fertilization may be achieved in a variety of ways. It may be accomplished *externally* (Figure 7.19), or *internally*, as is the case in the group of animals which has

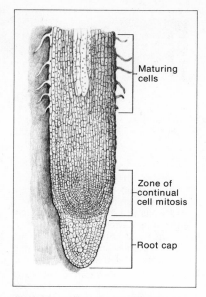

Figure 7.14 At the terminal portion of the root is the root cap. It protects the root tip as it pushes through the soil. The reproductive zone lies just behind the root cap. It is here that the mitotic process is carried on.

Figure 7.15 Lateral meristemic tissue. This is a cross-section through part of a stem. The thin cambial cells produce xylem on one side and phloem on the other.

Sclerenchyma
fibers

Collechyma
fibers

Figure 7.16 Two types of plant supportive tissue, sclerenchyma fibers and collenchyma fibers. The cell walls are thick, relatively rigid, and remain after the cell dies, thus giving structural support to the plants, and making wood a sought-after building material.

Figure 7.17 The germination and early growth of the common garden bean (top right). The stem and leaves arise from the epicotyl. The lower region of the hypocotyl gives rise to the root.

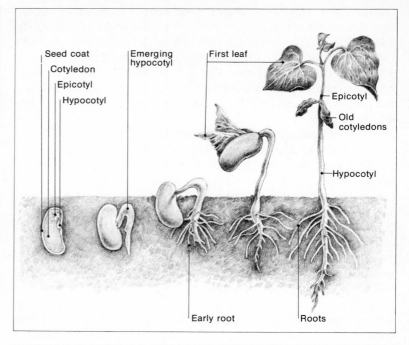

Seed coat
Cotyledon
Epicotyl
Hypocotyl

Emerging hypocotyl

First leaf

Epicotyl

Old cotyledons

Hypocotyl

Early root

Roots

Figure 7.18 A leaf cross-section. The epidermis is found on the upper and lower surfaces and protects the more delicate inner cells. It is usually one cell-layer thick. In the lower epidermis are openings called stomata through which carbon dioxide enters and the gaseous by-products of photosynthesis escape. The passage of gases through the stoma is regulated by the opening and closing of the guard cells. Most of the internal tissue is comprised of parenchyma cells which contain chloroplasts.

Chloroplast

Upper epidermis

Parenchyma cell

Lower epidermis

Intercellular spaces

Stoma

Guard cell

Phloem

hair and nurses its young—the *mammals*.

In mammals, fertilization is internal and is accomplished by the act of *copulation*. In most mammals copulation occurs only at specific periods when the female is physiologically and behaviorally receptive (in *estrus* or *heat*). One advantage of such timing is that it may increase the likelihood of young being born at the most opportune time of year, such as when food is most abundant. The receptivity of the female mammal is advertised by odors associated with vaginal discharge in many species, including dogs and horses. In certain primates, receptivity is communicated by a vaginal odor acting in coordination with a reddened and enlarged area around the rump. Receptive female chimpanzees are sometimes referred to as "pink ladies."

In humans, theoretically, there is no period when the female is not sexually receptive or when she is not sexually attractive to males, your personal experience notwithstanding. Women "stand" physiologically ready to copulate at almost any time and there is no particular signal which is involved in attracting males. Anthropologists have suggested that the continual receptivity of women may have evolved as a way to entice larger and stronger males to remain affiliated with the female which may be bearing his children. In such an association, the female might have greater reproductive success with the male's assistance, and the male might leave more offspring, as well, by assisting his weaker reproductive partner who is usually smaller and burdened with the responsibility of nursing the young. His affiliation might have been, and still might be in certain cultures, advantageous in terms of both food and protection. In time, a particular bonding, feeling, or attraction might have developed between reproductive partners which strengthened the union established on a purely physical basis. Perhaps this kind of bond provided the basis for what is called "love" or perhaps such bonding *is* love. It's hard to say since no one has been able to come up with an acceptable definition of the phenomenon. But it is known that the human often develops and, in fact, seems to have a need for strong attachments with another human. Such attraction, as we all know, may be long-term or short-term. Long-term relationships are sometimes formalized as "marriage," a system which in our culture discourages the breaking of the association. It should be pointed out that other species may build long-term bonds. For example, geese often mate for life, and a member of a pair might never mate again if one is lost.

In humans there is no specific point at which a male and female are likely to initiate copulation. The events preceding copulation may vary widely, even between established sexual partners. In certain other animals specific signals always precede the copulatory act, such as the female chimpanzee's "presentation" of her livid rear end to a male (although the blasé males may or may not

Figure 7.19 The male frog, in order to inseminate eggs, clasps the female behind her forelimbs and, as she extrudes the eggs from her body, he releases sperm over them. Since the fertilization takes place outside her body it is called external fertilization.

Vas deferens
(sperm duct)

Pubis

Urethra

Penis

Glans penis

Testis

Seminal vesicle

Bladder

Prostate gland

Rectum

Cowper's gland

Urethral bulb

Anus

Epididymis

Seminiferous
tubules

mount the soliciting temptress). Under normal conditions the physical prerequisites of copulation or *intercourse* in humans are simply an erect *penis* and a lubricated *vagina*. These conditions are usually achieved during the course of precopulatory sexual behavior, or *foreplay*. Foreplay, as well as the copulatory act itself, varies widely in its expression, between cultures, from one individual to the next, and even between established sexual partners from one time to the next.

The human *penis* is supplied with a great number of blood vessels, some of which open into large blood chambers. During sexual arousal, blood is restricted from leaving the penis by constriction of certain sphincters (rings of muscles) around the vessels at the base of the penis. Thus the penis becomes filled with blood. The result is that it grows stiff and its length increases, sometimes to a surprising degree—although, perhaps, more surprising to some than others. The penis stands erect at such times, and is said to be in the *tumescent* condition by the more delicate among us.

During foreplay, in the woman, cells in the vaginal wall usually secrete a lubricating substance which aids in the insertion of the penis. The *labia minum* and *clitoris* may become enlarged and reddened due to the rush of blood to those areas. The nipples may also harden and enlarge. If the foreplay has been sufficiently extensive, other changes may also occur, such as a mottling of the shoulders and chest and an enlargement of the breasts.

Copulation begins when the erect male penis is inserted into the female vagina. After a few or many pelvic thrusts (depending on the species, and in the case of humans, depending also on the individual, the degree of arousal, and one's schedule), *ejaculation* occurs in the male.

Ejaculation takes place in two stages. First there is a contraction of the entire genital tract, including the epididymis, sperm ducts, and seminal vesicles. Fluid from the seminal vesicles flows into the upper urethra (urinary tract), where repeated contractions of the prostate gland mix it with prostate fluid. As the urethra becomes filled with semen the urethral bulb at the base of the penis expands to accommodate the influx. The second stage begins as a sphincter muscle closes off the urethra at the bladder. The urethral sphincter at the base of the prostate then relaxes, allowing semen to move into the urethral bulb and into the penis. This is followed by contractions of the muscles in the area behind the testicles, which pump the accumulated fluid out through the penis. Several regular contractions may occur, but most of the semen is expelled in the first contraction. The semen may eject from the penis with considerable force in the first contractions and it is in this first

Figure 7.20 The male reproductive system includes the testes, scrotum, seminal vescicles, accessory glands, various ducts and the penis. The *testes* are the glandular organs in which sperm are produced. Before birth the testes descend from the abdominal cavity into a pouch behind the penis called the *scrotum*. One testicle (usually the left) hangs lower than the other so the two cannot be crushed together. There are two testes (or testicles) which contain hundreds of coiled *seminiferous tubules* in which the sperm develop. These tubules lead into the *epididymis* which appears to be a large, short tube but is actually a highly coiled, very thin tube about 20 feet long. The epididymis acts as a filter to remove any cell debris or pigment accompanying the sperm and to propel the sperm into the sperm duct. If more sperm are produced than can be utilized, the epididymis may "digest" some of them and return their nutrients to the body. Also the epididymis acts to increase the fertility of the sperm as they pass through it. The *sperm duct,* or *vas deferens,* transports the sperm over the pelvic bone to the back of the urinary bladder where it joins the sperm duct from the other testis. The *seminal vescicle* is a tightly coiled tube which lies in a pouch at about the point of juncture of the seminal ducts. Fluid from the seminal vescicle is discharged at ejaculation. The *prostate gland* lies just below the bladder and secretes *seminal fluid*, which together with fluid from the seminal vesicle, activates and maintains the sperm. Below the prostate gland the *Cowper's gland* or *bulbourethra,* also contributes a fluid to the semen. The *penis* is composed of three long areas of spongy tissue bound together by fibrous tissue. Two of these spongy *cavernous* areas are attached to the base of the Cowper's gland and at the penis tip, to form the *glans penis*. This structure may be covered by the foreskin, the fold which is removed at circumcision.

Figure 7.21 The reproductive system of the human female includes the ovaries, oviducts, uterus, vagina and external genitals. The *ovaries* are located at the sides of the abdominal cavity and are supported by ligaments. Eggs released from the ovaries move into the tubelike *oviducts* which extend to join to either side of the *uterus,* where the embryo develops. The oviduct is lined with cilia which sweep the egg toward the uterus. The uterus is pear-shaped, muscular and thick-walled, with the lower end opening into the vagina. The inner lining or *endometrium* of the uterus is glandular and is the site where the fertilized egg will develop. The lower part of the uterus, just above the vagina, is called the *cervix.* The vagina is a muscular tube, 3 or 4 inches long, which leads from the exterior to the cervix. The vaginal muscles are thin-walled and the epithelial lining may have a corrugated appearance. The external genitals, or *vulva,* is comprised of the *mons veneris,* a fatty mound which becomes covered with hair at puberty. The *labia majum* are thick folds which lie on either side of the vagina and which are also covered with hair. The *labia minum* are smaller folds of skin which lie between the labia majum and the vaginal orifice. Where the labia minum meet above the vaginal orifice is the *clitoris,* a small erectile organ which is highly sensitive and which corresponds developmentally to the penis of the male. The urethra (urinary tract) opens between the clitoris and the vagina. There may be a mucous membrane stretched across the vaginal opening, called the *hymen.*

fluid that the greater part of the sperm is concentrated.

Ejaculation is ordinarily accompanied by *climax* or *orgasm,* which is the culmination of the copulatory act. Accompanying physiological changes include an increase in blood pressure and a quickened pulse. The postclimactic period is marked by a feeling of relaxation, sometimes to the point of drowsiness. The sphincter muscles in the male relax and the penis again becomes flaccid or limp.

Orgasm in women is much less well documented. In fact, the very existence of the female orgasm was argued until 1966, when a pioneer study of human sexual response was published by the physician-psychologist team of Masters and Johnson. On the basis of many carefully monitored observations of sexual activities of all sorts, they demonstrated that the female orgasm is little different from the male orgasm. However, it is not so universal a response, and there is a much wider range of individual differences. The response is rarely or never achieved in some women, while others may reach orgasm frequently and easily. The inability to reach orgasm, however, has no effect on the ability to reproduce.

It is not known whether females of other species experience climax. However, Jane van Lawick-Goodall, in her continuing study of the wild chimpanzees of the Gombe Stream Preserve, has witnessed apparent nymphomania in one adolescent female, a behavior which would be difficult to explain in the absence of sensory reinforcement. Also, physiological tests on females of other species have demonstrated behavioral responses in copulation which are similar to those occurring in human females at orgasm.

The topic of copulation has proven a particularly thorny issue in many cultures, including our own. It has become interwoven with religious, legal, and social restrictions which are imposed to different degrees from time to time, even within the same culture. In our own culture morality has come to refer primarily to sexual matters. To many in our society an "immoral" person is one who is sexually promiscuous, *period.*

To some extent this attitude stems from the concept that childbearing is the only basis for copulation. It has been argued that, in nature, copulation is practiced solely for the purpose of procreation, and that copulation for other purposes is "unnatural," and therefore wrong. Logic aside, the premise is erroneous. Actually, copulation has many functions in nature. Among rhesus monkeys, subordinate males often present their rears to dominant males as an act of submission, an act which has the effect of assuaging aggression on the part of the dominant individual. The dominant male may follow up with a demonstration of his authority by

Ovary

Mons veneris

Bladder

Pubis

Urethra

Vagina

Clitoris

Fallopian tube (oviduct)

Uterus

Cervix

Glands of Bartholin

Rectum

Labium minum

Anus

Labium majum

mounting the subordinate and even making several pelvic thrusts, although there is no actual penetration. Among chimpanzees a pink lady may copulate with every adult male in the group, as well as some adolescent ones. Since only one male is needed to impregnate the female, the act of copulation is believed to be a means of developing social bonds within the group. Among baboons the females copulate during the earlier phase of their estrus period with males of any rank, but at the time they are most likely to conceive, they will accept only high-ranking males. Baboons are also highly social, and in this case, too, the earlier acts of copulation may serve to reinforce group bonds.

Humans are highly complex and can only be viewed with what must be a high degree of subjectivity by other humans. It is therefore difficult to define the bonding quality of copulation in our own species. It seems likely on an intuitive basis, however, that copulation may be an important bond builder. If this is so, it would help to explain our caution regarding what seems, on the surface, to be a rather mechanical process. In other words, it may not be advantageous to build bonds indiscriminately.

An indication that, in humans, copulation may have functions other than procreation is provided by evidence that whereas women may be prepared to copulate at any time, pregnancy can occur only at a specific time in the *menstrual cycle.*

THE MENSTRUAL CYCLE

The menstrual cycle begins in females at puberty, ordinarily around the age of twelve to fourteen, although the onset has become progressively earlier in American girls, probably as a result of increasingly good health. The cycles will continue for about the next thirty to forty years of the woman's life, usually interrupted only by pregnancies, and terminated finally by menopause.

Menstruation is the result of the shedding of the blood-rich *endometrium*, the lining of the uterus, when the egg released from the ovary has failed to be fertilized. The endometrium had become engorged with blood, blood carrying life-sustaining food and oxygen, in preparation for receiving a fertilized egg. But if the egg reaches the uterus unfertilized, the endometrial preparations have been for nil and the uterus reverts to its previous condition. The old endometrial lining breaks loose and slides out the vaginal opening, accompanied by blood from ruptured vessels.

An egg is released from an ovary about every 27 to 30 days. If it is not fertilized, menstruation will begin about 14 to 16 days after that and the period will last from 3 to 6 days. However, there is wide variation in the length of both the cycle and the menstrual

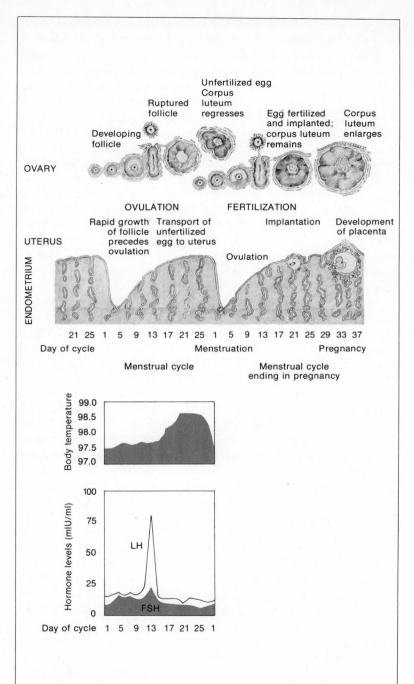

Figure 7.22 The time relationship of the ovarian cycle, the uterine cycle and temperature cycle to fluctuations in the leutenizing hormone (LH) and the follicle-stimulating hormone (FSH).

period among women and any pattern is considered normal as long as it recurs on a rather regular basis. Figure 7.22 describes the relationship of the ovarian, uterine and hormonal cycles when pregnancy does not occur, and when it does. *Hormones* are "chemical messengers," produced by various glands in the body, which act to regulate the course of events in other parts of the body.

At the end of a menstrual period, the cycle begins anew as deep within an ovary other follicles are developing. These are rounded bodies, each containing an egg which protrudes inward from the follicle wall into a large fluid-filled space. As the follicles mature, they grow larger and migrate to the surface of the ovary. Their development is stimulated by rising blood levels of a hormone called simply the *follicle-stimulating hormone* (FSH), interacting with another hormone called the *luteinizing hormone* (LH). Both of these hormones are produced in a small but rather incredible structure called the *pituitary gland*, which is located at the base of the brain. (We'll look more closely at this important gland later.)

As the follicles mature, they begin to secrete a hormone of their own called *estrogen*, which induces a thickening of the uterus lining, the endometrium. For some reason, one bubble-like follicle begins to outgrow the others, and as soon as its ascendancy is established, the others stop developing. This larger follicle then moves to a point just below the surface of the ovary. In response to the pituitary's luteinizing hormone, the follicle ruptures on about the thirteenth or fourteenth day after the follicle-stimulating hormone had initiated its growth. The egg is released, leaving behind the old yellowed body of the follicle, now called the *corpus luteum.* However, the corpus luteum is far from being a corpse, for it begins to secrete another hormone, *progesterone*, which stimulates the uterus to prepare for a fertilized egg. The endometrium thickens and becomes engorged with tiny blood vessels which will carry food and oxygen to the embryo in the event that fertilization occurs. The interaction of the progesterone and estrogen also inhibits the production of the pituitary's follicle-stimulating hormone at this time, to prevent the start of a new cycle. These inhibiting properties of estrogen and progesterone are the basis for their use in birth-control pills, which we will discuss later.

If the copulation has been successful (depending on your point of view), and the egg which reaches the uterus is fertilized, the zygote will implant in the uterine wall and begin its growth, taking its necessities from its mother's blood in parasitic fashion. (Although in some cases this may be only the beginning of a long parasitic existence.) However, when fertilization does not occur, the

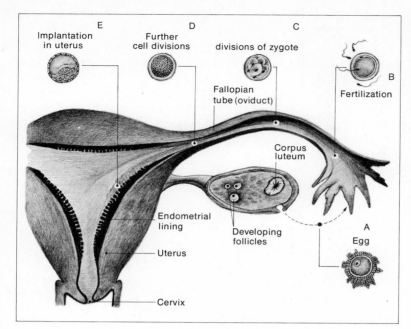

Figure 7.23 A section through the fallopian tube and uterus showing the sequence from fertilization to implantation. The egg cell(A) bursts from the ovary, surrounded by supporting cells, passes into the oviduct where fertilization occurs(B), and begins to undergo mitosis(C). During its preparation for implantation (a period of six days) rapid cell division(D) results in the formation of a hollow ball of cells, the blastocyst(E). The thickened inner region will become the embryo, while the thin layer around the sphere will give rise to the chorion, a forerunner to the placenta.

Labels in figure:
E — Implantation in uterus
D — Further cell divisions
C — divisions of zygote
B — Fertilization
A — Egg
Fallopian tube (oviduct)
Corpus luteum
Endometrial lining
Developing follicles
Uterus
Cervix

corpus luteum stops its secretion of progesterone, and the swollen, life-sustaining uterus lining begins to shrink and break down. The connecting blood vessels rupture, and this blood carries the bits of cell tissue out of the uterus and through the vaginal opening during menstruation. The pressure of blood-engorged tissue often causes discomfort and cramping as materials build up and are then dislodged. In some cases the changes in hormone balance at this point cause accompanying emotional depression or despondency. As soon as the uterus has returned to its original state, the pituitary stimulates the development of more follicles, and the cycle, well, the cycle begins again.

CONCEPTION

Although only one egg is produced during each menstrual cycle, the semen released at each ejaculation may contain millions of sperm, each with its own unique genetic makeup. There is no way to predict which sperm will fertilize the egg. During copulation the sperm are usually ejaculated into the upper reaches of the vagina, near the cervix. Each sperm cell has a whiplike "tail" (Figure 7.24). which propels it along at a rate of about ½ centimeter per minute. Surprisingly enough, however, sperm may reach the cervix only 1½ to 3 minutes after ejaculation. This is probably because their own propulsion is further assisted by ciliary action and currents

Figure 7.24 The human sperm.

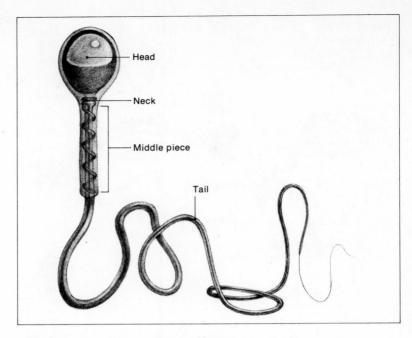

which are set up by contractions of the vaginal muscles.

The sheer number of the sperm in each ejaculation is an adaptation to the hazards they face in making their way to the egg. Some may die as a result of their own physiological changes ("natural causes"). Others may be rendered immobile by the chemical environment of the vaginal tract or devoured by roaming white blood cells of the woman which treat them as if they were foreign bacteria. The semen in which the sperm are carried is alkaline, or basic. The fluids in the vagina, however, are slightly acid, and sufficient exposure causes the sperm to become inactive. When the sperm are first activated at the time of ejaculation, they thrash violently and spasmodically, and some make it into the friendly alkaline environment of the uterus. Then, through ways which remain a mystery, a few manage to find their way through the uterus, into the tiny pore of the oviduct (also called the *Fallopian tube*), and up the tube nearly to the ovary, where they encounter the egg on its way down. Conception actually takes place, then, not in the uterus, but far up in the oviduct.

Several sperm may reach the egg at once, but for some reason only one will penetrate the egg's membrane and effect fertilization. The mechanism that prevents fertilization by all but one of the sperm is still unknown.

Once conception has taken place, the zygote will continue

Box 7.1 The Theory of Recapitulation Human embryos, in the course of their development, go through stages when aspects of their appearance very strongly resemble characteristics found in the embryos of other species of animals. For example, human embryos develop tails, which usually disappear before birth. (In a small percentage of births, a rudimentary tail must be surgically removed.) Human embryos also develop gill-like pouches which disappear or are altered to produce other structures such as the ear canal. Gill pouches of shark and man are compared in the art below. Below that you can see the strong similarities in the circulatory systems of the embryos of sharks and humans (Adapted from Simpson, Pittendrigh and Tiffany 1957-Harcourt, p. 353.)

In its earliest stages the human embryo is very similar to a starfish embryo. In later stages it bears strong resemblances to fish, amphibian and reptile embryos. Finally, however, it develops characters that are unmistakably human. E. E. Haeckel (1834–1919) summed up this phenomenon, which had been noted earlier, in his statement: "Ontogeny recapitulates phylogeny". This statement is misleading since it assumes that human embryos go through stages in which they resemble adults of lower life forms. Instead, the phenomenon makes more sense to us when we understand that the developmental mechanism of our fish-like ancestors could have been inherited by us. This helps to explain why it is that early embryos of many modern species are so similar. But the parallels cannot be drawn out too specifically, because even in these early embryos characteristics are found which are specific to only one species.

MAN

Gill pouches
Arm limb bud
Heart
Umbilicus
Leg limb bud
Tail

SHARK

Gill pouches

Arterial arches
Heart
Veins

Box 7.2 Why Your Children Will Not Look Exactly Alike In the developing female embryo, there are about 7 million eggs in the ovaries by 20th week. By puberty many of these have been lost (though the means of selection is unknown) so that there are by then only about 400,000 eggs. At about the age of sexual maturity (usually around 12 to 16) there may only be 300,000 eggs. Of these, only one will escape a ruptured follicle every 28 days. This goes on for about 40 years, so a total of about 400 eggs are produced. Each woman could actually produce a maximum of about 40 children in a lifetime (the record is 69). So, of millions of potential ova, only 25 to 40 can contribute their genes to the next generation. If no type of developing germ cell has an advantage, the probability of any one developing to maturity is remote and the probability of ova with identical sets of genes approaches nil.

Males, after puberty, normally produce millions of sperm each day and the production may continue throughout life. The chromosomes of each sperm and egg have undergone meiosis and crossing over and there is almost no likelihood of two identical sperm being produced. Of millions of sperm in each ejaculation, only one will fertilize an egg. One reason so many sperm are necessary is that the female reproductive tract is essentially a hostile environment and most sperm will quickly perish there. Also, a mucuos plug may block the cervix and enzymes carried by sperm are necessary to break it down to open the way for other sperm. These same enzymes act to break down the follicle cells surrounding the egg so that a sperm can get through to fertilize it. If no particular type of sperm (in terms of genetic constituents) has an advantage, you can see that the likelihood of identical individuals being produced from the same parents is very slim. In fact, the odds of your having children with identical genetic constitutions have been calculated at about 1 in 14 trillion.

down through the oviduct to be received by the ready uterus. As soon as the zygote implants itself in the uterine wall, it will begin to receive sustenance as oxygen and nutrients filter through from the blood of the mother. Then, given the conditions for its continued development, in another 266 days a new individual will join the United States' population.

There has never been another individual quite like this one, and there won't be others to follow. Not exactly like this one. The genetic combination cannot be duplicated and neither can the experiences which will be superimposed on that constitution. In some cases its existence will be brief, the time counted in days or even minutes. In other cases it will join the world's population. where it may become anything from a Sikh to a Republican. Whatever it will call itself, though, it will be unique.

CONTRACEPTION

Now let us consider a problem at the other end of the reproductive spectrum: the avoidance of parenthood.

WHY NOT CONCEIVE?

Why would anyone want to avoid bringing a baby into the world? You might immediately think of a reason. But actually the rationales are varied. In fact, more couples are making the decision to remain childless or to adopt children which are already here. Adoption as a way of becoming a parent has no risk in terms of the health of either the wife or the child, but there is an element of "chance" in the genetic makeup of the child. Also, as an ego trip, it has less going for it. Couples who have no children are citing reasons such as personal freedom, financial well-being, fewer worries with resulting better health, and more time to be with each other. They are, strangely, sometimes labeled as "selfish" by people who want children and who, therefore, have them.

One of the most frequently cited reasons for avoiding reproduction involves overpopulation. There are, by almost any measure, too many humans on our limited earth. Worse yet, it seems our numbers are increasing daily to the point that our population can only be stabilized by very drastic methods. It is doubtful that many couples are aware or concerned enough about overpopulation to willingly forego having children. However, one of the reasons for having children has been that society seemed to expect, and even demand, it. In recent years, as more Americans become aware of the population problem, it has become more socially acceptable to forego children. Thus, with the relaxation of public pressure and the generation of more public discussion of alternatives, perhaps some couples forego children because of overpopulation without really knowing it. Even those who have children are having fewer, for now—at least in the United States (which has about 6 percent of the world's people). We cannot assume that Americans are copulating less so we must assume contraception is being more effectively practiced. How?

METHODS OF CONTRACEPTION

Contraception, or avoidance of pregnancy, can be achieved in many ways. Some means are used by the female, some by the male, and some by both partners. Unfortunately, their common bond is that none, except abstinence, is 100 percent effective.

Historical Methods

Historically, conception has been prevented by a number of means.

The ancient Egyptians blocked the cervix with leaves, cotton, or cloth. The encouragement of homosexuality among the ancient Greeks probably reduced the rate of conception in that culture. In the middle ages *condoms* to fit over the penis were fashioned from such materials as linen and fish skins. Douching has long been practiced as a means of flushing sperm out of the vagina. The finger has also been used to remove semen after it has developed a stringy quality through being exposed to the vaginal environment. The Old Testament refers to *coitus interruptus*, or the removing of the penis just prior to ejaculation. The most effective method, of course, has always been total abstention, a method which has met with little applause, except in certain circles. There are a number of "folk" devices which are employed even today. These range from regulation of intercourse according to the phase of the moon, and stepping over graves, to using cellophane sandwich-wrapping as condoms and douching with soft drinks (the carbonic acid and sugar are believed to be spermicidal and a shaken drink provides the propulsion for the agents to reach far into the vagina).

The effectiveness of most folk methods is unknown, so let us consider other means whose effectiveness or ineffectiveness has been investigated a bit more carefully.

The Rhythm Method
It has been jokingly said that people who practice the *rhythm method* of contraception are called "parents," although the joke is funnier to some than others. Basically, rhythm involves periodic abstention. It is theorized that the time during which a woman may become pregnant can be predicted, and by refraining from intercourse during this time, pregnancy cannot occur. The couple must therefore avoid intercourse at least two days before and one day after ovulation, although, for safety, this interval is usually increased. Interestingly enough, although the method involves couples refraining from intercourse at the very time when the woman may be psychologically and emotionally most receptive, and although the system requires the careful use of thermometers, calendars, paper and pencils, certain groups claim that this is the only "natural" means of birth control. Unfortunately, in general, the method is of doubtful reliability, partly because the menstrual periods of many women are highly irregular.

Coitus Interruptus
In spite of the obvious inconvenience of withdrawing the penis from the vagina immediately before ejaculation, coitus interruptus

Figure 7.25 Different types of contraceptives: spermicide jelly and diaphragm (lower left), spermicide foam (upper left), prophylactics or "rubbers" (upper middle), spermicide cream (upper right), birth control pills (lower right), and one type of intrauterine device—IUD (lower middle).

is one of the most widely used contraceptive practices in Europe today. Again, there are little good data on the effectiveness of this method. It is likely of low reliability because the man must do just the opposite of what he is very highly motivated to do. The practice is risky, also, because sperm may leave the penis long before ejaculation occurs. Because of residual semen in the penis after the first ejaculation, there is higher risk of conception for those who are able, and have the time, to immediately copulate again.

The Condom

The *condom*, or rubber, is one of the most widely used contraceptive devices in the United States. Basically, it is a balloonlike sheath made of rubber, or other material, which fits tightly over the penis and traps the sperm. There are a number of grades of condoms ranging from the two-for-a-quarter specials sold in restroom vending machines to more expensive types made from animal membranes.

The best condoms are only 85 percent effective (partly because they are often incorrectly used). This means that out of 100 couples who depend on them for a year, 15 of the women may become pregnant. Disadvantages of the method include a reduced level of sensation on the part of the man and the interruption while the condom is frantically located and put on. One important reason for its popularity, however, is the protection it affords against the transmission of venereal disease.

The Diaphragm

The *diaphragm* is a dome-shaped rubber device with a flexible steel spring enclosed in the rim. It is inserted into the vagina and fitted over the cervix to prevent sperm from entering the uterus. Unlike the condom, the diaphragm must be individually fitted. It should be coated with a spermicidal cream or foam and left in place for several hours after intercourse. It is about 90 percent effective if well-fitted and used correctly. With a certain element of anticipation on the part of the woman, it need not interrupt the sexual activities in order to be inserted.

The Cervical Cap

The *cervical cap* functions similarly to the diaphragm. It is a small plastic or metal cap which fits tightly over the tip of the cervix. One of its advantages is that it can be left in place instead of having to be reinserted each time. In fact, it is sometimes removed only for menstruation. Its effectiveness may be somewhat higher than that of the diaphragm, since it ordinarily cannot develop leaks. However, its loss is also less likely to be noticed.

Spermicides

A number of sperm-killing foams, jellies, aerosols, suppositories, and creams are available. They are usually quickly and easily applied inside the vagina and require neither fitting nor prescription by a doctor. They are less effective than the mechanical devices, but their effectiveness increases when they are used with the diaphragm or cervical cap.

The Intrauterine Device (IUD)

The *intrauterine device*, or *IUD*, has been known as an effective contraceptive for many years. It is reported that Arab camel drivers placed fruit seeds in the uteruses of their camels to prevent pregnancies during the long, hard caravan marches. Just how the practice started is not known; there must have been better places to store seeds. Today IUDs of a variety of shapes and materials are in use. However, beyond the fact that the presence of a foreign body in the uterus somehow prevents pregnancy, no one knows exactly how the IUD works. It is theorized that the IUD may prevent implantation of the zygote, or that it may cause the egg to move too rapidly through the oviduct, or even that it may alter the condition of the uterine endometrium.

The IUD has several major advantages. Once it has been installed in the uterus, it requires no further attention. The expense of the device is minimal, and its effectiveness is about 97 percent.

One problem with IUDs is that in some women the uterus rejects them. As a result, the device may be expelled from the uterus, sometimes without being noticed. Some IUDs have a string that hangs into the vagina, so that there is a means of checking to ensure that the device is in position. For some reason IUDs are most likely to be rejected by the uterus in women who have never been pregnant. There may also be side effects such as bleeding and pain just after insertion, and if these effects persist, the IUD must be removed.

The Birth-Control Pill

The *birth-control pill* is an oral contraceptive. It is composed of the hormone estrogen, either natural or synthetic, and an "artificial hormone," similar to progesterone, called *progestin*. This combination apparently inhibits the formation of egg follicles in the ovary. A pill is taken each day for 20 or 21 days, beginning on the fifth day after the onset of menstruation. The next menstrual period begins when this series of pills is finished. Regular usage of the pill on this basis results in a menstrual period precisely every 28 days. Apart from the convenience of knowing even the day of the week on which menstrual periods will begin, in many cases this regulation may alleviate the pain accompanying menstruation. The greatest advantage of the pill is that when it is used correctly it is probably over 99 percent effective. There have also been reports that regular use may delay menopause, and increase the libido of the middle-aged user, although the bases for these effects are not known.

Birth-control pills may cause undesirable side effects, especially during the first few months of use. In many ways these side effects are similar to the symptoms experienced during pregnancy. These may include weight gain, fluid retention, nausea, headaches, depression, and irritability. However, another such effect is enlargement of the breasts. Some or all of these symptoms may diminish after the first few months of use.

There may also be more serious problems associated with birth-control pills. You have probably heard reports that the pill may be associated with cancer. Early results of a study sponsored by Planned Parenthood of New York were inconclusive; although such studies indicate a higher incidence of "precancerous" areas in the cervix among women who use the pill, these areas do not

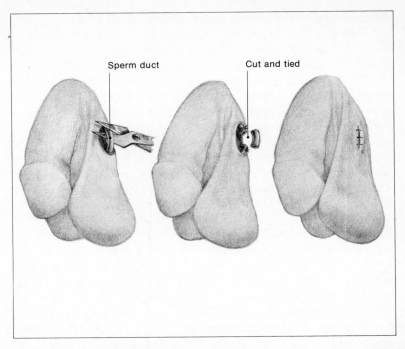

Figure 7.26 The procedure in a vasectomy. This operation does *not* impair or interfere with sexual performance.

Sperm duct

Cut and tied

invariably develop into cancer and are easily and totally curable.

Another possible risk may be thromboembolism, or blood clots, including thrombophlebitis (inflammation and clotting in the veins) and pulmonary embolism (blood clots in the vessels of the lungs). According to a British report in 1968 on women over thirty-five those using the pill had a higher chance of dying from thromboembolism (8 in 200,000) than those who were not using the pill (1 in 200,000). Women under thirty-five had half the risk in both cases. However, about 450 women in 200,000 between the ages of twenty and thirty-five die as a result of pregnancies, so from this standpoint the pill is still far safer than pregnancy.

In any case, although the pill has produced no serious medical problems for the overwhelming majority of its users, we still have no information on its long-term use and research must be rigorously continued in this area.

STERILIZATION

Sterilization is 100 percent effective in ensuring that pregnancy will not occur. Either the male or female can be sterilized, although it is a much simpler matter in the male.

A *vasectomy* for a male normally takes only about 15 minutes and can be performed in a doctor's office. A small incision is made in the side of the scrotum, a short section is cut from the seminal duct (the vas deferens), and the incision is then closed (Figure 7.26). The operation can sometimes be reversed to restore fertility; this has been successful in 50 to 80 percent of the cases in those areas where doctors have had the greatest experience with this operation. Another method, in which a removable plastic plug is inserted in the seminal duct, shows promise of being reversible in virtually all cases.

Sperm continue to be produced after vasectomy, but instead of being ejaculated, they are reabsorbed by the body. Seminal fluid is otherwise ejaculated as before; it simply does not contain sperm. As a form of "insurance" for men who feel they may want to have children later, it is possible to have sperm frozen and kept in a sperm bank for as long as ten years. If children are desired later, the woman can be impregnated by artificial insemination.

Vasectomy is, in itself, a simple procedure, but there are other factors that must be taken into account. For example, in the United States, as well as in some other countries, men often identify strongly with the "male role." The realization that they are no longer able to impregnate women may have a rather marked psychological effect in some individuals. For this reason the procedure is ordinarily advisable only for those men who are fairly secure in their own sexual identity. Also, in some areas it may be difficult to find a doctor who will perform a vasectomy. Although there were thirty-eight vasectomy clinics operating in the United States in 1971, the reaction of the medical profession has, as a rule, and as usual, been very conservative.

Sterilization in women is usually accomplished by *tubal ligation* (Figure 7.27). The oviduct is cut and both ends are tied back. This operation is much more complicated than a vasectomy, since the abdominal wall must be opened. A method which is still in the experimental stages involves entry through the vaginal opening and, when perfected, may make the ligation a far simpler procedure.

As with vasectomy, tubal ligation causes no change in hormone production, sex drive, or sexual performance.

ABORTION

Abortion, the artificial termination of a pregnancy, has been one of the more common forms of birth control throughout recorded history. In most modern societies the practice is tightly regulated, but the regulations vary widely. In the United States there has been a recent relaxation of restrictions on abortion which has paved the way for new programs facilitating abortion. This policy does

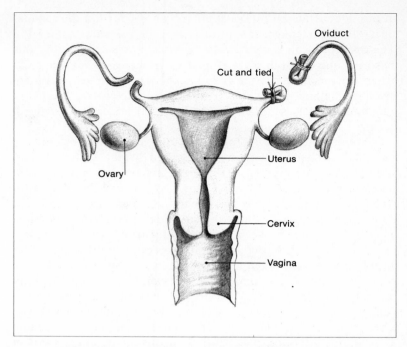

not necessarily mean that there are more abortions now than there were before; rather, it means that *illegal* abortions may become less common. When restrictions were in effect estimates ran as high as one illegal abortion for every two live births in the United States. Interestingly, illegal abortions are most common where laws are most restrictive. In Italy, for example, where even contraceptives have been banned, the abortion rate is estimated to be equal to the birth rate.

Illegal surgical abortions and do-it-yourself methods have resulted in innumerable deaths and permanent damage. Some of the methods employed have been as barbaric as they were hazardous. And even then, they have not always accomplished the intended result. In fact, a woman is almost always safer having a baby than attempting to abort the fetus herself without proper medical help.

When performed properly, under medical supervision, an abortion during the first three months of pregnancy is a simple procedure which entails no more risk than a tonsillectomy. The most common method is one called *dilation and curettage,* or *D and C.* With this method the vagina and cervix are dilated and the lining of the uterus is scraped with a *curette,* a steel loop at the end of a handle. Another method coming into increasing use is *vacuum curettage*, in which a suction device is used to pull the embryo from the uterine wall. Another increasingly common procedure involves the injection of salt solutions into the uterine cavity.

Abortion became illegal in the United States in the nineteenth century, when it was ruled that the practice was dangerous to the prospective mother. The recent court rulings in the United States which uphold the right of a woman to undergo abortion on request have generated heated debates. One argument has revolved around the point at which a developing embryo becomes a "person." There are those who feel that the zygote becomes a human being at the moment of conception and thus has a "right" to be born. Others argue that abortion does not constitute the destruction of a "life" until some specific later point, such as at the time of the first heartbeat, or the point at which the embryo first begins to stir or "quicken," or the point at which it is capable of surviving after birth. There are many puzzles and paradoxes in such arguments, especially since the sacredness of human life often depends on whose life we are considering. Many who oppose abortion on this ground have no qualms about supporting capital punishment or the mass destruction of life during wars. Obviously, the question of the "morality" of abortion is not likely to be resolved on a scientific basis.

Abortion has also been opposed on the basis that it encourages sexual freedom, a possibility that is viewed with more alarm in some quarters than in others. One of the arguments heard in more than one statehouse across the land is that the woman must "pay the piper." She has had her fun, so now let her accept the consequences in the form of "compulsory pregnancy." Apart from the assumption that moral responsibility should be expected only of women, the idea that a child is considered a form of punishment is an interesting one, rarely acknowledged by its *de facto* proponents. Of course, these are only a few of the arguments which are being heard regarding abortion, but they should provide a starting point for discussion. If you are interested in hearing some other arguments, bring up the subject at your parents' next party.

As a final point of consideration, in Sweden a study was made to see what happened to children whose mothers were denied abortion. They were compared to a matched set of children who were wanted. It was found that twice as many of the unwanted ones grew up as "illegitimate," or in broken homes, or in institutions. Twice as many had records of delinquency, and twice as many were declared unfit for military service. Twice as many had required psychiatric care, and five times as many had been on social-welfare programs during their teens.

NEW DEVELOPMENTS IN CONTRACEPTION

World population pressures, the desire of an increasing number of couples to have children by intent rather than by chance, and the wish of many to entirely avoid the rigors of parenthood have encouraged vigorous new research in birth control. The resulting ideas have been rather varied and imaginative. As with the present contraceptive methods, however, most of the new ones are for use by women.

Progestins are artificial hormones which are administered in low continuous doses. They may be administered as a "minipill" (daily, with no breaks), as an injection that lasts three months, or by the implantation of a "time capsule" beneath the skin. The capsule is made of silicone rubber and releases progestins into the blood stream at a regular rate for perhaps as long as twenty-five or thirty years.

A progestin alone does not suppress follicle development, as it does when taken with estrogen in the form of "the pill." It is not known precisely how progestins work, but they may alter the cervical mucous, preventing the sperm from entering the uterus. It has also been suggested that they may cause the egg to move through the oviduct too fast, or that they may prevent implantation or the formation of essential hormones.

Progestins are not subject to some of the risks suspected of the present birth-control pill, such as thromboembolism; however, they may cause headaches, dizziness, irregular bleeding, or the absence of menstruation. They are also reported to cause nodules in the breasts of beagles, so beagles probably shouldn't take them. They have currently been withdrawn from clinical testing, but if they are perfected, they offer hope of a simple and effective contraceptive.

Prostaglandins are hormonelike substances similar to fatty acids which are found in many kinds of animal tissues. They somehow seem to regulate hormones at their target sites in cells and organs. Prostaglandins also show promise in the treatment of a variety of ailments, from nasal congestion to hypertension. They have been used to treat both male sterility and painful menstruation, and they can induce labor at any stage of pregnancy. In controlling pregnancy, they seem to cause regression of the corpus luteum, which in turn prevents the maintenance of the pregnancy. Testing is still in the early stages, but prostaglandins show promise of providing a safe, self-administered contraceptive which could be applied only a few times a year.

The *"morning-after" pill* is a heavy dose of estrogens, to be taken by the woman for several days after copulation. It will prevent pregnancy, but the fact that it usually induces nausea for one or two days has prompted the irreverent suggestion that its principle mechanism lies in negative conditioning. These pills are now available in some American medical centers.

Stevia rebaudiana is a South American weed. The Indian women of Paraguay drink a cup of tea made from the plant each day as a contraceptive. The powdered weed has been found to reduce pregnancies in rats, and this method, as well as certain other folk remedies, are being seriously investigated in the search for the ideal contraceptive.

In the wide-ranging search for better or more effective contraceptives, a variety of other methods is under consideration. These include long-lived spermicides, removable oviduct plugs, vaginal rings which release progestins or spermicides, IUDs which release spermicides or hormones, immunization of a woman against her husband's sperm, immunization of a man against his own sperm, a nonsteroid by-product of dynamite manufacture to be administered to men (this one stirs the imagination), and clips and plugs for the vas deferens.

All these methods, of course, must undergo rigorous testing by such agencies as the Food and Drug Administration. In the United States clinical testing on humans may be started shortly after tests on other animals are completed, but a minimum of eight to ten years of such testing is required before a drug will be released for general use. These rigorous controls have led some researchers to other countries in which the controls are less stringent. The need for birth control is increasingly urgent, but obviously no method should be extensively employed without maximum assurance of its safety.

8
Development

The story of development concerns the events befalling that single fertilized egg after conception—how it grows and changes. Some of these events are programmed into the zygote and are highly predictable. For example, the first cell divisions occur precisely on time, and any embryologist can describe in detail the appearance of a chick embryo seventy-two hours after fertilization. In its broadest sense, the story of development might also include the learning experiences of an organism as it interacts with its environment, its aging and its death. In this chapter, however, we will concern ourselves with what happens between fertilization and birth in animals. Again, the focus will be primarily on our own species, and we will consider further certain ethical and philosophical factors associated with the production of new humans. In addition, we will discuss the motivation for having children, from a biological point of view. But let's begin by noting that animal eggs are basically of three types.

TYPES OF EGGS

Animals produce eggs of three basic types, roughly categorized according to the amount of yolk they contain. Since yolk is a food store, the embryos of those animals which derive their nutrients mainly from their mother's blood, as is the case with humans, need little yolk; perhaps just enough to last the zygote until it has implanted in the wall of the uterus. In contrast, since birds leave their mother's body at a very early developmental stage, they must carry their entire prehatching food supply with them. So, whereas a human egg is smaller than the period at the end of this sentence, you would be hard put to hide an ostrich egg with this whole book, although the differences in the size of both the young and the adults of these two species are not particulary great.

In the species producing a third type of egg, the new organism begins to acquire its own food at a stage some time after fertilization, but before it has arrived at what will be its final body organization. Thus, the frog egg has just enough yolk to get the developing

FROG EMBRYO

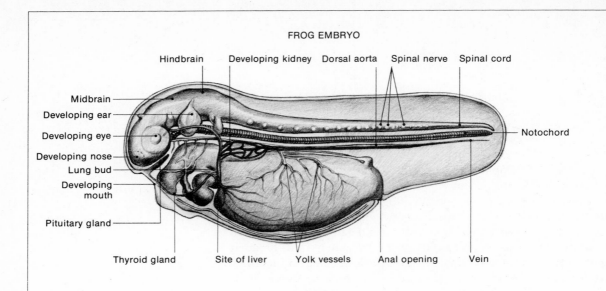

Midbrain
Developing ear
Developing eye
Developing nose
Lung bud
Developing mouth
Pituitary gland

Hindbrain
Developing kidney
Dorsal aorta
Spinal nerve
Spinal cord

Notochord

Thyroid gland
Site of liver
Yolk vessels
Anal opening
Vein

BIRD EMBRYO

Midbrain

Developing eye
Developing nose

Allantoic sac

Hindbrain
Cranial nerves
Developing ear
Atrium
Dorsal aorta
Pulmonary artery
Developing lung
Ventricle
Yolk stalk (to yolk)
Spinal cord
Stomach
Spinal ganglion

HUMAN EMBRYO

Hindbrain   Developing ear

Midbrain

Forebrain

Developing eye

Developing eye lens

Developing nose

To placenta

Ventricle

Anal opening

Spinal cord

Lung

Atrium

Stomach

Liver

Intestine

Dorsal aorta

Tail bud   Umbilical cord

embryo to the tadpole stage, and after that the tadpole survives on a combination of the food stored in its tail and what it can find to eat. The frog zygote, then, has a yolk supply intermediate to that of mammals and birds.

Figure 8.1 Frog, bird (opposite page) and human embryos at comparative stages.

EARLY DEVELOPMENT IN THE FROG

Recall that at fertilization any organism receives all the genetic information it is ever going to get. No matter what differences arise in the cells that make up the various structures of the adult, they all derive from this single cell, with its directing set of chromosomes.

THE FIRST CELL DIVISIONS

The first step in the process that will result in all the highly differentiated tissues and organs of the adult is the first mitotic cleavage of that primal cell, the zygote. The genetic material doubles, virtually identical chromosomes move to opposite poles, and the cell cleaves into two daughter cells. Shortly afterward, each of these

Figure 8.2 Early development in the frog. The top row shows the fertilized egg and the first three cleavages; The middle row, blastula stage (left, cutaway) and the early gastrula stage (center, surface view and right). The bottom row shows the gastrula in later stages of development. Individual cells have by now become so small in relation to the embryo that they are no longer shown. The neural tube is the early spinal cord, the area which will become the brain. Why is it essential that the brain tissue develop so early?

daughter cells divides again.

The first two cell divisions are usually at right angles to each other, but along the same vertical axis; the third line of division is perpendicular to these (Figure 8.2). In the case of the frog this third cleavage is displaced toward one hemisphere called the *animal pole*. The reason for this displacement is that the relatively inert yolk material is concentrated in the other hemisphere, the *vegetal pole*. The cytoplasm in the area of the animal pole is less encumbered by yolk, and hence more active. The axis and polarity of these cleavages are determined by the point at which the sperm penetrated the egg at fertilization.

As the cells of the embryo continue to divide, they form a cluster, or ball of cells. These divisions are accompanied by new kinds of changes which produce a hollow cavity inside this ball, now called a *blastula*. The unequal distribution of the yolk has resulted in smaller cells of more nearly equal size having formed at the animal pole.

GASTRULATION

Early in the second day after fertilization, a process called *gastrulation* begins in the frog embryo. At this stage, the embryo, logically enough, is called a *gastrula*. Gastrulation involves the migration of certain cells toward an area in the vegetal hemisphere, where they converge and begin to roll under to form an opening called the *blastopore*. The rolling under takes place along a curved line, at what will be the back, or *dorsal* side of the yolk area. The process of involution extends outward from the edges of the crescent, until finally the crescent becomes a circle. The vegetal area inside this circle is called the *yolk plug*, and this will eventually be enclosed by the expanding layer of cells around it.

At this stage three types of cells may be distinguished. The ectoderm lies on the outside, an inner *endoderm* is formed from cells of the yolky vegetal area, and the cells which had rolled under from the outside and have come to lie between the ectoderm and endoderm become the *mesoderm*. Each of these cell layers, or *germ layers*, will give rise to very specific structures. For example, the ectoderm will form the outermost layer of the skin and certain other structures including the sense organs, parts of the skeleton in various animals in the head and neck and, importantly, the nervous system. The ectoderm will also form hair and feathers. The mesoderm will form tissue associated with support, movement, transport, reproduction and excretion (for example muscle, bone, cartilage, blood, heart, blood vessels, gonads and kidneys). The endoderm will produce structures associated with breathing and digestion (including the lungs, liver, pancreas and other digestive glands).

REGULATION

Now, at this stage of development it is hard to see just how this peculiar ball is going to turn into a frog. The precise changes involved are difficult even for students of embryology to visualize so we won't try your patience with much detail. Some understanding of the mechanisms, though, should provide food for thought. You might be interested in knowing, for example, that even at this early stage (the second day) the fate of some cells has been sealed for hours. In other words, they are on their way to assuming their final role, and the changes which have occurred in their genetic

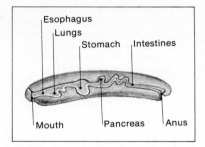

Figure 8.3 The basic tube-within-a-tube structure of advanced animals. (General body plan of vertebrates which is permitted by the events at gastrulation.)

Figure 8.4 At gastrulation the three primary germ layers of the body are established. While more than one layer may contribute to the development of a structure, some tissues are clearly derived largely from certain germ layers, as shown.

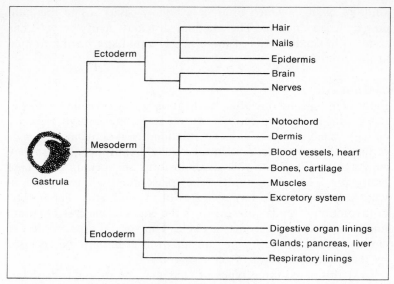

constitution are irreversible. The genes which have been shut down as the chromosomes direct the production of only certain kinds of *m*RNA can never be reactivated.

Other cells in the embryo have not yet specialized, and these cells can still theoretically become part of any number of organs or systems. In fact, if some of these generalized cells are transplanted to other parts of the embryo, they will give rise to the same types of cells as their new neighbors. Had they been left in their original location, they would have formed cells of an entirely different type. As each hour goes by, more and more cells become committed, as their changes take them past the point of no return.

The Flexible Fate

We know that in some species, such as the flatworm, the fate of most cells is never sealed; or how could regeneration take place when it is cut in half? Let's consider what happens to such cells in the process of regeneration. In some kinds of salamanders, when a leg is amputated the cells in the area of the cut despecialize and revert back to the more primitive state typical of embryos. Then they begin to specialize again, some taking entirely new developmental pathways, until finally the limb is restored.

If the leg of a mouse is amputated, however, it doesn't grow back. In these higher animals the cell differentiation is more permanent, and the cells are less able to resort to an earlier condition. Epidermal cells, it is true, may migrate to grow over and protect the cut area, and some cells may change to some degree because of the wound. But general deorganization and regeneration is not possible. We might well ask at this point what determines the form a cell will take. How is development regulated?

Sequencing

For the highly complex and integrated systems of an adult vertebrate to develop, an incredible array of regulatory mechanisms must be involved. These mechanisms must see to it not only that muscle cells develop in a limb area, but that muscles of very specific sizes, shapes, and attachments are formed.

The sequence of development is also important. For example, before the nerves appear which will regulate those first timorous palpitations in the incipient heart muscles, there must be some place for the blood to go. Thus the first step is the appearance of great channels, which will eventually become major vessels, in those looseknit, undifferentiated first-blood pools. Obviously, then, the sequencing of the tissues and organs is critical, but what regulates this sequence?

The exact mechanism of sequencing and other forms of regulation are still a mystery, but we can explain them conceptually in terms of the processes we already know something about. We know, for example, that certain areas of the chromosomes can become activated while others shut down, enabling them to produce cells of various types. Certain cells may also slow down or stop their reproductive activities while others undergo a burst of reproduction, accompanied by a surge in the growth of the tissue they comprise. And, of course, cells can die. It is essential, in fact, that some cells die, such as those which lie between what will be the fingers of a developing hand. Hence the highly organized changes that take place in the embryonic body may actually be "only" the sum of a few kinds of cellular changes we already know something about.

Other changes in the cells of developing embryos probably occur as a function of time. For example, a particular segment along the length of a chromosome may have a shorter active life than another segment. After an initial period of activity, then, one set of genes might cease its activity while another continues. The question here, of course, is what determines the active life span of any set of genes. Why do some of them stop making *m*RNA so soon? Biologists are searching for the answer.

Stress

Another regulatory factor is stress. Just as karate practitioners build up certain tissues around their knuckles by pounding makiwara

Figure 8.5 Spinal cord induction. the notochord induces the ectoderm to thicken and fold, thus forming the spinal cord. (This figure shows development in a chick embryo, chosen for clarity.)

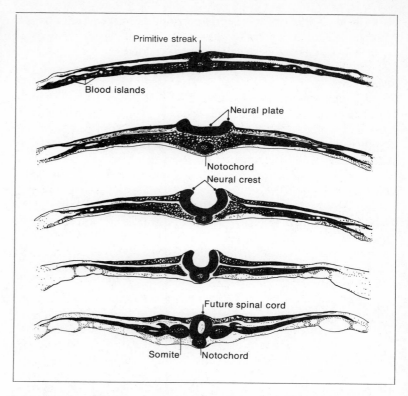

boards, so an embryo builds up tissue along its lines of stress. The bending and twisting of the developing embryo as it grows produces stress in certain areas. A type of mobile, undifferentiated tissue called *mesenchyme* then moves into these stressed areas, settles there, and begins reproducing and specializing as supportive tissue.

Induction

Regulation may also be accomplished by one type of tissue acting upon another. For example, after the mesodermal layer is formed, the tissue begins to thicken just under the area that will ultimately become the spinal cord. This thickened rod of mesoderm is called the *notochord*. The notochord itself doesn't form the spinal cord, but it influences the overlying ectoderm in a peculiar way. First, the ectoderm which lies over the notochord begins to thicken, then tissue starts to build up along each side of the long, thickened plaque (Figure 8.5). These two ridges, called the *neural folds*, enlarge and begin to fold toward each other until they join and fuse. The result is a hollow tube lying along the dorsal surface of the embryo. This is the early *spinal cord*. The spinal cord is then over-

Figure 8.6 In the developing vertebrate, outpocketings of the brain come to lie under the head ectoderm and induce it to thicken and form the lens of the eye. The brain will form the sensory layers of the eye.

grown by ectodermal cells from the surrounding area.

In most vertebrates, after the spinal cord forms, the underlying notochord that has induced its formation disappears, or at least diminishes greatly. Any supportive function it might have had is taken over by vertebrae which form from mesoderm lying alongside the spinal cord, eventually enclosing it. The role of the notochord as the inducer in this case has been demonstrated by transplanting segments of the notochord to other areas of the embryo. When this is done at a certain period, before the ectoderm has differentiated too far, incipient spinal cords can be made to develop almost anywhere over the surface of the embryo.

There are a number of inducers which operate in the developing embryo. In some cases tissues which have developed from the same germ layer may interact as inducer-inducee after they have differentiated to some extent. For example, the lens of the eye is induced in ectodermal cells by an underlying area of the brain, which has also formed from the ectodermal layer. Of course, these examples cover only a few aspects of the regulatory processes that govern the complex developmental changes in the embryo. However, they give you some idea of the role of regulation and a few of the ways in which it can work.

Now let's take a brief look at development in two other kinds of animals: birds and mammals. Keep in mind that although there are important differences between frogs, birds, and humans, many of the basic patterns of development are the same.

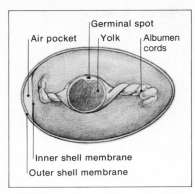

Air pocket | Germinal spot | Yolk | Albumen cords

Inner shell membrane
Outer shell membrane

Figure 8.7 A cross-section of a hen's egg. The embryo will take its food primarily from the yolk and the albumen. The albumen also acts as a shock absorber and the embryo is supported by twisted cords of albumen. Note that the white of the egg is on the outside of the zygote. The yolk will become incorporated into the body of the developing embryo.

DEVELOPING BIRDS AND THEIR MEMBRANES

Although birds' eggs are large, the early embryo is very small; in fact, the zygote is at first only a *germinal spot*. The life of a chicken, then, begins as a tiny disk on the yolk. And when cell division takes place, it is limited to that disk since cleavage of the great body of inert yolk is impossible. The blastopore in the chicken embryo does not develop as a crescent, but as a slit along what will be the body axis (Figure 8.8). As a result, the mesoderm is formed somewhat differently. The endoderm first splits away from the underside of the ectoderm, and cells from the upper ectodermal layer then roll under along the midline of the embryo to form the mesoderm.

After the spinal cord has formed, rapid growth in what will be the head and tail regions produces folds at either end of the embryonic axis. At each end of the body the ectoderm pockets in and the endoderm pockets out, bringing the two layers together. These thinned areas then rupture, forming the mouth and excretory openings, and the tube-within-a-tube structure, characteristic of vertebrates, is complete.

In the meantime, a membrane composed of endoderm plus mesoderm grows around the yolk. Blood vessels developing in the mesoderm are able then to carry food to the embryo. Another membrane, called the *allantois*, grows out of the rear area of the developing gut and forms a receptacle for nitrogenous wastes. This sac will be left behind when the chick hatches.

When the extra-embryonic membranes fuse over the embryo, the embryo is enclosed in a four-walled membrane. The inner two layers become the *amnion* and the outer two become the *chorion*, each consisting of a layer of mesoderm and a layer of endoderm.

The amnion will fill with fluid which acts as a shock absorber to protect the embryo from injury. This amniotic fluid also serves as a lubricant, so that as the limb buds and other projections develop they will not fuse to the body. The mesoderm layer on the outside of the allantoic membrane will later fuse with the mesoderm on the inside of the chorion to form a three-layered *chorioallantoic membrane*. This membrane will come to surround the albumen, and thus will be just under the porous shell. Its function, as you may guess, is to pick up oxygen, which diffuses in through the porous shell, and transport it to the embryo. It also provides a surface area through which carbon dioxide from the respiring embryo can leave the egg.

As the embryo continues to develop, it eventually separates from the yolk, except for a thin stalk through which food is carried. The anterior, or head, part of the spinal cord begins to bulge and form the lobes of the brain. The mesoderm on either side of the embryo segments into blocks, or *somites*, which will soon form the vertebrae and large trunk muscles. The mesodermal membranes we have been discussing have formed from lateral extensions of these somites.

A large blood vessel underneath the embryo begins to pulsate, and later it loops and fuses together to form the heart. Tiny flaps appear, which then develop into the limbs. The brain outpockets on either side to induce the great orbs which will be the eyes. Meanwhile, the endoderm also outpockets and proliferates in various places to form the highly complex glands and organs associated with digestion.

Many other barely perceptible changes occur as the embryo continues its transformation. At about the eleventh day there will be a great eruption of enzymes, and the various systems of the organism will begin to function in even greater harmony. After about three weeks the chick will begin its first feeble pecks to break free of its protective shell. When it finally emerges, it will begin to register strange and fascinating new experiences which further alter the arrangements of its constituents. Then someone will eat it.

HUMAN DEVELOPMENT

Now that we have some idea of certain general principles of development, let us consider the human animal. And, as you are doubtless aware, it is all too easy to find oneself plunged into controversy by this simple shift in species. When considering our own group there is an understandable tendency to seek "specialness" and to minimize the similarities with other, less "noble," species of the earth.

As an example of one controversy, there is an aged but growing argument over the period at which "life" begins for humans. There are those who see great importance in the zygote who would never argue that every gamete had the right to impose its genes on the world. Others would argue that a human embryo is not sacred or "one of us" until implantation occurs, or until the heart starts beating, or until the embryo begins kicking and moving (quickening), or until it draws its first independent breath. Others argue that it is *never* sacred and can be removed from the population at any point in its postnatal period without need for particular concern. Yet others argue that the sacredness of human life is somehow related to political or religious beliefs. It seems to be true that a life which, as a result of its experiences, has come to agree with us, is more important than one which has reached other conclusions. A prominent Southeast Asian leader recently noted that

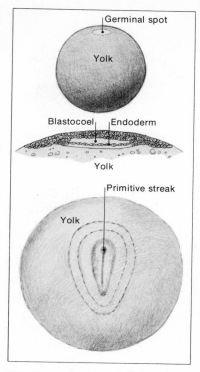

Figure 8.8 The early development of the chick. Cells along the primitive streak (an elongate blastopore) are rolling under. Some of these cells will form the rodlike notochord.

Figure 8.9 Later development of the chick embryo. Note the development of the membrane system, allantois and the erosion of the yolk mass. Gray areas show developing embryo.

Americans cannot be moved by the deaths of Asians. We are chagrined when trainloads of bombs, intended to be rained on people whom we have never seen, explode prematurely and kill Americans. There is no way to resolve such arguments on a scientific basis. We must establish our values and arbitrarily draw lines. Since some individuals find information useful in forming their opinions in such matters, we have an added incentive in considering the development of the human embryo.

Descriptions in bus-station novels notwithstanding, fertilization occurs with the mother-to-be totally unaware of the event. If there are sperm cells thrashing around in the genital tract at any time within forty-eight hours before ovulation to about twelve hours after, the odds are very good that pregnancy will occur. As soon as the egg is touched by the head of a sperm, it undergoes violent pulsating movements which unite the twenty-three chromosomes of the sperm with its own genetic complement. From this single cell, about $\frac{1}{175}$ of an inch in diameter, a baby weighing several pounds and composed of trillions of cells will be delivered about 266 days later.

For convenience, we will divide the 266 days, or nine months, into three periods of three months each. We can consider these *trimesters* separately, since each is characterized by different sorts of events.

THE FIRST TRIMESTER

In the first trimester the embryo begins the delicate structural differentiations that will lead to its final form. It is therefore particularly susceptible during this period to any of a number of factors that might influence its development. In fact the embryo often fails to survive this stage.

The first cell divisions result in cells that all look about alike and have roughly the same potentials. In other words, at this stage the cells are, theoretically anyway, interchangeable. Seventy-two hours after fertilization, the embryo will consist of sixteen such cells. (So, how many divisions will have taken place?) Each cell will divide before it reaches the size of the cell that has produced it; hence the cells will become progressively smaller with each division. By the end of the first month the embryo will have reached a length of only $\frac{1}{8}$ inch, but it will consist of millions of cells.

The First Month

The first cleavages actually take place in the oviduct as the zygote moves toward the uterus, pushed along by the beating cilia that line the oviduct and by contractions in the oviduct walls. When

Figure 8.10 A 72-hour chick embryo. The embryo has rolled onto its side and flexed its back. The most prominent areas are swollen lobes of the developing brain which lie over the beating heart. (Courtesy: Grant Heilman, Lititz, Pa. Photo by Runk/Schoenberger)

the zygote reaches the uterus, it does not implant for about three days, although the uterine wall is already swollen and ready to receive it. In fact, by the time it implants it has already reached the hollow-ball blastula stage. One side of the blastula is thin-walled, and this portion, called the *trophoblast*, will give rise to the membranes that will come to surround the embryo. The thicker-walled side, called the *inner cell mass*, will become the embryo proper.

Figure 8.11 Relative sizes of the human embryo at various stages in its development.

14 days

18 days

24 days

4 weeks

6½ weeks

7½ weeks

9 weeks

11 weeks

15 weeks

The trophoblast secretes a *proteolytic* (digestive) enzyme which breaks down some of the blood vessels in the uterine wall beneath it. The embryo thus comes to lie in a pool of the mother's blood. The bleeding caused at this time by implantation is sometimes mistaken for menstruation. After about a week the embryo will have eaten its way deep into the uterine wall, where it will lie steeped in a bloody glycogen-laden fluid—a fluid which will sustain it during these early days. In about twelve days the injured uterine wall will have repaired itself, and the embryo, buried deep within the flesh of the uterine wall, will silently begin to change.

After implantation the trophoblast extends fingerlike projections called *chorionic villi* into the uterine wall. At the same time the mother's tissues undergo certain changes in the areas around the villi. Thus chorionic tissues of the embryo, together with these maternal tissues, form the *placenta*, which serves as the organ of transport between the embryo and mother. Through the placenta will pass food and oxygen from the mother and waste and carbon dioxide from the fetus.

At the end of a month the embryo is still smaller than a pea (Figure 8.11), but momentous changes have occurred within it. The anterior end of the spinal cord has developed bulges and lobes as it begins to form the brain. The mesoderm has divided into thirty-two pairs of dense blocks of somites. These will later give rise to much of the skin and the voluntary muscles. Three pairs of arches have also appeared in the neck region, through which arterial blood is pumped. In the human embryo these will later disappear, but in fish they become the gills. The opening which will be the mouth has broken through, although no food will enter here for some time yet. Even large visionless eyes have begun to appear.

Also during this first month roaming cells in the yolk sac and chorion change in appearance and become blood cells. These, by day 24, will be pumped by a tubular incipient heart through unfinished blood vessels and will begin picking up oxygen which has diffused across the placenta from the mother's blood. Near the end of the first month primitive kidneys form, but these will function in removing nitrogenous waste from the blood for only a few days. They will be replaced by a more specialized kidney later. The first cells of the endocrine glands are now apparent. And the tiny limb buds have appeared which will eventually form the arms and legs.

The Second Month
In the second month the features of the embryo become more recognizable. Bone begins to form throughout the body, primarily

in the jaw and shoulder areas. The head and brain are developing at a much faster rate than the rest of the body, so that at this point the ears appear and open, lidless eyes stare blankly into the amniotic fluid. The circulatory system is developing and blood is pumped through the umbilical cord out to the chorion, where it receives life-sustaining nutrients and deposits the poisons it has removed from the developing embryo. The nitrogenous wastes and carbon dioxide filter into the mother's bloodstream, where they will be circulated to her own kidneys and lungs for removal. At about day 46 the primordial reproductive organs begin to form, either as testes or ovaries, and it is now, for the first time, that the sex of the embryo becomes apparent. Near the end of the second month fingers and toes begin to appear on the flattened paddles which have formed from the limb buds. By this time the embryo is about two inches long and is more or less human in appearance; it is now called a *fetus*.

The Third Month
Growth and differentiation continue during the third month, but now the fetus begins to move. It breathes the amniotic fluid in and out of bulblike lungs and swallowing motions become distinct. At this point individual differences can be distinguished in the behavior of fetuses. The clearest differences are in their facial expressions. Some frown a lot; others smile or grimace. It would be interesting to correlate this early behavior with the personality traits that develop after birth.

At the end of the first trimester there are pronounced changes in the reproductive and excretory organs of the embryo, and its urine begins to appear in the amniotic fluid—to be filtered into the mother's blood where it is removed by her kidneys. Bone continues to form throughout the body. Each bone arises independently, usually in an area that will be near the middle of its mass; it then grows outward to meet other bones at what will be the joints. Although all the organ systems have been formed by this time, if the fetus is removed from its mother it will die.

THE SECOND TRIMESTER
In the second trimester the fetus grows rapidly, and by the end of the sixth month it may be about a foot long, although it will weigh only about a pound and a half. Whereas the predominant growth of the fetus during the first trimester was in the head and brain areas, during the second trimester the body grows at a much faster relative rate than the brain and begins to catch up in size with the head.

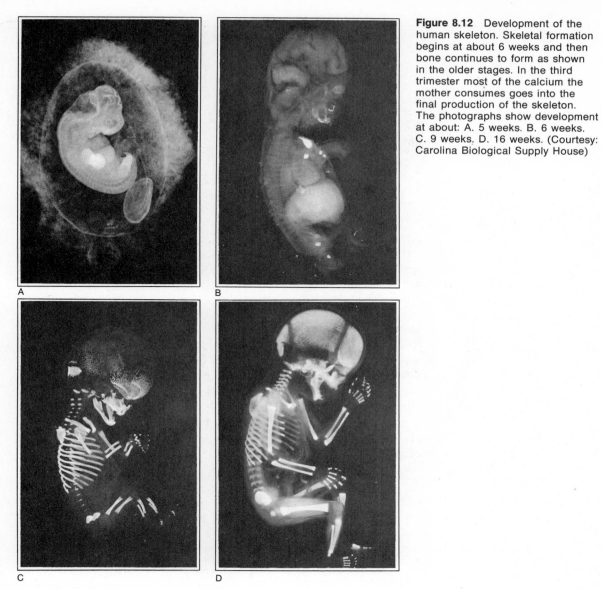

Figure 8.12 Development of the human skeleton. Skeletal formation begins at about 6 weeks and then bone continues to form as shown in the older stages. In the third trimester most of the calcium the mother consumes goes into the final production of the skeleton. The photographs show development at about: A. 5 weeks. B. 6 weeks. C. 9 weeks. D. 16 weeks. (Courtesy: Carolina Biological Supply House)

The Fourth Month

In the fourth month the back becomes less curved, and the abdominal muscles grow to enclose the viscera. The muscles become much stronger and the fetus is able to hold up its head. Fingerprints are beginning to appear as an adaptation for government employment. If the fetus is stimulated during this period, it will react at first with very general bodily movements, but later in the

Chorionic villi

Uterine lining

Placenta

Remnant of yolk sac

Amnion

Amniotic fluid

Umbilical cord

Chorion

Uterine smooth muscle

Uterine cavity

Cervical plug

Vagina

fourth month the fetus will respond with specific reflexes.

During the first trimester blood cells were formed mainly in the liver and spleen. In the second trimester the bone marrow begins to form blood cells. The fetal heart is now pumping strongly, and the heartbeats can be heard with a stethoscope.

The Fifth Month

The fetus is by this time behaving more vigorously. It is able to move freely within its sea of amniotic fluid and the delighted mother can feel it kicking and thrashing about. Interestingly, the fetus must sleep now, so there are periods when it is inactive. It is capable of reacting to more types of stimuli as time passes. For example, by the fifth month the eyes are sensitive to light, although there is still no sensitivity to sound. Other organs seem to be complete, but remain nonfunctional. For example, the lungs are developed, but they cannot exchange oxygen. The digestive organs are present, but they cannot digest food. Even the skin is not prepared to cope with the temperature changes in the outside world. In fact, at the end of the fifth month the skin is covered by a protective cheesy paste consisting of wax and sweatlike secretions mixed with loosened skin cells (*vernix caseosa*). The fetus is still incapable in nearly all instances of surviving alone.

The skeleton has been developing rapidly during the second trimester, with some bones arising anew from undifferentiated embryonic cells, and others forming through the gradual replacement of cartilage cells by bone cells. At this stage the mother must supply large amounts of calcium and other bone constituents in order to provide building materials for the fetal skeleton.

The Sixth Month

By the sixth month the fetus is kicking and turning so constantly that the mother often must time her own sleep periods to coincide with her baby's. The distracting effect has been described as similar to being continually tapped on the shoulder, but not exactly. The fetus moves with such vigor that its movements are not only felt from the inside, but can be seen clearly from the outside. To add to the mother's distraction, the fetus may even have periods of hiccups. By this stage it is so large and demanding that it places a tremendous drain on the mother's reserves.

At the end of the second trimester the fetus has the unmistakable appearance of a human baby (or a very old person, since its skin is loose and wrinkled at this stage). In the event of a premature birth around the end of this trimester, the fetus may be able to survive.

Figure 8.13 A human fetus suspended in its protective amniotic fluid. Note the relationship of the various tissues associated with pregnancy.

Figure 8.14 Motherhood may be considered an extreme form of altruism. As the infant places greater demands on the mother's body, she is placed at a severe physical disadvantage. Birth itself is a risky business and then only signals the start of years of self-sacrifice for many mothers. (Courtesy: Photo by Erika)

THE THIRD TRIMESTER

During the third trimester the fetus grows until it is no longer floating free in its amniotic pool. It now fills the abdominal area of the mother. The fetus is crowded so tightly into the greatly enlarged uterus that its movement is greatly restricted. In these last three months the mother's abdomen becomes greatly distended and heavy, and her posture and gait may be noticeably altered in response to the shift in her center of gravity. The mass of tissue and amniotic fluid that accompanies the fetus ordinarily weighs almost twice as much as the fetus itself. Toward the end of this period, milk begins to form in the mother's mammary glands, which in the previous trimester may have undergone a sudden surge of growth.

At this time, the mother is at a great disadvantage in several ways in terms of her physical well-being. About 85 percent of the calcium she eats goes to the fetal skeleton, and about the same percentage of her iron intake goes to the fetal blood cells. Of the protein she eats, much of the nitrogen goes to the brain and other nerve tissues of the fetus.

Some interesting questions arise here. If a woman is unable to afford expensive protein-rich foods during the third trimester, what is the probability of a lowered I.Q. in her offspring? On the average the poorer people in this country show lower I.Q. scores. Are they poor because their I.Q.'s are low, or are I.Q.'s low because they are poor? Is there a self-perpetuating nature about either of these alternatives?

In the third trimester, the fetus is large. It requires increasingly greater amounts of food, and each day it produces more poisonous wastes for the mother's body to carry away. Her heart must work harder to provide food and oxygen for two bodies. She must breathe, now, for two individuals. Her blood pressure and heart rate rise. The fetus and the tissues maintaining it form a large mass which crowds the internal organs of the mother. In fact, the crowding of the fetus against the mother's diaphragm may make breathing difficult for her in these months. Several weeks before delivery, however, the fetus will change its position, dropping lower in the pelvis (called *"lightening"*) and thus relieve the pressure against the mother's lungs.

There are important changes occurring in the fetus in these last three months, and some of these are not very well understood. The effects of these changes, however, are reflected in the survival rate of babies delivered by Caesarian section (an incision through the mother's side). In the seventh month, only 10 percent survive; in the eighth month, 70 percent; and in the ninth, 95 percent survive.

Interestingly, there is another change in the relationship of the fetus and mother at this time. Whereas measles and certain other infectious diseases would have affected the embryo during the first trimester of pregnancy, at this stage the mother's antibodies confer an immunity to the fetus, a protection which may last through the first few weeks of infancy.

At some point about 255 to 265 days from the time of conception the life-sustaining placenta begins to break down. Certain parts shrink, the tissue structure begins changing, and the capillaries begin to disintegrate. The result is a less hospitable environment for the fetus, and premature births at this time are not unusual. At about this time the fetus slows its growth, and drops into position with its head toward the bottom of the uterus. Meanwhile, the internal organs undergo the final changes which will enable the newborn to survive in an entirely different kind of world. Its home has been warm, rather constant in its qualities, protected, and confining. It is not likely to encounter anything quite so secure again.

BIRTH

The signal that there will soon be a new member of the earth's most dominant species is the onset of *labor*, a series of uterine contractions which usually begin at about half-hour intervals and gradually increase in frequency. Meanwhile, the sphincter muscle around the cervix dilates, and as the periodic contractions become stronger, the baby's head pushes through the extended cervical canal to the opening of the vagina. The infant is finally about to emerge into its new environment, one which, in time, may give it the chance to propel its own genes into the gene pool of the species.

Once the baby's head emerges, the pattern of uterine contractions changes. The contractions become milder and more frequent. After the head gradually emerges through the vaginal opening, the smaller shoulders and the body appear. Then with a rush the baby slips into a new world. As soon as the baby has emerged, the umbilicus by which it is attached to the placenta is tied off and cut. The placenta is expelled by further contractions as the *afterbirth*. The mother recovers surprisingly rapidly. In other species, which deliver their young unaided, the mother immediately chews through the umbilicus and eats the afterbirth so that it will not advertise to predators the presence of a helpless newborn. Fortunately, the behavior never became popular in our own species.

The cutting of the umbilicus stops the only source of oxygen the infant has known. There is a resulting rapid buildup of carbon dioxide in the blood which affects a breathing center in the brain.

An impulse is fired to the diaphragm, and the baby gasps its first breath. Its exhaling cry signals that it is breathing on its own.

In American hospitals the newborn is then given the first series of the many tests it will encounter during its lifetime. This one is called the *Apgar test series*, in which muscle tone, breathing, reflexes, and heart rate are evaluated. The obstetrician then checks for skin lesions and evidence of hernias. If the infant is a boy, it is checked to see whether the testes have properly descended into the scrotum. A footprint is then recorded as a means of identification, since the new individual, despite the protestations of proud parents, does not yet have many other distinctive features that would be apparent to the casual observer. And there have been more than a few cases of accidental baby-switching.

MISCARRIAGE

Of course, not all pregnancies are terminated by delivery of a healthy offspring. The environment in which the fetus develops is optimal in terms of security and comfort; however, the route between conception and birth is fraught with risk. Part of the risk stems from the very complexity of living things. There is a phenomenal array of physical and chemical interactions going on within the body of a developing organism. If, at any point along the line, a component fails in its function or its timing, the embryo may be lost. If this happens very early, the embryo may be broken down—digested—and reabsorbed by the mother's body, perhaps wthout her ever knowing she was pregnant. If the embryo or fetus should fail at some period later in its development, the body of the mother often mysteriously is able to recognize that something has gone "wrong," and the fetus is aborted as a "miscarriage." Actually, about half of all miscarriages involve fetuses which are structurally or physically abnormal.

Miscarriages may be caused by a variety of factors. For example, an embryo may not implant correctly if it invades the uterine wall too soon in its development. Alcoholism or heavy smoking by the mother may cause premature birth of otherwise healthy babies. A weak cervix may not be able to support a fetus and thus the offspring may be lost. In fact, so many things *can* go wrong that the wonder is that healthy babies are ever delivered—or so it seems when one reviews the long list of hazards which threaten embryos.

WHY HAVE CHILDREN?

It may be useful at this point to consider why people have babies at all, since childbearing certainly weakens and occasionally kills the mothers, and since there is a risk of having abnormal offspring which will never achieve independence. As you are undoubtedly aware from your own experience or from that of friends, people have children for a number of reasons—reasons of various levels of merit. A couple may feel that it might help to save an otherwise shaky or unwise marriage, or a man may feel it will focus a clinging wife's attention on something else. In some cases it may be a response to questions such as, "When are you going to start your family?" (this, often, to a blissfully happy and devoted couple who thought they had), or "When are you going to give us some grand-children?" or even, "Why don't you have any children?" In our society it is generally more permissible to ask a couple this last question than to ask why they have so many children. In early America, and in many agricultural societies today, children were a "free" labor force as well as a form of social security. Even today there are rumors of Americans having children for tax or welfare advantages, but this is harder to understand considering the expense of children. Many people have children probably because they simply had never considered any alternative. It seems that, often, cultural traditions persist long past the circumstances which gave rise to them. We are taught from Dick and Jane books onward that "life" involves growing up within a family, enduring adolescence, getting married, having children and working hard to be a productive and accumulating member of the community all the while. It is unquestionably hard to break such cultural patterns, difficult both psychologically and sociologically, since aberrant behavior often goes unrewarded, or even punished to some degree, in our society. Benevolent smiles and well-wishes are usually reserved for those who go along.

There are also other reasons for having children, as you are well aware. Many people seem to have an innate affection for children, simply because they are children. Others seem to reserve their love for those children who are members of their own race or socioeconomic group—although such people may think of any child as "cute" (even those recovering from bombings which these people have helped to finance).

It is often said that children contribute to feelings of fulfillment for the parents. The "fulfilling" nature of children is an interesting one. Let's consider it briefly from a biological point of view. First, we must remember that we are all the result of millions of generations of successful reproducers. If our forebears had not reproduced we would not be here. In fact, we are likely to be the offspring of the *most* hard-driving reproducers of past generations. Those individuals of any generation who reproduced with less than maximum effort left fewer genes in the following generation; those more devoted to reproduction left a greater percentage. Hence, each generation would tend to consist increasingly of the offspring of the most successful reproducers, as the proportion of the genes of the less successful ones progressively diminished in the population through time.

Now, if the motivation to reproduce (or to do the things which result in reproduction) is an innate response to the influence of genes, we, as the offspring of the most successful reproducers, can be expected to feel rather strongly about having children ourselves and caring for them intensely. Such a feeling may take a variety of forms, from a peculiar longing to see one's "own" children and to interact with them, to a general positive response or protective feeling toward any baby or child.

Since we can expect variation between individuals for any genetically determined character, we can expect to find people among us who are less fascinated by babies, or who have less longing to see their own. What will happen to the genes of these people in the next generation? (Hint: if your parents didn't have any children you probably won't either.) Instead, the individuals in the *next* generation will be composed primarily of the offspring of what sort of people in *this* generation?

It might be mentioned that reproductive ability doesn't say much about other qualities an individual may possess. There is no evidence that Leonardo da Vinci left his peculiar genes or combinations of genes to subsequent generations. Jonathan Livingston Seagull may have been a remarkable flier and philosopher, but the next gull population would certainly consist of the offspring of humdrum, 9-to-5, hardworking, child-rearing gulls. Jonathan's genes would have died with him.

Of course there are innumerable social influences which affect our attitudes toward children, or anything else for that matter, and there is no intent to say here that any such complex creature as man is blindly obeying the callings of his genes. As we learn more about the concepts of modern biology, however, it is sometimes fun, and possibly important, to speculate on the influences of our ancestry. Which traits are influenced genetically and which are simply acquired? Have we made the conscious decision to love these individuals who are so helpless, demanding, inexperienced, inarticulate, small, dirty and broke? Are we *taught*, somehow, to love them? Or are we responding, as the descendants of the most successful reproducers, to the inner, compelling call of our heritage?

The question is not a simple one as you can see, although, if you broach the subject with many people you will get a string

of responses off the tops of their heads. Part of the problem is that we tend to consider social customs as being acquired culturally or artificially through conscious decisions made in some sort of a historical vacuum, especially if these customs appear in writing in the form of codes, laws or religious dogma. Perhaps, however, the "vacuum" is a myth. Any genetically influenced behavior will be the result of untold generations of development. As we become aware of the presence of that behavior in our society we might begin to consider it as normal or proper since it obviously works, and, in time, it might even come to be written into our traditions, religions or laws. Then, at some point we might note that there is some enforcement of the behavior, that people who, perhaps, do not wish to adopt the behavior are constrained, in some sense, to do so. As the environment changes, there may come a time when the old codes, which have worked so well and which have come to be embraced in our traditions, are no longer necessary. Then it is an easy step for those interested in the roots of our customs to believe the behavior exists as a dogmatic cultural phenomenon, and to change it we need only to reeducate society, change our institutions, and/or rewrite our laws. We can see the problems inherent in this position in light of biological principle. Possibly our attitudes toward children, mates, society, war, competition or territory could be more readily modified to fit our new kind of world if we could begin to realize that we are the products of evolution. Love of children, jealousy of mate, defense of territory and aggression against competitors brought our species to whatever success we enjoy today. If we have changed our world so that these qualities are unimportant (or threaten us), we might deal more successfully with these relic tendencies by recognizing their origins. Such recognition does not imply that they cannot be overcome. After all, man is a brainy and determined creature. We have been able to overcome gravity, but we did not do it by denying its existence. Instead, we recognized it for what it was and dealt with it accordingly.

9
Systems and Their Control

Have you ever wondered why, on your way to school, you have never encountered a 7-ton amoeba in the parking lot? There are probably a lot of good reasons. For one thing, an amoeba is a *protozoan,* a one-celled animal, and as you know, the volume of an object increases at a proportionately greater rate than its surface area. Thus a 7-ton amoeba would have an unsatisfactory surface-to-volume ratio. In other words, its regulatory surface membrane would be too small to accommodate its mass. Second, the unsupported weight would create such great internal pressures that the animal would probably rupture, causing a temporary parking inconvenience. Third, oxygen and carbon dioxide can move only across wet membranes, and the amoeba would have no way of keeping its membrane wet. Moreover, even if oxygen were able to move in the amoeba's body, once it crossed the membrane the only way it could reach the inner cytoplasm would be by diffusion. aided by some circulation resulting from the animal's sluggish movements. Thus the movement of carbon dioxide and oxygen through the body would be very slow and inefficient. Since amoebae have these problems, then, they must remain small and live in water.

There are 7-ton animals that live on dry land, although fortunately we rarely encounter them. These animals have had to solve all the attendant problems of large size and terrestrial life. For example, they have developed *multicellularity.* This permits groups of cells to act together to perform special tasks, such as supporting weight, circulating fluids, and gaining oxygen. These groups of cells are called *tissues.* Groups of tissues which have the same functions are called *organs.* The tissues and organs associated with a particular life process are loosely termed as *systems.* Thus we can talk about supportive, contractile, and integumentary systems, respiratory, circulatory, and digestive systems, or the hormonal and nervous systems. In this chapter and the next we will

Figure 9.1 A group of sponges (left), a section through a simple sponge (right) and a spicule being formed from mesenchyme cells (lower).

Figure 9.2 A molting crayfish. (Courtesy: John H. Gerard from National Audubon Society)

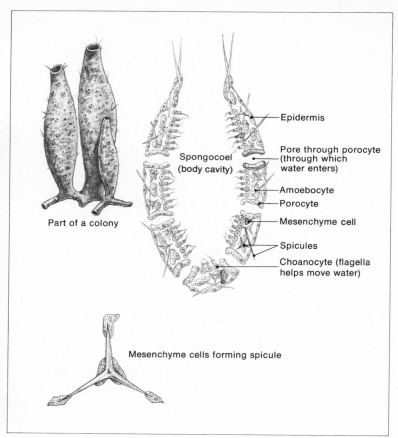

Part of a colony

Spongocoel (body cavity)

Epidermis

Pore through porocyte (through which water enters)

Amoebocyte

Porocyte

Mesenchyme cell

Spicules

Choanocyte (flagella helps move water)

Mesenchyme cells forming spicule

consider the various systems of multicellular organisms. We don't have the space, or for that matter, the energy, to consider the systems of all the phyla, so we will give attention to only a few groups whose body organizations illustrate the evolutionary descent of the organ systems.

SUPPORTIVE SYSTEMS

We have noted that the great stems and trunks of plants acquire their structural strength from the combined strength of many tiny cell walls. Animal cells, unlike plant cells however, do not have cell walls, and animals' bodies are instead supported by various *skeletal* structures consisting of material that has been secreted by cells and then hardened to some degree. Such materials include bone and cartilage in animals that have internal skeletons and *chitin*, the major component of the outer skeletons of insects. In certain soft-bodied animals such as hydra, flatworms, and earthworms, a certain rigidity of the body results from the compression of cell fluids, by muscular contraction forming a type of hydrostatic skeleton.

TYPES OF SKELETAL SYSTEMS

One of the most primitive skeletal systems is that of the sponge (Figure 9.1). In these animals the body is largely supported by the water in which they live, so they have developed only weak skeletons. In fact, sponge skeletons consist only of tiny *spicules* scattered through the tissue. These spicules are formed from unspecialized mesenchyme cells, but only particular kinds of spicules develop. The 10,000 species of sponges are classified according to the type of spicules that comprise their skeletons. The various classes have either calcium carbonate spicules, six-rayed glassy spicules, or fibrous or glassy spicules without six rays.

Some kinds of animals have skeletons outside their bodies, called *exoskeletons*. These include such species as lobsters, snails, insects, and oysters. Such skeletons serve well in protecting internal parts (as do our skull and rib cage), but animals with exoskeletons may have a problem with growth. Snails and clams simply secrete extensions to their shells as they grow, so that their shells grow with them. Lobsters, crabs, and crayfish, however, have to discard their old shells and grow new, larger ones. They do this by first withdrawing the valuable minerals from their shells, which causes the shell to soften. The shell then splits down the back. The soft, vulnerable animal crawls out, swells by absorbing quantities of water, grows rapidly for a time, and then secretes a new shell into which it redeposits the minerals.

Vertebrate animals, including man, have an *endoskeleton*, a skeleton inside their bodies. In the most primitive vertebrates, such as sharks and rays, the skeleton consists entirely of cartilage. In most other vertebrates, however, the cartilage that forms in the developmental stages is replaced by bone. Also, some parts of the skeleton form directly as bone.

TYPES OF CONNECTIVE TISSUE

Cartilage and bone, as well as tendons, ligaments, and fibrous connective tissue, are all referred to as *connective tissue*. Generally, cells comprise only a small part of connective tissue; most of the mass actually consists of substances secreted by the cells. These substances are nonliving and are called *matrix*. The specific nature

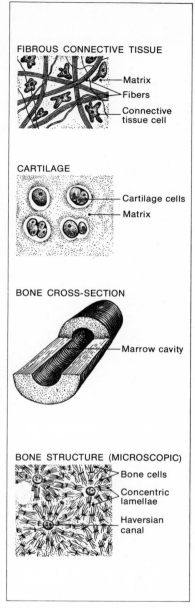

Figure 9.3 Connective tissue types.

of each type of connective tissue is largely dependent on the qualities of the matrix (Figure 9.3).

The matrix of bone contains calcium salts, which make the bone hard, and a protein called *collagen*, which gives the bone some degree of flexibility. Because of its high density, bone tends to be heavy, but this weight is offset by the fact that most larger bones are hollow. The tissue inside these hollow cavities is called *marrow*. *Yellow marrow* consists mostly of fat cells, while *red marrow* is the site of the formation of red blood cells and some of the body's white blood cells. Running throughout the bony matrix are *Haversian canals*, small channels encompassing the blood vessels and nerves which serve the bone cells. The bone cells and the Haversian canals are connected by little canals called the *canaliculi*, through which the bone cells receive food and oxygen and eliminate waste.

The matrix of bone also contains cells which have the peculiar ability to withdraw calcium salts from the matrix so that the shape of a bone can change in response to the stresses placed on it. We see such changes in certain African cultures in which the practice of binding the woman's head in a tight metal ring results in a change in head shape.

In *fibrous connective tissue* the matrix consists of a heavy, interlaced mat of microscopic fibers which are secreted by the connective tissue cells (Figure 9.3). Such tissue is found throughout the body as a binding material, holding skin to muscles, holding glands in position, and binding bones together.

Tendons and *ligaments* are specialized types of fibrous connective tissue. Ligaments are fairly elastic and serve to bind one bone to another. Tendons are inelastic but flexible cords which connect muscle to muscle or muscle to bone. (The fibrous connective tissue just below the skin, when it is tanned, becomes leather.)

THE HUMAN SKELETON

The adult human skeleton consists of about 200 bones (the number changes as some bones fuse with age). In humans, as well as in other vertebrates, these bones make up two basic skeletal parts called the axial skeleton and the appendicular skeleton (Figures 9.4, 9.5).

The *axial skeleton* is that part of the skeleton which runs approximately along the axis of the body—that is, in the direction of the backbone. Thus it includes the skull, the spinal column, the *sternum* (breastbone), and the ribs. The skull consists of the fused bones of the braincase, or *cranium*, and the bones of the face. The backbone is made up of thirty-three separate *vertebrae*, which differ

Skull

Clavicle

Scapula

Sternum

Humerus

Rib cage

Vertebra

Radius

Pelvic girdle

Ulna

Carpals

Metacarpals

Phalanges

Femur

Patella

Fibula

Tibia

Tarsals

Metatarsals

Phalanges

AXIAL SKELETON (Dark)   APPENDICULAR SKELETON (White.)

Figure 9.5 The axial (dark) and appendicular (light) skeleton in frogs.

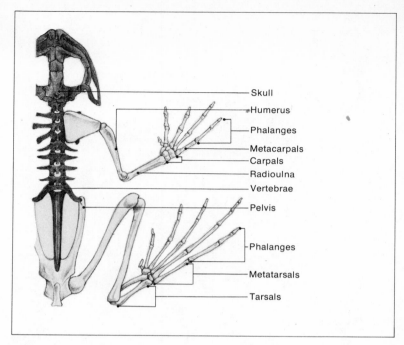

- Skull
- Humerus
- Phalanges
- Metacarpals
- Carpals
- Radioulna
- Vertebrae
- Pelvis
- Phalanges
- Metatarsals
- Tarsals

Figure 9.6 A lumbar (lower back) vertebra showing the centrum which articulates with vertebra on either side, the neural arch through which the spinal cord passes, and the various projections for the attachment of ribs and muscles.

- Neural spine
- Neural arch
- Transverse process
- Centrum

somewhat according to their position along the length of the spine (Figure 9.6).

The ribs are attached dorsally (toward the back of the body) to the vertebral column, and ventrally (toward the belly surface) to the sternum or to cartilage attached to the sternum, except for the last pair, called *floating ribs*, which are not attached ventrally. The rib cage is flexible and slightly moveable so that it expands during breathing and can absorb blows in the protection of the heart, lungs and other vital organs.

The appendicular skeleton is comprised of the arms and legs and two girdle complexes (Figure 9.7). The pectoral girdle consists of two collar bones (*clavicles*) and two shoulder blades (*scapulas*). The pelvic girdle is composed of three bones fused to form the hips. You may find it surprising to learn that fish have rather primitive hip and shoulder girdles. On the other hand, fossil evidence indicates that the arms and legs of land animals evolved from the fins of early fish, so perhaps, all things considered, the girdles in fish are not so surprising.

The upper parts of both the arms and legs are comprised of single bones which are joined to two bones of the lower limbs (Figure 9.8). Both the arms and legs terminate in a complex of several bones. The arms end with the wrists (*carpals*) and the legs with the ankles (*tarsals*), which slide past each other to permit a rather wide range of movement for the hands and feet. The hands and feet are made up of bones called the *metacarpals* and *metatarsals*, respectively. The hands and feet each terminate in five digits, the fingers and toes. One of the identifying characteristics of humans is a highly moveable and specialized hand existing on the same body with a much less talented foot. Interestingly enough, both the human arms and legs are rather primitive structures as vertebrate limbs go.

Figure 9.7 The pectoral (A) and pelvic girdles (B). In humans, the bones of the pectoral girdle are loosely joined and can be fixed. In leaping animals, such as cats, the collarbone is greatly reduced so as not to interfere with the forward movement of the forelegs.

TYPES OF JOINTS

Some bones move freely, while others are only slightly moveable, and yet others are solidly fused together. The freedom of movement of any bone depends on its relationship to other parts of the body. The most moveable bones are those which form points of *articulation,* or joints, with other bones (Figure 9.9). The *ball-and-socket joints* that connect the humerus to the scapula at the shoulder and the femur to the pelvis at the hip permit great freedom of movement. The hinge points at the knee and the elbow also provide free movement, but only in very restricted directions, back and forth. A joint such as that at the pubic symphysis permits only slight flexibility, as in the separation of the hips when giving birth. The bones of the skull, which arise separately and later fuse along *suture lines*, are completely immobile—thus the difficulty of flexing one's head.

Clavicle

Scapula

Humerus

Radius

Ulna

Carpals

Metacarpals

Phalanges

Arm and Leg

Hip bone

Femur

Tibia

Fibula

Tarsals

Metatarsals

Phalanges

Now let us consider how all these bones are able to move with such precision that a cigarette can be rolled with one hand, and such strength that, in the next instant, tall fences can be leaped with a single bound.

CONTRACTILE SYSTEMS

TYPES OF MUSCLE

Muscles are made up of *contractile* fibers and may be divided into three types, according to their structure and the nerve stimulus that activates them. *Smooth muscle* is found in the walls of the digestive tract, and around some blood vessels and certain internal organs. *Cardiac muscle* comprises the walls of the heart. Both these types of muscles have long been termed "involuntary," since they function without conscious control. Thus you don't have to lie awake nights keeping your heart beating or remember to keep breathing in and out. There is evidence that some degree of control over such "involuntary" responses can, in fact, be learned. We'll discuss some of the research in this area in the next chapter. Meanwhile, what we are really interested in here is a third kind of muscle, the voluntary muscles, *skeletal muscles*. These are the muscles that enable you to move your body.

Both the skeletal muscles and the cardiac muscles have several nuclei in each fiber, or cell, whereas smooth-muscle fibers have only one nucleus each (Figure 9.10). However, in the skeletal muscles these nuclei lie just under the membrane of the fiber, instead of being scattered throughout the cell. In addition, the cells of skeletal muscles may be very long, perhaps even extending the entire length of the muscle. The microscope reveals that skeletal and cardiac muscles have alternate light and dark bands, called *striations*, about which we will have more to say later. The characteristics of the three muscle types are summarized in Table 9.1.

SKELETAL MUSCLES

Typically, a skeletal muscle is a bundle of millions of individual muscle fibers bound together by connective tissue. (It is the connective tissue which makes some meat so tough.) The muscle is surrounded by a very tough sheet of whitish connective tissue called the *fascia* (those people who study it are not called fascists!). The slippery fascia permits the muscles to move past each other with a minimum of friction.

Usually, the ends of a muscle are attached to two different bones by tendons. When the muscle is contracted, it becomes shorter and thicker, drawing the bones closer together. Ordinarily one of the bones moves less than the other or is completely sta-

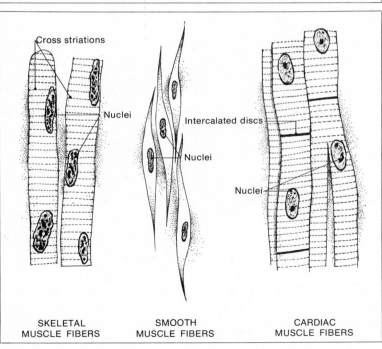

Figure 9.9 Representative joints in the human body.

Sutures (immovable)

CRANIUM

Pubic symphysis (slightly movable)

PELVIS

Ball and socket joint (free movement)

Hinge joint (moderate movement)

KNEE

HIP

Figure 9.10 The three basic types of muscle tissue in humans.

Cross striations

Nuclei

Intercalated discs

Nuclei

Nuclei

SKELETAL MUSCLE FIBERS

SMOOTH MUSCLE FIBERS

CARDIAC MUSCLE FIBERS

Figure 9.8 The human arm (opposite page) and leg are very similar in their basic design. What are their obvious differences? Are there any differences in moveability?

Systems and Their Control 219

Table 9.1—Comparison of Muscle Types

	Skeletal	Smooth	Cardiac
Location	Attached to skeleton	Walls of viscera, around arteries	Wall of heart
Shape of fiber	Elongate, cylindrical	Elongated, spindle-shaped	Elongated, cylindrical, branched
Number of nuclei	Many	One	Many
Position of nuclei	Peripheral	Central	Central
Cross striations	Present	Absent	Present
Speed of contraction	Rapid	Slow	Intermediate
Ability to remain contracted	Least	Most	Intermediate
Type of control	Voluntary	Involuntary	Involuntary

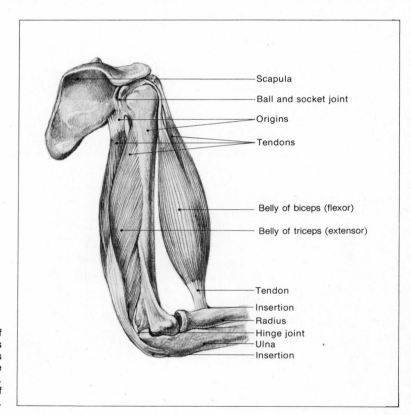

Scapula
Ball and socket joint
Origins
Tendons
Belly of biceps (flexor)
Belly of triceps (extensor)
Tendon
Insertion
Radius
Hinge joint
Ulna
Insertion

Figure 9.11 The attachments of the biceps brachii muscle and its antagonistic muscle, the triceps brachii. Note what effect the contraction of either would have. Belly simply refers to the body of the muscle.

tionary. The bone which moves less provides the area of attachment for the end of the muscle called the *origin*; the moveable bone provides the surface for the muscle's *insertion*.

Now let's consider an example. If you ask a fellow to show you his muscle, if he is a gentleman he will display his *biceps brachii* (Figure 9.11). Note that the origin of this muscle is at the joint of the humerus and scapula (*biceps* means "two-headed," so note also the double origin). The muscle inserts on the radius, the arm bone which terminates behind the thumb. As the forearm is rotated, the radius describes an arc, hence its name. As the biceps is *flexed*, the lower arm is drawn toward the upper arm. The maximum circumference of the biceps can then be measured by those crass souls interested in such things.

Antagonistic Muscles

On the other side of the humerus from the biceps brachii there is another large muscle. This one originates at the juncture of three bones and is called the *triceps brachii*. It inserts on the elbow, or *olecranon process*, the part of the ulna which extends past the elbow joint. As the biceps is flexed, consider what happens to the triceps. Obviously, it must relax and lengthen. Then, when the triceps is flexed, the arm straightens as the biceps lengthens. Thus muscles usually work in opposing pairs in an *antagonistic* system.

In some cases antagonistic muscles must contract simultaneously. For example, simultaneous contractions of opposing leg muscles permit us to stand. The *flexor* and *extensor muscles* of the leg pull against each other and lock the knee in place. The tension, of course, must be highly regulated. Otherwise, we would be leaping up from chairs and collapsing all over sidewalks. Walking also demands an extremely high degree of coordination between opposing muscles as the tension changes and weight shifts from one leg to the other.

Types of Movement

Whereas the contraction of the biceps functions in flexing the arm, other muscles have different sorts of actions. *Adductors* and *abductors* move parts of the body toward or away from the central axis of the body, as when the arm is pressed against the body or swung outward. *Levators* and *depressors* raise or lower parts of the body, as in shrugging or lowering the shoulders. *Pronators* rotate body parts downward, as when you turn your palm downward to look at your watch, and *supinators* rotate those parts upward, as when accepting M&Ms. *Sphincters* and *dilators* are rings of muscle which open and close body openings, such as the muscles of the lips. However, most muscles of this kind are involuntary, like those at the juncture of the stomach and small intestine or between the urinary bladder and the urethra.

The superficial muscles of the human body are shown in Figure 9.12. You need not memorize these, but look them over carefully to get a good working idea of how you're put together.

MUSCLE CONTRACTION

The striated muscle fiber, or cell, has several nuclei, as we have noted, and is surrounded by a rather tough cell membrane. Upon closer examination, we see that each fiber is made up of small threads of protein called *myofibrils*. It is the cross banding or striations in the myofibrils which gives the fiber its banded appearance. Muscle tissue is highly active, so you shouldn't be surprised to find that it contains a rather large complement of mitochondria, which, as you recall, are energy-producing organelles. The mitochondria lie snugly against the contractile units, the myofibrils.

As shown in Figure 9.13, the tiny myofibrils are made up of still smaller units, called *myofilaments*. There are two kinds of myofilaments; the thicker ones are about 100 Ångstroms in diameter and the thinner ones are about half that size. The banded appearance of the myofibril is due to the arrangement of the two kinds of myofilament in each thread.

The myofilaments consist primarily of two proteins, called *actin* and *myosin*. It is believed that the thinner filaments are composed of actin and the thicker ones are composed of myosin, since the dark bands formed by the thicker filaments disappear when the myosin is extracted from the tissue. The thick myosin filaments are found only in this darker A band; the I band contains only the thin actin filaments, although these may extend into the A band.

During contraction the A band remains constant in length while the I band shortens, and the length of the H zone within the A band also decreases. It has been suggested that the filaments themselves do not actually contract, but instead slide past each other. In other words, during contraction the actin filaments extend further into the A band, so that the H zone and the I band become narrower as the ends of the myosin filaments approach the Z band. Got that?

It is not known how the filaments are able to slide along each other or to maintain a state of contraction. It may be that the bridges along the myosin fibers (believed to be able to attach to the actin fibers) break and rejoin somewhere else along the length of the actin. Apparently the energy provided by ATP is utilized in the formation of these bridges. In fact it is theorized that one ATP → ADP conversion takes place for each bridge formed.

CONTRACTILE SYSTEMS IN OTHER ANIMALS

Because the skeleton and muscle systems are so strongly inter-

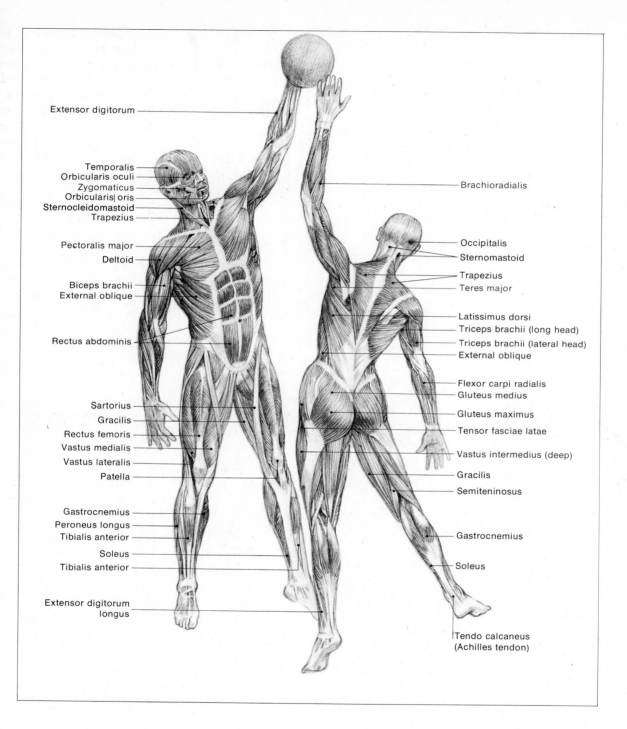

Extensor digitorum

Temporalis
Orbicularis oculi
Zygomaticus
Orbicularis oris
Sternocleidomastoid
Trapezius

Pectoralis major
Deltoid

Biceps brachii
External oblique

Rectus abdominis

Sartorius
Gracilis
Rectus femoris
Vastus medialis
Vastus lateralis
Patella

Gastrocnemius
Peroneus longus
Tibialis anterior

Soleus
Tibialis anterior

Extensor digitorum
longus

Brachioradialis

Occipitalis
Sternomastoid

Trapezius
Teres major

Latissimus dorsi
Triceps brachii (long head)
Triceps brachii (lateral head)
External oblique

Flexor carpi radialis
Gluteus medius

Gluteus maximus

Tensor fasciae latae

Vastus intermedius (deep)

Gracilis

Semiteninosus

Gastrocnemius

Soleus

Tendo calcaneus
(Achilles tendon)

Superficial frog muscles

Gastrocnemius muscle

Individual muscle fibers

Fibril

Nucleus

Z line

Fiber

Z line
I band
H zone
A band
I band
Z line

Banding pattern of filaments in fibril

Z line
I band
H zone
I band
Z line

Bridges

A band
Myosin
Actin

Actin
Myosin

A

B

C

D

related, we have considered the two systems in humans together. However, we should also learn something about the contractile systems of other organisms before we move on.

Plants are capable of movement, as almost everyone knows. The movement, however, is usually restricted to *tropic* responses, such as a leaning toward light, or the opening and closing of flowers or stomates. Such activities are usually responses to differences in growth rate or fluid pressure within parts of the plant. It should also be pointed out that in some plant species the organism, or its gametes, can travel about quite freely under its own power.

Coelenterates are animals which show *radial symmetry* and which are organized basically as hollow sacs. Animals in this group include the jellyfish, hydra, and sea anemone (Figure 9.14). Coelenterates never develop mesoderm and therefore cannot have true muscles. However they have, in their double-walled bodies, a system of contractile fibers. The fibers in the outer wall (formed from ectoderm) run lengthwise and those in the inner wall (formed from endoderm) run circularly. Since these animals are unrestricted by a rigid skeleton, they are able to perform some rather remarkable contortions.

The basic system of internal, longitudinally arranged, contractile fibers and external, circular fibers is also common in certain phyla of worms which have mesoderm, and hence, true musculature. Included here are roundworms (phylum *Nemathelminthes)*, (Figure 9.15), flatworms (phylum *Platyhelminthes)*, and segmented worms (phylum *Annelida)*. In addition to longitudinal and circular fibers, these more "advanced" phyla may also have additional muscle arrangements, such as diagonal or dorsoventral fibers.

In arthropods, the joint-legged group which includes the insects, the exoskeleton demands a different sort of muscle system. In this case the muscles, of course, lie within the skeleton. The joints are also unusual, in that they are simply thin, flexible places in the exoskeleton. Muscles may extend across a joint so that their contraction causes the joint to bend. Or a muscle may lie entirely within a single segment of an appendage, with its origin on the inner wall of the exoskeleton and its insertion on an *apodeme*—a long, thin, tough part of the exoskeleton that extends from one section of the body into the next (Figure 9.16).

The arrangement of muscles within an exoskeleton imparts remarkable strength to arthropods. You have undoubtedly noticed tiny ants carrying loads of food many times their weight or dragging along the corpses of huge insects such as crickets.

This muscle-skeleton arrangement is also responsible for the much-loved whine of the mosquito. The mechanism of the in-

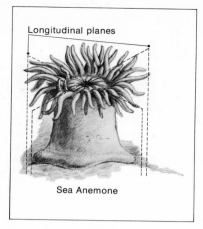

Longitudinal planes

Sea Anemone

Figure 9.14 Coelenterates, such as the sea anemone, show radial symmetry. In other words, they don't have a left and right side. Why do you suppose bilateral symmetry evolved—what advantage does it bestow?

Figure 9.13 How muscles are believed to work (left). The gastrocnemius, or "calf" muscle of a frog is used to illustrate the "sliding filament" hypothesis. In the lower row, A shows a section of a myofibril with its light and dark bands. B shows the molecular structure of a relaxed muscle. In C, the muscle is contracted. D shows a cross-section through C. Relate these areas to the description in the text.

Box 9.1 Classification Living things are divided, for convenience, into smaller groups. One such division separates plants and animals into two *kingdoms*. Thus, an organism in the animal kingdom could be expected to have more in common with another organism in its own kingdom than one from the plant kingdom. But among animals, certain ones share a number of traits in common, and these may be placed together into a *phylum*. Thus, all animals which have notochords at some period in their development are placed in one phylum, chordata. As finer and finer divisions are made among plants and animals, one finally arrives at a group of organisms which have so many things in common that they normally interbreed. These organisms are said to be of the same *species*. The taxonomic breakdown is:

Kingdom
  Phylum
    Class
      Order
        Family
          Genus
            Species

Members of two genera (plural of genus), then, won't share enough things in common to be able to interbreed, but they are all within the same family. And members of one family have more things in common with other members of the order than with families from another order.

Definitely not in our family!

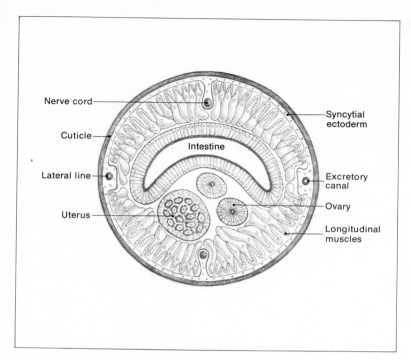

Figure 9.15 Cross-section of a roundworm.

Nerve cord

Cuticle

Lateral line

Uterus

Syncytial ectoderm

Intestine

Excretory canal

Ovary

Longitudinal muscles

Figure 9.16 A comparison of the musculature of an insect and man. Note, particularly, the relationship of the muscles and joints.

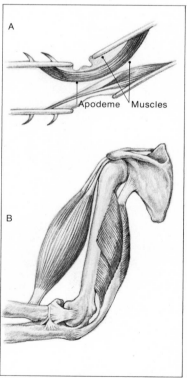

A

Apodeme   Muscles

B

credibly rapid contractions which move the wing once puzzled physiologists, but it has been found that only a single muscular contraction can set up a reverberation in the insect's exoskeleton. As the thoracic section vibrates, it causes the wings to move, thus a single muscle contraction may produce several wingbeats.

Most of the energy for muscle contraction in active animals involves the use of oxygen. So let's consider some of the ways living things move oxygen from the environment into their bodies.

EXTERNAL RESPIRATION

External respiration involves getting oxygen from the environment. Internal respiration, as you recall, involves intracellular metabolism and thus includes the use to which oxygen is put.

During the day the tissues of green plants may be steeped in oxygen which the plants have manufactured themselves. But at night when photosynthesis ceases, they may have to acquire oxygen from the environment. In the case of many plants, opened stomates allow oxygen to enter the body of the plant, where it can then diffuse across the membranes of thin-walled interior cells.

There are also other ways plants may gain oxygen. For example, in very thin blades of aquatic plants, oxygen may diffuse

Figure 9.17 The typical bony fish gill. The lamellar projections serve to increase the surface area of the gill. Note the blood vessels within each lamella. Blood passes from the gill arch into the lamellae of the gill filaments. Here gases are exchanged. The oxygenated blood then returns to the gill filaments. Here gases are exchanged. The oxygenated blood then returns to the gill arch and back to the body.

Figure 9.18 The external structure of the chiton, dorsal and ventral views. The animal attaches to rocks by suction produced by muscles of the foot. It feeds by rasping algae off rocks with a horny toothed structure called a radula. Between the foot and the upper body is a concavity in which lie 11 to 26 leaf-shaped gills. From the gills extend tiny cilia which, by their beating, keep oxygen-laden water flowing past them.

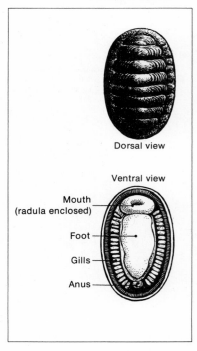

across the outer cell walls to permeate the plant's tissue. Many types of smaller aquatic animals acquire their oxygen by a similar method. Planarian flatworms have no organs for taking in oxygen, but their thin, flattened shape provides them with a large surface area in relation to their mass. Diffusion of oxygen from the water into their bodies thus gives them all the oxygen they need for their slow-paced lives.

Larger aquatic animals, even though they lead no more active or exciting lives than the flatworm, have proportionately less surface area exposed to oxygen-laden water; hence they need some more specialized system of obtaining oxygen. Consider the chiton, a mollusk which is common to both the Atlantic and Pacific coasts (Figure 9.18). The Atlantic species may be less than an inch long; the Pacific species run as long as 10 inches. You have probably noticed chitons clinging tenaciously to rocks along the shoreline and may have commented on their apparent lack of personality. Gills project from a concavity around their bodies, and each gill is covered with tiny cilia whose beating circulates oxygen-laden water past the gill. The gills, as would be expected, are very thin-walled and full of tiny blood vessels, so that oxygen can enter the bloodstream easily and be carried quickly to the rest of the body.

Figure 9.19 This battle-scarred shark lacks a water-pumping system and must stay in motion to keep oxygen-laden water flowing over its gills. (Courtesy: Elgin Ciampi from National Audubon Society)

Although fish are much more complex creatures than chitons, their means of acquiring oxygen are basically the same. The gills in fish are essentially protrusions which come into contact with the water (Figure 9.17). The cell membranes at this interface are very thin and readily permit the passage of gases across them. The gills themselves are laced with tiny, thin-walled capillaries which carry carbon dioxide out of the bloodstream and transport oxygen from the water to the rest of the body. Remember that there are normally higher carbon dioxide levels within the body, and higher oxygen levels outside it, and it is these differences which determine the direction of flow of the dissolved gases. Thus the higher level of carbon dioxide within the body causes it to move outward, and the lower level of oxygen in the body causes oxygen to move inward from the surrounding water.

Just as the chiton brings oxygenated water into contact with its gills by the movement of cilia, the fish washes water past its gills by taking in water through the mouth and pumping it out through the gill covering. Some fish, such as sharks, lack the muscles for pumping water and must swim constantly to keep water flowing over their gills. If they stop swimming they will suffocate.

Insects generally solve the oxygen problem in a different way. They breathe through a *tracheal system* of tiny tubes permeating their body tissue. These tubes open to the surface at tiny holes called *spiracles*. As shown in Figure 9.20, oxygen enters the tracheal system directly from the air and, aided by the insect's bodily move-

Figure 9.20 The tracheal system in a generalized insect. Air moves through a diffuse tubular system which penetrates almost every part of the body.

Longitudinal trachea   Spiracle

Figure 9.21 Seaweed has no need for a circulatory system since it is thin-bodied, with a large surface area exposed to the oxygen-laden environment. (Courtesy: C. G. Maxwell from National Audobon Society.)

Figure 9.22 The human respiratory system (right). The contraction of the diaphragm creates a partial vacuum in the chest and causes oxygen-filled air to rush in to fill the void. The air is warmed and moistened as it travels through the upper air passages. The air passes through increasingly smaller passages until it reaches the very thin-walled sac-like alvioli which allows oxygen to pass into the bloodstream and carbon dioxide to pass into the lungs from where it is expelled.

ment, moves through the tracheae deep into the body tissue, where it enters directly into the cellular environment. Thus a circulatory system is not very important for external respiration in insects. (But insects do have circulatory systems. Why is this true of insects?)

The insect's system of getting oxygen probably places an inherent limitation on its size, since a tracheal system could be efficient only for organisms below a certain size or body mass. Thus an evolutionary direction taken eons ago may be responsible for the fact that you don't have to contend with 800-pound mosquitoes or 600-pound crickets.

Among those hairy milk-givers called mammals, the group which includes humans, external respiration takes place by means of special pouches called *lungs*. Oxygen-laden air is brought into the lungs by *ventilation*, or breathing. The surface of the lungs is moist, and it is this thin film of water that allows oxygen from the air to dissolve into a solution so that it can diffuse across the thin cell membranes and enter the bloodstream. The blood transports the oxygen to the tissues where it diffuses into the cells. In the cells the oxygen picks up spent hydrogen atoms from the electron transport chain and is thus converted into water.

Meanwhile, the carbon dioxide that has been produced in metabolic reactions diffuses from the cells into the capillaries, which carry it to the lungs. In the tiny, thin-walled sacs of the lungs it then diffuses across the cell membranes into the lung cavity and from there it is exhaled out into the environment. So, the study of external respiration is simply the consideration of how oxygen is brought in to pick up used hydrogens, and how carbon dioxide, which is of no direct use to animals, is discarded. Figure 9.22 describes the human respiratory system.

CIRCULATORY SYSTEMS

We will now consider a few selected species to see how oxygen and carbon dioxide, as well as food and metabolic waste products, get from one place to another within living things.

You will recall that in the vascular system of higher plants, water enters the root hairs from the soil and is carried upward through the xylem. Food is transported throughout the plant from the leaves by the phloem. The problem of circulation in algae that live in water is solved by the fact that almost the entire plant is flattened, so that every cell can exchange materials with the surrounding water or with nearby cells which are in contact with water. Thus there is no need for a circulatory system, and this is why you don't see "veins" in the seaweed you find on the beach.

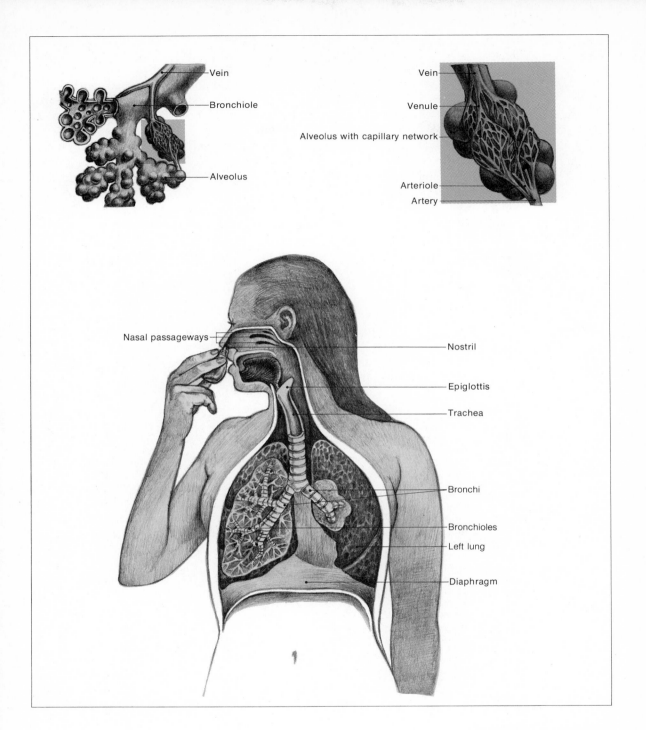

Vein

Bronchiole

Alveolus

Vein

Venule

Alveolus with capillary network

Arteriole

Artery

Nasal passageways

Nostril

Epiglottis

Trachea

Bronchi

Bronchioles

Left lung

Diaphragm

Dorsal | Ventral | Hearts | Capillaries
blood | blood
vessel | vessel

Figure 9.23 The closed circulatory system of the earthworm. The tubular rings which make up the hearts force blood backward through the ventral vessel and forward through the dorsal vessel. Note the branches of the major vessels which lead into other parts of the body. The blood is continuously enclosed in vessels, thus the system is "closed."

TYPES OF CIRCULATORY SYSTEMS

Size may be an important factor in the development of circulatory systems. In one-celled organisms and small multicellular species there may be no need for a circulatory system, since materials can easily diffuse from one part of the body to another and a large part of the total membrane area of the body is exposed to the outside environment.

Coelenterates have an unusual kind of circulatory system. Some of the cells which line the body cavity (or the inner layer of the two-layered animal) are *amoeboid.* That is, they change shape as they crawl and squirm their way through the body in amoeba fashion. When they reach the inner layer of the pouchlike animal, they may surround and engulf food particles just as an amoeba does. After engulfing food particles at the surface of the body cavity, they withdraw into the body with the nutrients and distribute them to other cells.

Flatworms also have saclike guts, with a single opening which serves both for food intake and elimination of waste. Since the body is flattened, no cell is very far from the gut, where digestion takes place, and thus nutrients can easily diffuse to the areas where they will be utilized. More complex animals, however, have specialized circulatory systems with some kind of device which pumps fluid throughout the body.

Circulatory systems may be classified according to whether they are *open* or *closed,* the open systems being more primitive. In mollusks, such as clams and snails, and in arthropods, such as insects and crayfish, a tubular heart pumps blood through vessels for a way, but then it empties into open cavities, or *sinuses.* From there it seeps or percolates through tissues until it begins to collect again in a large sinus around the heart. When the heart relaxes, blood flows in through tiny, one-way openings called *ostia* in the heart wall. When the heart contracts the ostia close, and blood is pumped out again through the vessels. Open circulatory systems, then, are those in which the blood is not continuously enclosed in vessels.

Among the most primitive of the closed systems are those of the annelids, or segmented worms. In the earthworm, for example, the heart consists of a series of five tubular rings which force blood backward through one large ventral vessel and forward through another large dorsal vessel (Figure 9.23).

Vertebrates also have a closed circulatory system, as you probably know. Actually, the term is somewhat relative, since blood may leave vessels and flow through *sinusoids,* the small open cavities of the liver and certain bones. Man is not at all unusual among

Box 9.2 Blood   Among invertebrates the circulatory fluid, or blood, is quite diverse. It may be a colorless fluid or it may be green (copper-containing), or even red (iron-containing). In vertebrates, the *erythrocytes,* or red blood cells, float in a colorless fluid called *plasma*. Other things floating in the plasma include proteins and various ions, as well as food and waste material. In addition to erythrocytes, the plasma may carry other kinds of cells such as leucocytes or *white blood cells,* which function mainly in combating invasions or infections, and cellular fragments called *platelets,* which function in clotting.

The red blood cells of vertebrates contain a protein-iron compound called *hemoglobin* which is what makes blood red. Hemoglobin is able to combine reversibly with oxygen. The result is that human blood is able to carry about 60 times more oxygen than could be dissolved in plasma. In body parts where the oxygen concentration is high (such as in the lungs), hemoglobin combines readily with it. Where oxygen is low (such as in the tissues), hemoglobin tends to give up its oxygen.

Hemoglobin is comprised of four subunits, each of which has a protein chain (globin) and *heme* (a complex carbon ring structure which contains iron). Each of these subunits can hold one oxygen molecule.

vertebrates in having a large muscular heart that pumps blood into large muscular *arteries*, which in turn branch into increasingly smaller *arterioles*, and finally into *capillaries*. The walls of the capillaries are so thin that oxygen, carbon dioxide, nutrients, and metabolic wastes are able to pass through them, to or from the tissues.

From the capillary beds the waste-laden blood flows into larger vessels called *venules*, and then into increasingly larger *veins*, which return the blood to the heart. The problem the body faces, of course, is how to get rid of the wastes it has picked up, primarily carbon dioxide and various nitrogen compounds.

Dissolved carbon dioxide gases are in equilibrium in the blood with carbonic acid and ionic bicarbonate. That is,

$$CO_2 + H_2O \quad \leftrightarrow \quad H_2CO_3 \quad \leftrightarrow \quad H^+ + HCO_3^-$$

| Dissolved gaseous carbon dioxide | Carbonic acid | Ionic bicarbonate |

The problem here is to make sure that the waste produced by the cells is in the form of carbon dioxide when it reaches the lungs, since it is this form which passes through membranes most easily. The waste, in fact, is likely to reach the lungs as carbon dioxide because of the presence of an enzyme in the blood, called carbonic anhydrase, which moves the reactions to the left.

The nitrogenous waste products come from the body's breakdown of proteins, which leaves the unused components of the large molecules behind. Among these bits of waste matter are excess nitrogens, which can be extremely poisonous to the system. So how are the nitrogenous waste products gotten rid of? It may be in poor taste to tell you the end of an exciting story first, but you may as well know: the nitrogenous waste products are excreted in the kidneys.

THE HEART

The muscular pump called the *heart* varies widely in invertebrates according to the life-style of the organism, but in the vertebrates there is a rather regular progression from simple to complex.

In fish the heart is two-chambered, one chamber being a thin-walled *atrium* which accepts blood after it has been circulated through the body. When the atrium contracts, blood is forced into the *ventricle*. The atrium doesn't have to be very strong, because it doesn't take much work to pump blood only to the next chamber. The ventricle, however, must exert tremendous pressure, since it has to force the blood through the large artery called the *aorta*

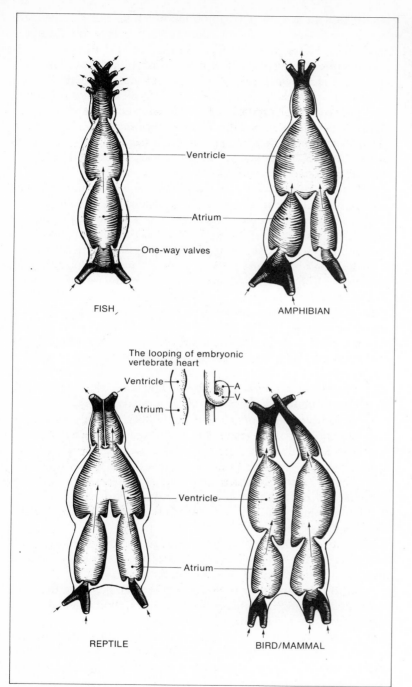

Figure 9.24 The progression of the vertebrate heart. Note that the heart of birds and mammals is in essence, a divided fish heart. In the development of the embryo the bird and mammal hearts double back on themselves. The result is that the atria end up on top. In this diagram however, the heart appears as it would if no looping occurred.

FISH

AMPHIBIAN

The looping of embryonic vertebrate heart

REPTILE

BIRD/MAMMAL

Figure 9.25 Frogs are "cold-blooded," or *ectothermic*. In other words, their body temperature is largely dependent on the temperature of their environment. Since the system doesn't utilize a lot of energy in maintaining a constant temperature, they can afford a rather inefficient circulatory system. (Courtesy: Jack Dermid from National Audubon Society)

into the tiny mesh of the gill capillaries, where gases are exchanged, then through larger vessels into the capillary beds of the rest of the body, from where it reenters the veins. The fish ventricle, then, must be able to pump blood through two capillary beds. Not surprisingly, the fish ventricle is thick-walled and muscular.

The amphibian heart is a bit more complex in that it has two atria. One receives oxygenated blood from the lungs, and the other receives deoxygenated blood from the rest of the body. The two are mixed in the single ventricle, so the system is not particularly efficient; but it is efficient enough for the "cold-blooded" and not particulary active frog.

Reptiles go one step better and have two atria and two ventricles, although in most species the ventricles are not completely divided (exceptions include crocodiles and alligators). Because of the incomplete ventricular septum, there is some mixing of blood from the two sides of the heart. Still, this system is more efficient than the frog's, since one side of the heart pumps blood through the lungs and the other side pumps it through the rest of the body.

In birds and mammals the right and left sides of the heart are completely separated. The result is much greater efficiency. Birds and mammals are "warm-blooded" animals. That is, in most cases they maintain a rather constant internal body temperature. Just as it takes energy to maintain the temperature of your apartment, energy must also be expended in order for a warm-blooded animal to maintain its temperature. Thus, the metabolic rate of birds and mammals is higher than that of other animals.

The development of efficient circulatory systems and warm-bloodedness has rather interesting evolutionary implications. You may have noticed that on a cold day you can catch sluggish lizards rather easily as their metabolism drops with the temperature and they become inactive. On the other hand, it is no easier to run down a rabbit on a cold day than on a warm day. The rabbit's metabolism is relatively independent of the outside temperature. As warm-bloodedness developed in some lines of prehistoric animals, then, they would have had a distinct advantage over their cold-blooded cousins. This advantage is cited as one of the major reasons mammals eventually replaced the dinosaurs.

But let's see what the separation of the two halves of the heart has to do with efficiency. When waste-loaded blood enters the right atrium from the large veins (the *superior vena cava* and the *inferior vena cava*), it is then pumped by contraction of the atrium through a valve to the right ventricle. The right ventricle sends the blood, which is now low in oxygen, through the *pulmonary arteries* to the lungs. Here the blood picks up oxygen and then returns to the left

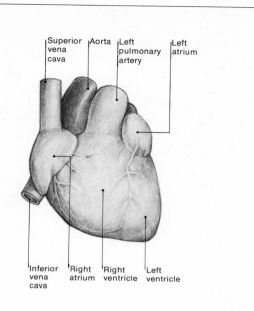

Figure 9.26 The external and internal appearance of the human heart. Note the size of the atrial and ventricular walls. The right side of the heart pumps blood to the lungs where carbon dioxide and oxygen are exchanged. The oxygenated blood then enters the powerful left side of the heart where it is then pumped to the rest of the body. Arrows show the direction of muscle movement. The heavy lines show the direction of blood flow. The atrial and ventricular contractions are shown by both for the right side.

Figure 9.27 The major blood vessels of the human body.

atrium via the *pulmonary veins*. The left atrium pumps the blood into the left ventricle, which then contracts to send the blood on its long journey throughout the body. The left ventricle, then, must be very thick-walled and muscular; hence, the left side of the heart is larger. For this reason you may have the notion that your heart is on the left side. It isn't. It's right in the middle of your thoracic cavity and is about the size of your fist. Thus when you pledge allegiance to the flag, your hat is actually over your left lung.

The heart is supplied by nerves from special centers in the brain. These centers respond to a variety of stimuli. They increase or decrease heart rate according to whether arteries are being stretched (which decreases the heart rate) or veins are being stretched (which increases it) and according to the relative levels of carbon dioxide and oxygen in the blood, as we will discuss later. The system is not a simple one. For example, the diameter of specific arteries is controlled by the nervous system, so that after a meal more blood is shunted to the stomach and intestines to aid in digestion. But the sight of a shadowy figure in an alleyway as you leave a restaurant can cause blood to be shunted back to the muscles.

Figure 9.27 shows the major blood vessels of the human body. Where are the pulmonary vessels? What is the large network below the heart on the right side of the body? You might now wonder why the good guys survive so many shoulder wounds in so many bad Westerns. According to the figure, where is a better place to be shot? Why aren't the veins on the back of your hand in the same place as those of the person sitting next to you? (You might refer to circulatory system development in Chapter 8.)

THE LYMPHATIC SYSTEM

Before we leave the topic of circulation we should take note of another, often-neglected, circulatory system, the *lymphatic system* (Figure 9.28). The lymphatic system circulates *lymph* through the body outside blood vessels and returns this fluid to the blood, where it is then called *plasma*. Blood plasma, minus the dissolved proteins, filters out into the tissues through tiny capillaries, bathes the cells, moves between them and finally winds up in small channels in the tissue called *lymph capillaries*. The lymph fluid may then move into larger channels and finally reach any of various lymph nodes, where it is filtered, thus removing any cellular debris or bacteria. Frogs and some other vertebrates have "lymph hearts" which move the lymph along, but mammals rely chiefly on muscular movements to circulate the fluid.

White blood cells called *lymphocytes* are found in great num-

Superior sagittal sinus

Transverse sinus

Carotid artery

Brachial artery

Cephalic vein

Basilic vein

Renal artery

Inferior vena cava

Femoral artery
Femoral vein

Plantar arteries

Jugular veins

Superior vena cava

Aorta

Pulmonary artery

Heart

Portal vein

Iliac artery

Saphenous vein and tributaries

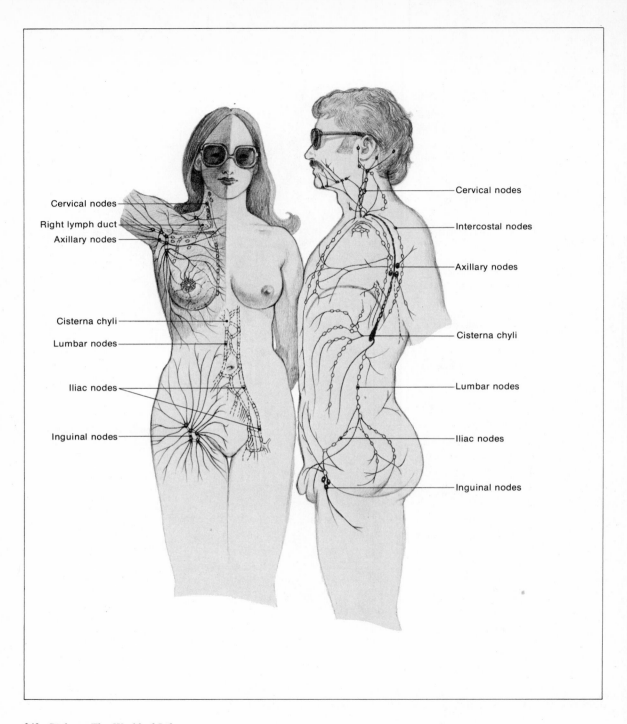

Cervical nodes

Right lymph duct

Axillary nodes

Cisterna chyli

Lumbar nodes

Iliac nodes

Inguinal nodes

Cervical nodes

Intercostal nodes

Axillary nodes

Cisterna chyli

Lumbar nodes

Iliac nodes

Inguinal nodes

bers in the lymph nodes. If bacteria have entered the system, they will probably eventually find themselves at one of these nodes, where they will be destroyed by the lymphocytes. The lymph flows into increasingly larger channels, or ducts, until finally, it empties into large veins near the heart.

DIGESTIVE SYSTEMS

The tired old notion of our bodies being internal-combustion engines, with our food the fuel that runs them, can still be squeezed for elements of truth. The idea is that work must be done to minimize body entropy, and the energy for this work must be released from the food in which it is bound. That energy is released through first breaking the food down into its constituents (digestion), transporting the product to the cells (circulation), and enabling the cells—which are actually the engines, if the analogy is to be valid at all—to burn the food for energy (metabolism). Of course, not all ingested material is "burned." Some of it is used as building blocks for the repair and maintenance of the machinery.

DIGESTIVE ARRANGEMENTS

The basic scheme for digestion is the same in all animals, except for the details. However, these details vary widely, as you can see in Figure 9.29. An amoeba, for example, simply flows around a food particle and engulfs it, so that the particle is held in a *food vacuole*. It then pumps acidic digestive enzymes into the vacuole until the food is digested. Indigestible particles are brought to the surface of the animal and squeezed out through an opening which appears in the membrane.

The hydra and the flatworm both have pouchlike digestive systems. Whereas the hydra stings its prey with special cells on its tentacles and then hauls the prey into its digestive cavity, the flatworm is able to extrude its *pharynx*, or "throat," to capture food. In both these animals digestion begins in the *gastric cavity*, or "stomach," but before the process is completed, the particles of partly digested food are engulfed by cells of the body wall and digestion is completed within these cells. So these animals digest food by both extracellular and intracellular processes. (Which of these two processes typifies human digestion?)

In most more complex animals the digestive system is a tube rather than a pouch, with one opening through which food enters and another through which it exits—a much more civilized arrangement.

Although both humans and earthworms have tubelike digestive systems, you will notice in the earthworm several structures

Figure 9.28 The human lymphatic system. The lymph nodes may trap bacteria and other foreign matter. A type of white blood cell is abundant in the lymph nodes. The lymph nodes often become involved in cancer of the breast, creating serious surgical problems. At present the most prominent treatment for breast cancer is removal of the breast, sometimes with axillary lymph nodes and underlying muscles.

Flagellate

Nucleus

Enzymes released

Food

Undigested remains

FOOD VACUOLE FORMING IN AN AMOEBA

Wastes Food

Mouth
Ectoderm
Food
Endoderm secretes enzymes and absorbs

HYDRA

Ectoderm
Mesoderm
Endoderm secretes enzymes
Absorbed food
Stomach
Pharynx
Mouth
Food
Wastes

FLATWORM

Food Pharynx Crop (stores food) Intestine

Mouth Esophagus Gizzard (grinds food) Enzymes secreted and food absorbed in intestine Anus

EARTHWORM

Mouth Esophagus Stomach Intestine

Liver Pancreas Anus

SALAMANDER

which are absent in man, such as the crop and gizzard. The *crop* is where food is stored, and the muscular *gizzard* is where food is ground against small stones which have been ingested. So there is now no reason for you to lie awake night after night wondering how an earthworm chews. Birds, which also lack teeth, have a similar digestive arrangement.

As another fascinating aside, earthworms have amoeboid cells in their bodies which can ingest particles that move through the digestive tract as the worm literally eats its way through the soil. These cells then migrate through the body and come to rest just under the skin, giving the earthworm a color similar to the soil in which it lives, causing the early bird to come up short.

In vertebrates, such as the salamander, food enters a well-defined mouth, is swallowed by voluntary action, and then moves down the smooth-walled *esophagus* by involuntary muscle contractions. The food then enters the stomach, where digestion begins. It then passes on to the intestine, where it is further broken down by the action of secretions from the pancreas and liver. When the food is finally digested, it moves through the intestinal wall into blood vessels which lie in the wall. The nutrients are then carried by the blood to the tissues where they will be used. The indigestible particles move on through the intestine, to be eliminated through the *anus* or the *cloaca*. (The *cloaca* is a common opening for the intestine, kidneys, and reproductive organs found in amphibians, reptiles, and birds—a primitive plumbing arrangement which has been "improved" upon in mammals.)

The most important chemical process in digestion is hydrolysis. Hydrolysis, as we saw in Chapter 3, is the breaking down of substrate matter by the chemical addition of water. Large molecules of proteins, fats, and carbohydrates are thus broken down into amino acids, glucose, glycerol, and fatty acids, which are absorbed through the intestinal wall and enter the bloodstream. Our concern here, then, is the conversion of larger food particles to smaller ones.

THE HUMAN DIGESTIVE SYSTEM

In the human digestive system, shown in Figure 9.30, the breakdown of food particles begins in the mouth, where large particles are physically broken into smaller ones by the action of chewing. Chewing is especially important to animals which eat green plants, since the cellulose cell walls cannot be broken down by digestive enzymes. In contrast, animal tissue *can* be entirely broken down by digestive enzymes. This is the reason you never see cats chewing their cud. In the mouth, the food is lubricated by saliva (this is not my favorite part of biology, either). Also, some digestion begins

Figure 9.29 A comparison of the digestive mechanisms of the amoeba, flatworm, Hydra, earthworm and the salamander.

Figure 9.30 The human digestive system.

Parotid gland

Tongue

Sublingual gland

Submaxillary gland

Esophagus

Liver

Gall bladder

Stomach

Spleen

Transverse colon

Descending colon

Ascending colon

Small intestine

Cecum

Sigmoid colon

Appendix

Rectum

in the mouth as starches are changed to disaccharides, or double sugars. The food then moves to the stomach through the smooth-walled esophagus, with food further moistened by glands in the esophagus.

The *stomach* is a muscular sac which churns the food as it mixes it with various secretions from the stomach wall, such as mucous, hydrochloric acid, and enzymes. The food is sealed in the stomach through all this process by two sphincters or rings of muscles, one at either end of the stomach. After the mixing is completed, the lower sphincter opens, and the stomach squeezes the food into the small intestine.

The *small intestine* is a long, convoluted tube in which digestion is completed and through which the nutrients enter the bloodstream. The inner surface of the small intestine is covered with tiny, fingerlike projections called *villi*, which greatly increase the surface area of the intestinal lining. The surface area of the villi themselves is further increased by tiny projections called *microvilli*. Each villus may have about 3,000 of these microvilli. Within the villi are numerous tiny lymph vessels, as well as tiny branches of the mesenteric arteries and veins, the blood vessels which supply much of the viscera. Some of the products of digestion move into these blood vessels.

Interestingly, fat doesn't have to be entirely broken down before it enters the circulatory system. Smaller fatty particles enter the blood vessels, but larger ones may pass directly into the lymph vessels in the villi. It is interesting and almost weird that the products of fat digestion are so quickly reconstituted to fat in the intestinal mucosal cells that after a meal high in fat content these particles may appear in the bloodstream and give the blood itself a milky appearance.

The process of absorption is accomplished in part by passive *diffusion* across the membranes of the intestinal lining, but also in part by *active transport*, a process in which energy is expended to do the work.

Digestion in Humans

The events of digestion are complex and numerous, but we might briefly review the process in humans. The first ten inches of the small intestine comprises the *duodenum*. The enzymes produced by the pancreas enter the gut in this area. The pancreas secretes powerful *proteolytic* enzymes, which break down proteins. These are powerful agents, so why doesn't the pancreas digest itself? the sharp-witted student will ask. The agile-minded physiologist will quickly respond that this is prevented by the fact that the enzymes

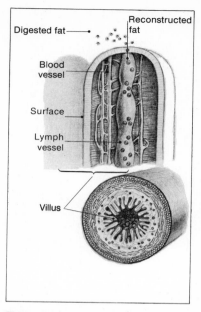

Figure 9.31 A cross-section of the villus from the digestive system.

are stored in an inactive form in the pancreas and are activated by the duodenal environment. Well, then why isn't the duodenum digested? As a matter of fact, it would be if it were not "alive," with a membrane system, which actively excludes harmful substances such as digestive enzymes. This is also why, if you are unfortunate enough to be "wormy," when a worm dies, it is immediately digested. But I suppose such ultimate vengeance provides slim consolation to the host. In addition to this inherent protection, the intestine secretes a protective shield of mucus, and the high level of sodium bicarbonate from the pancreas also helps to neutralize the acids of the stomach juices.

Another substance secreted into the duodenum is *bile,* which is stored in the gall bladder after it is produced by the liver. Bile contains certain sodium salts which act as detergents (the biodegradable kind) breaking up the large fat droplets, so that the pancreatic enzymes can work on them.

Nutrient-laden blood from the intestine is carried into the *hepatic portal vein* which leads to the liver (a *portal* vessel is one that lies between two capillary beds). The blood then filters through tiny vessels and sinuses in the liver. If it contains excess nutrients, some are removed by the liver and stored. If it contains too few nutrients, the liver responds by releasing some of its stored nutrients. Thus proper glucose and amino acid levels are maintained in the blood, despite variations in food intake. This is one more example of the amazing internal regulation, or *homeostasis,* accomplished by the body. As the body's resources of glucose drop below the amount needed to maintain a constant blood level, we have the sensation of hunger and are motivated to go out and find some glucose (in one form or another).

Food passes from the duodenum through the next two segments of the small intestine, the *jejunum* and the *ileum,* where further digestion and absorption occurs. The remaining waste products then move on into the *large intestine.* The large intestine consists of two portions, the *colon* and the *rectum.* The inner surface of the large intestine is rather smooth, and no digestion occurs here. However, the colon is a veritable hotbed of a bacterium called *Escherichia coli,* which provides some vitamins for the body in the form of its own waste product. (The presence of these bacteria in the environment is an indication of fecal contamination, an important factor in the spread of some diseases, such as infectious hepatitis.) The large intestine also extracts water from the solid waste products, or *feces,* with constipation resulting if you put off nature's call (more "relevance"!). The feces are stored in the rectum before elimination through the anus.

THE EXCRETORY SYSTEM

In addition to ridding themselves of solid wastes, animals have the problem of getting rid of metabolic wastes, or "cell garbage." In the amoeba and some other protozoans metabolic wastes simply diffuse across the cell membrane and out into the surrounding water. However, protozoa which live in fresh water have another problem as well. Since protoplasm has a higher concentration of particles in solution than pond water does, water constantly enters the body by osmosis, creating the danger that the body will swell and rupture. Some protozoa solve this problem with a *contractile vacuole*, a small vesicle which continually pumps water out from the cell interior. So in the case of fresh-water protozoa the problems of metabolic waste and internal water regulation are solved by different mechanisms. In other animals this may not be the case, and the two problems may be closely interrelated.

In earthworms, for example, each segment has a pair of coiled tubes called *nephridia*, which connect the body cavity to the outside (Figure 9.32). The end of the tube which opens into the body cavity is rimmed with beating cilia which circulate the fluids within the body cavity. Capillaries lie tightly coiled around the outside of the nephridia, and metabolic wastes from the blood can thus enter the tube which leads to the outside. Water, glucose, and other useful substances are reabsorbed from the body fluids back into the capillaries as the cilia-driven fluid passes out. The amount of water leaving the animal depends on how much water has moved into the body cavity. Worm urine, in case you've ever wondered, is very dilute and is excreted at a rate of about 60 percent of the worm's body weight each day.

SOLUTIONS TO THE WATER PROBLEM

The problems of ridding the body of metabolic wastes and maintaining proper internal water levels are solved in vertebrates in a number of ways. Fresh-water fish, like fresh-water protozoa, live in a *hypotonic* medium; that is, the concentration of particles in their body fluids is higher than that in the surrounding water. Hence water tends to move into their bodies by osmosis. One means by which they get rid of excess water is through a greater number of tiny clumps of blood vessels called *glomeruli* in their kidneys. We'll consider the structure of the glomerulus below, but in general this means that there is a larger surface through which fluids can leave the blood to be collected in the kidney. The result is that the metabolic waste in fresh-water fish is highly dilute.

Salt-water animals, interestingly enough, have the same problem as desert animals: conserving water. Seawater is a *hyper-*

Figure 9.32 A cross-section of the earthworm showing the paired nephridia. Each tube has one end opening into the body cavity and the other end opening to the outside.

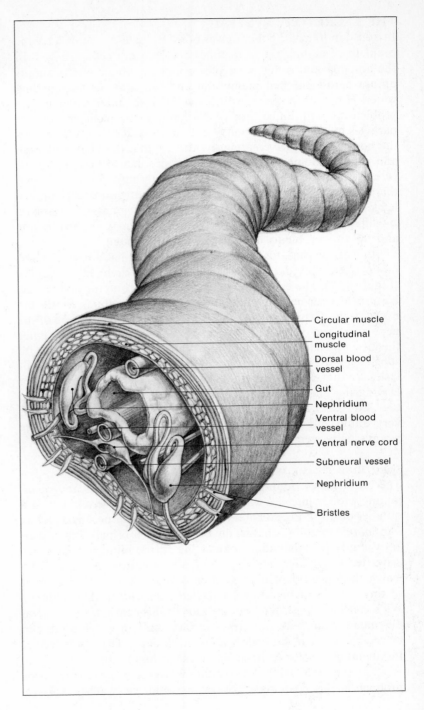

Circular muscle

Longitudinal muscle

Dorsal blood vessel

Gut

Nephridium

Ventral blood vessel

Ventral nerve cord

Subneural vessel

Nephridium

Bristles

Box 9.3 Special Solutions to the Water Problem   Some insects are able to survive for years enclosed in jars of dry pepper. They do this by utilizing metabolic water and excreting almost dry uric acid.

A camel can tolerate dry conditions by adopting a number of tactics. At night it drops its body temperature several degrees so that bodily processes slow down. In the heat of the day it doesn't begin to sweat until its body temperature reaches about 105° F. In addition, the camel can lose twice as much of its body water (40%) without ill effect than can most other mammals. Peculiarly, the thick coat of a camel acts as insulation to keep the heat out. Also when the camel does drink, it may hold prodigious amounts of water.

Some seabirds and turtles have salt removal glands. The huge tears which may appear in the eyes of sea turtles have nothing to do with the turtle's realization that we are poisoning the oceans. They are merely the result of specialized salt removing glands. These glands, however, remove only sodium chloride from the blood.

The tiny kangaroo rat doesn't drink at all and it lives on dry plant material. It survives because it doesn't sweat and is active only in the cool of night. Its feces are dry and its urine is highly concentrated. Most of its water loss is through the lungs. Fats produce more metabolic water than other foods, so it prefers fatty foods. On the other hand, soybeans are high in protein and thus produce a lot of nitrogenous waste for which a lot of water is needed—so if a kangaroo rat is fed only soybeans it will die of thirst.

tonic medium, that is, it has a higher concentration of solutes than the marine animal's body fluids; hence they tend to lose water by osmosis. As a result they have developed smaller glomeruli, so that less water filters out through the kidney. This mechanism may be augmented in some species by special glands which excrete the salt taken in with the seawater the animals drink. In other species the gills may function in eliminating nitrogenous waste as the blood passes through them. Most sharks and rays have a different strategy. They retain some urea, or waste, in their blood, and thus raise the osmotic pressure of their body fluid to more nearly that of seawater.

All terrestrial mammals, including man, also face the problem of how to flush nitrogenous wastes out of their bodies without losing too much water. A large part of human food is protein (for those who can still afford it), and protein is, paradoxically, at the source of man's excretory problems. In digestion, as we know, proteins are hydrolyzed into amino acids, which are transported to the cells where they will be utilized. Some of the amino acids are used as building blocks for proteins and other molecules, with no great

Figure 9.33 Ways of regulating water and salt balances in a fresh water fish, a salt water fish, a shark, and a cat (a terrestrial mammal).

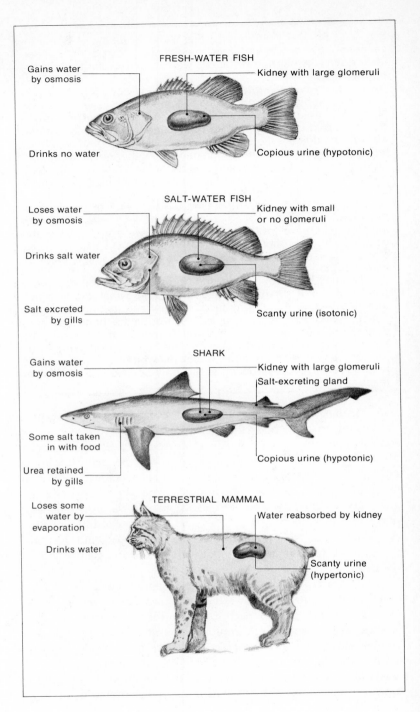

FRESH-WATER FISH

Gains water by osmosis

Kidney with large glomeruli

Drinks no water

Copious urine (hypotonic)

SALT-WATER FISH

Loses water by osmosis

Kidney with small or no glomeruli

Drinks salt water

Salt excreted by gills

Scanty urine (isotonic)

SHARK

Gains water by osmosis

Kidney with large glomeruli

Salt-excreting gland

Some salt taken in with food

Urea retained by gills

Copious urine (hypotonic)

TERRESTRIAL MAMMAL

Loses some water by evaporation

Water reabsorbed by kidney

Drinks water

Scanty urine (hypertonic)

change in their structure being required. Most amino acids, however, are broken down further by the liver for storage as an energy source.

Now, the first step in deriving energy from an amino acid is to get rid of the nitrogens, since no nitrogen enters into either the anaerobic or aerobic energy cycles. The nitrogens which have been stripped from the proteins usually end up with three hydrogens— and NH_3 is ammonia, a deadly poison. So, the problem involves how to dispose of the ammonia. It can't pass out through the lungs like carbon dioxide, so it must be excreted by the kidneys. However, ammonia is so poisonous that it must be highly diluted to be handled by the body, and thus its excretion would entail the loss of too much water. So carbon dioxide is added to the ammonia to form *urea*, which the body can handle in higher concentrations without adverse effects. The moderately dilute urea is then passed out of the body as *urine*.

Since fresh-water fish have water to spare, they excrete dilute ammonia directly. However, some other organisms, including many egg-layers such as insects and birds, convert the ammonia to an insoluble *uric acid*, which is then excreted as a paste, with most of the precious water extracted and conserved by the body. The chemistry of the waste products of metabolism as they generally occur in fish (ammonia), birds (uric acid), and mammals (urea) are shown below.

$$NH_3$$
Ammonia

$$\begin{array}{c} NH_2 \\ | \\ C=O \\ | \\ NH_2 \end{array}$$
Urea

Uric Acid

Many evolutionists believe that vertebrates evolved in fresh-water habitats (contrary to popular assumption), so the transition to land must have involved the development of new ways to handle the waste problem. The fresh-water animal had the problem of

Nephron

Renal artery

Renal vein

Ureter

Bowman's capsule

Branch of renal artery

Glomerulus

Branch of renal vein

Proximal convoluted tubule

Distal convoluted tubule

Capillaries

Collecting tubule

Henle's loop

eliminating the excess water which was continuously entering its body, and one solution was probably to allow great quantities of water to pass out through its extensive glomeruli. On land, however, the problem changed. There wasn't much water, so what there was had to be retained. In large part, the problem was apparently solved by a decrease in the size of the glomeruli.

THE HUMAN KIDNEY

Humans, along with other mammals, have two kidneys (although we can live with only one—nature is often redundant). The kidneys lie in the dorsal area (at the back) and extend slightly below the protective rib cage (this is why blows to the kidney area are so dangerous). Each kidney contains about a million *nephric units*, consisting of a glomerulus, a Bowman's capsule, and the connecting tubules. The structure of these units is shown, along with the kidney itself, in Figure 9.34.

The exit from the glomerulus is smaller than the entrance, so that the glomerular blood is placed under high pressure as it enters from the renal artery. As the blood, under intense pressure, wends its way through the circuitous and tortuous route of the glomerulus, much of its fluids move out through the thin-walled vessels and into the cuplike *Bowman's capsule*. Solid particles, such as blood cells and certain proteins, are too large to pass through the glomerular walls.

The filtrate received by the Bowman's capsule contains a high concentration of waste products, but also some good stuff, including water, salts, and nutrients. So as the filtrate begins to move through the tubules on its way to the urinary bladder, the blood vessels continuing on from the glomerulus head off the filtrate at the tubules and the *loop of Henle*. Here they wind around the loop and retrieve many of the usable products, both through diffusion and by active transport, allowing the urea and some water to continue on their way.

The remaining fluid in the tubules proceeds to a larger collecting tubule, and then to the *ureter*, which leads from each kidney to the *urinary bladder*. From the urinary bladder a single opening, controlled by sphincter muscles, allows the urine to pass into the *urethra*, which then leads to the outside.

Have you ever noticed that after drinking a lot of beer or other brews you have to urinate quite frequently? You may have also noticed that the quantity going out seems to be more than you thought had come in. Furthermore, you may have noticed that after a particularly eventful evening you are thirsty the next day. The reason for this is that reabsorption of water at the tubules is con-

Figure 9.34 The human kidney (left) as seen in Captain Kidney. See the text for an explanation of the kidney mechanism.

Box 9.4 Pressure Gradients in the Kidney Tubule   Let's look at the mechanism behind how hypertonic urine is formed in the human nephric unit. First, body fluids filter through the capillary wall into Bowman's capsule and from there through the proximal convoluted tubule (here straightened), the long loop of Henle, the distal convoluted tubule and then into the collecting duct. As the urine passes through the ascending loop of Henle (on the right), cells lining the loop pump sodium out of the urine into the tissue surrounding the loop. These sodium ions then diffuse over to the descending loop (on the left) where they enter the loop. From there they flow to the ascending loop where they are pumped out again. This constant circulation means that the loop of Henle and the adjacent collecting duct are always bathed in a salt solution and it is this solution that sets up the osmotic gradient which withdraws water and solutes from the urine. The ascending loop is apparently impermeable to water since no water passes out with the sodium. The collecting duct, however, is permeable to water and water freely flows out in response to the high salt concentrations in the surrounding tissue. This water is picked up by the blood vessels which come from the glomerular region and the hypertonic urine passes to the bladder.

trolled by an *antidiuretic hormone* (ADH), which is secreted by the posterior lobe of the pituitary. Alcohol suppresses the secretion of this hormone, so that less water is reabsorbed; the urine thus becomes hypotonic, or more dilute than the body fluids, and the body actually becomes somewhat dehydrated. Relevance strikes again! On the other hand, ADH secretion also decreases when you drink a lot of water since the osmotic concentration of the blood decreases (drinking a lot of water can temporarily "thin" the blood). This also results in the formation of a urine that is hypotonic to the blood.

In case you are planning to be shipwrecked sometime, you may also be interested in another bit of practical physiology. A person on a life raft cannot get the water he needs by eating fresh fish, contrary to popular rumor. The reason is that human urine cannot exceed a salt concentration higher than about 2.2 percent. Fish fluids are less concentrated than this, it is true; but fish are high in protein, so to get rid of all the nitrogen atoms, you would have to excrete a lot more water than you would be getting from the fish. You would do better eating fish than drinking seawater, of course, since the salt concentration in seawater is about 3.5 percent. Some shipwreck survivors claim that the spinal fluid of fish is a good supply of water, but all things considered, you would be better off to take some water with you.

This chapter has covered a lot of material, so let's consider the somewhat special cases of the endocrine and nervous systems in a chapter by themselves. You should continue to notice the myriad ways that living things adjust themselves, so that the bodily processes function between certain set limits—some wider than others, depending on the process and on the organism.

10

Hormones and Nerves

(Courtesy: Wide World Photos, Art Zeller, American Stock Photo/Harold Lambert Studios, Inc.)

In this chapter we will look at the hormone and nervous systems together because both are intimately involved in the processes of internal regulation, and also because in some cases they act together. Regulation, as we have already seen, is a many-faceted phenomenon. Consider the regulatory function of DNA in making protein, or the regulatory function of the kidney in maintaining water balance. Regulation in its broadest sense serves to keep bodily processes within allowable limits and coordinates those processes so that the organism can function more efficiently in the part of the world it has available to it. Hormones and nerves, as we will see, are involved in regulating and coordinating an entire orchestra of bodily processes so as to minimize "dissonance," or *entropy*. In the next chapter we will deal with behavior, so it may also be good to set the stage by considering the broader role of the nervous system. Let's begin with hormones.

HORMONAL REGULATION

First, what is a hormone? (Ready with your markers?) A *hormone* is a chemical that is produced in one part of the body and carried by the blood to another part of the body, where it ultimately influences some process or activity there. Hormones are produced by *endocrine*, or *ductless*, *glands* which are found in various parts of the body.

Hormones perform regulatory functions in invertebrates as well as vertebrates; in fact, much important hormone research has been carried out on various arthropods, such as insects and crayfish. Hormones have been found to regulate the changes in body organization which many insects undergo as they pass through their larval and pupal stages. A typical experiment to show the importance of hormones in such development might entail depriving part of an insect body of a hormone, noting the lack of development in that body part as the rest of the body matures normally, and then injecting the proper hormone into the inhibited part and observing the subsequent normal growth.

Figure 10.1 The major endocrine glands in the human. Because of the immense regulatory power of small amounts of these gland secretions and because of their often intricate interaction, these secretions must be very precisely regulated in amount and in time.

Pituitary gland

Thyroid gland

Adrenal glands

Pancreas

Testicles (♂)
Ovaries (♀)

One of the greatest problems in arranging such an experiment is acquiring the hormones. Animal bodies actually contain very little hormone substances, since only tiny amounts in the body are necessary to produce great changes. For example, a woman produces only about a teaspoonful of estrogens in an entire lifetime. To extract only a few milligrams of another of the estrogens, estradiol, more than *two tons* of pig ovaries were required.

"Chemical messengers," as hormones are sometimes called, may function in keeping variable bodily processes within their proper limits, as does the antidiuretic hormone, or they may direct permanent changes, such as growth or sexual maturation. They may also function in emergency situations by causing the body to prepare for stress. And they may help the organism to survive better in its environment in other ways, as when amphibians or reptiles change colors to match their surroundings. Hormone action is also a part of nature's chemical warfare arsenal. Some plants, for example, manufacture substances which mimic the growth hormones of insects and thus they disrupt the normal growth of insect larvae which feed on them.

THE HUMAN ENDOCRINE SYSTEM

In the human endocrine system the *pituitary* has been called the "master gland," but this term isn't really an apt description, since many of its activities are regulated by the hypothalamus (described later). The pituitary does have a number of important functions, however, and it also exerts an effect on certain other glands. It is composed of two major parts, the *anterior* and *posterior lobes*, with a central *midlobe* area. The anterior and posterior lobes actually develop from different embryological tissue. The posterior lobe develops from nerve tissue and the anterior lobe from an outpocketing of the embryonic mouth. (What germ layers are involved?) The area between these, the midlobe, even has a function of its own.

The pituitary is about the size of a bean and is located at the geometric center of the skull, where it lies shielded by heavy layers of surrounding bones. Its small size belies its physiological complexity. In all, the pituitary secretes nine known hormones. Three of the hormones secreted by the anterior pituitary are called *tropic hormones* because they regulate the secretions of other glands—the thyroid, the adrenals, and the gonads, or sex glands.

The anterior pituitary also secretes a nontropic growth hormone called *somatotropin*, which controls bone and muscle growth. If too little somatotropin is present during childhood, the result is a *pituitary dwarf*, or *midget*. Too much somatotropin results in a

Figure 10.2 Insect development.

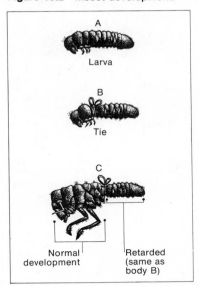

Figure 10.3 An experiment which illustrates the importance of hormones in insect development. When growth hormones produced in the head are kept from reaching other parts of the body, those parts do not mature.

Figure 10.4 This giant is 9 feet, 3½ inches tall and the midget is 3½ feet tall. The size of both the giant and the midget are the result of abnormal pituitary secretions. (Courtesy: Wide World Photos)

Figure 10.5 The secondary sex characteristics in the male and female are influenced by the sex hormones testosterone and estrogens. (Courtesy: Arnold Schwarzennegger. Photo by Art Zeller; American Stock Photo and Harold M. Lambert Studios, Inc.)

giant. Excessive somatotropin in an adult increases the size of only certain bones which are still able to respond to the hormone—the jaw and the bones in the hands and feet. This condition is called *acromegaly.*

When the *thyroid* is stimulated by the pituitary, it produces *thyroxine,* the hormone which controls the body's metabolic rate. Deficiency of thyroxine in childhood results in a characteristic physical appearance and mental retardation called *cretinism.* It is now possible to recognize thyroid deficiency early in life, so that thyroxine can be administered artificially to promote normal development.

The gonads, the testes and the ovaries, predominately produce *testosterone* and *estrogen,* respectively, hormones which are responsible for those secondary sex characteristics we have all come to love and admire. In males these characteristics include a deepening of the voice, broadened shoulders, hairy chests, facial hair (beards and mustaches), muscular development, sperm production, and increased production of red-blood cells (which might explain the term "red-blooded he-man"). In females estrogen causes changes in the distribution of body fat, of body hair, and development of the breasts. *Progesterone,* produced by the corpus luteum, stimulates the development of the uterine lining and causes the further breast development during pregnancy.

Ad- means "upon" and *renal* refers to the kidney, so guess where the *adrenal glands* are located. The adrenals have two distinct parts, the outer *cortex* and the inner *medulla.* About fifty steroids have been isolated from the adrenal cortex of various mammals, and it is believed some of these may be used in the production of different hormones, but it is not yet known just what role, if any, these steroids play. One hormone produced in the cortex is *cortisone* (manufactured when the cortex is stimulated by a secretion, called ACTH, from the anterior pituitary). *Aldosterone* is also produced in the adrenal cortex and functions in the regulation of salt and water in the body. It works by influencing the cells of the kidney's distal convoluted tubule to increase their rate of sodium reabsorption. When sodium is reabsorbed, water must follow because of the osmotic gradient which is set up. The renal cortex also produces some male sex hormones. This system is usually shut down in women, but if it should be activated, for example, by a tumor, it will cause such characteristic masculine qualities as beard growth.

The adrenal medulla secretes the hormones *norepinephrine* and *epinephrine,* which for our purposes will be referred to simply as adrenaline (which you will find out about shortly). As you will see, these hormones may produce effects similar to those caused by certain emotions. Other endocrine glands such as the thyroid, the hormones they produce, and their principal effects are listed in Table 10.1.

HOW HORMONES WORK

Actually, we don't know how hormones work. Although we have known the composition of some hormones for a long time, we still aren't sure how any of them work. One hypothesis, however, states that when hormones come in contact with the membranes of "target cells," they alter certain sites on the membranes of these cells and trigger the production of a group of substances called *prostaglandins* within the membrane (see Chapter 6). Prostaglandins are involved in the production of the enzyme *adenyl cyclase*, which helps to change ATP to a form of adenosine monophosphate, cyclic AMP, in the cytoplasm. The cyclic AMP may then do any of a number of things. For example, it may alter enzyme activity (think of the implications) or it may stimulate the production of *m*RNA. Changes such as these could theoretically produce all the effects which are attributed to hormone action.

FEEDBACK SYSTEMS

The regulation of hormone production is achieved, in part, through a negative feedback system. As an example, consider the interaction of the tropic hormones and their target endocrine glands. The anterior pituitary gland secretes a thyroid-stimulating hormone (TSH) which causes the thyroid gland to secrete thyroxine. As the thyroid responds and the thyroxine level rises in the blood, the thyroxine itself has a depressing effect on the secretion of TSH, resulting in an eventual lowering of thyroxine and a release of more TSH. The levels of both hormones are thus kept within tight limits, because of the influence they exert on each other. Such feedback mechanisms are common throughout the endocrine system and furnish some of the clearest and most fascinating examples of homeostasis in living things.

THE RELATIONSHIP OF HORMONAL AND NEURAL CONTROL

Nerves may regulate bodily processes in much the same way hormones do. In some cases they may even duplicate the work of hormones. In other cases the nerves act in coordination with hormones or independently of them. Whether such regulation is hormonal or neural (pertaining to nerves), it is accomplished *chemically*.

Table 10.1—Major Vertebrate Endocrine Glands and Their Hormones

Gland	Hormone	Principal Action
Anterior lobe of pituitary	Thyrotropic hormone (TS)	Stimulates thyroid
	Follicle-stimulating hormone (FSH)	Stimulates ovarian follicle
	Luteinizing hormone (LH)	Stimulates testes in male, corpus luteum in female
	Growth hormone	Stimulates growth of bones and muscles
	Adrenocorticotropic hormone (ACTH)	Stimulates adrenal cortex
	Prolactin	Stimulates secretion of milk, parental behavior (such as nest building in birds and fish)
	Interstitial cell-stimulating hormone	Stimulates testosterone production
Midlobe of pituitary	Melanocyte-stimulating hormone	Regulates color of skin in reptiles and amphibians
Posterior lobe of pituitary	Oxytocin	Stimulates contractions of uterine muscle, milk release
	ADH or vasopressin	Controls water reabsorption
Thyroid	Thyroxine	Controls metabolism
Parathyroid	Parathyroid hormone	Controls calcium metabolism
Testes	Testosterone	Stimulates production of sperm and secondary sex characteristics
Ovary, follicle	Estrogens	Stimulates female secondary sex characteristics
Ovary, corpus luteum	Progesterone	Stimulates growth of uterus
Adrenal medulla	Epinephrine, norepinephrine (adrenalin)	Activates sympathetic nervous system
Adrenal cortex	Cortisone, cortisone-like hormones, aldosterone	Controls carbohydrate metabolism, salt, water and sugar
Pancreas	Glucagon	Stimulates breakdown of glycogen into glucose
	Insulin	Lowers blood sugar levels, increases formation and storage of glycogen

The dividing line between regulation by nerves and hormones is not clear-cut. For example, the pituitary gland itself is partly formed from embryonic neural tissue. Also, it lies directly under the part of the brain called the *hypothalamus.* The anterior pituitary is at least partly under the control of substances released from the hypothalamus. The posterior lobe of the pituitary contains the ends of nerves which originate in the hypothalamus. Also, two hormones associated with the posterior pituitary are known to be actually manufactured in the hypothalamus and only stored in the posterior pituitary.

The adrenal glands provide other examples of close nerve-hormone interaction. *Adrenaline* is the hormone responsible for those astonishing physical feats we hear of people performing in emergencies (the story about a little old lady lifting a car off her trapped husband crops up periodically). Adrenaline is released as a result of neural stimulation of the adrenal medulla. Neural stimulation also causes the pituitary to manufacture more ACTH and thus to boost the production of cortisonelike hormones by the adrenal cortex.

THE NERVOUS SYSTEM

In considering the nervous system we will begin by learning something about that mysterious structure called the brain and the great nerve to which it is connected. But first let's see how this complex system arose by tracing some nervous systems from simple to more complex forms. We could say from "lower" to "higher" forms, except that the term *higher* is admittedly prejudiced. What is usually meant by the word is any characteristic similar to those of humans, such as a large "thinking" center. The implication of this usage is that all species tend to evolve toward humanlike characteristics. Nothing could be further from the truth.

THE EVOLUTION OF THE CENTRAL NERVOUS SYSTEM

Clues to the evolutionary development of nervous systems can be gleaned from a cross-species survey, a review of animal nervous systems from the simple to the more complex (Figure 10.6). One of the simplest nervous systems is that of the fresh-water hydra. It consists simply of a two-dimensional net of interconnecting *neurons*, or nerve cells, spread throughout the outer body layer. The entire surface of the animal is about equally covered. There is no part which controls the rest, no nerve center which functions in the regulation or coordination of the nerve net.

The flatworm has a somewhat more specialized nervous system. The neurons are arranged in two longitudinal *nerves* (a nerve is composed of many neurons) connected by *transverse commissures*. Your keen eye will undoubtedly also have noted the aggregation of nerves in the head region. These are the *cephalic ganglia*. A *ganglion* is a clump of neural cell bodies.

Have you ever wondered why the brain is located in the head in most animals? There have been exceptions, such as the huge, herbivorous dinosaur, the *Brontosaurus*, which had a second "brain" at the base of its tail to help direct its immense body as it browsed in prehistoric lakes. But the evolutionary reason for nerve centers in the *anterior*, or head, region may have been to permit

Figure 10.6 Examples of animal nervous systems. It is likely that the evolutionary route of vertebrate nervous systems was much the same. A, the *Hydra,* with its nerve net composed of connecting neurons; B, the planarian, or flatworm, with its longitudinal nerves connected by transverse nerves and its nerve concentration in the head; C, the earthworm, with its single ventral nerve cord and well-defined cerebral ganglia and; D the frog, with its dorsal hollow nerve cord and the well-developed brain protected by bone.

A

Nerve net

B

Nerve branches
Transverse nerves
Head ganglia
Nerve cord

Ganglion

Nerve branches
Cerebral ganglion

D

Brain
Spinal cord
Nerve branches

quick analysis of the environment into which the animal would be moving. These centers would in all probability have come to be associated with the organs of perception, the specialized receptors we refer to as the *senses*. And as we know, the senses of sight, sound, smell, and hearing are commonly located in the head region. Thus, the environment into which the animal is moving may be quickly assessed. If the brain and these special receptors were located at the posterior end instead, the animal might find itself in an inhospitable environment by the time it realized its predicament.

The earthworm has only a single longitudinal nerve, but even this one nerve shows indications of a paired arrangement, in that it is two-lobed, much like two cords pressed together. In the earthworm the nerve is ventral, with the heart and digestive tract lying dorsal to it. Note the distinctness of the cerebral ganglia and the obvious nodes of the nerve cord, each with paired nerves reaching into the body. The nodes correspond with the segments of the earthworm's body, each node innervating a particular segment.

The frog represents a relatively primitive vertebrate system. In vertebrates the longitudinal nerve is dorsal, and it is hollow (remember how it develops), filled with fluid, and protected by bone. At the anterior end it swells into a large specialized organ which is more than an enlarged ganglion, although it is probably derived from such nerve clumps through evolutionary changes. The vertebrate brain shows marked specialization, since different parts of it and structures within it are intimately concerned with very specific functions. As in all vertebrates, there is the basic paired and segmented arrangement indicative of a common annelidlike ancestry. These characteristics appear in the two-lobed structure of the spinal cord in cross section and in the paired nerves which branch off from the spinal cord and brain. The branching is no longer so regular and apparent because of specialization along the spinal cord as the vertebrate body plan became more complex.

THE VERTEBRATE BRAIN

Actually, there is no such thing as "the vertebrate brain," because the vertebrates include widely diverse groups, each of which is highly specialized and distinctive. In the midst of such diversity, however, it is possible to detect trends which give some clues to the general pattern of brain development in animals with backbones. Generally, those traits which are most like the hypothetical or presumed ancestral forms are called *primitive*; those which are more unlike this early condition are called *advanced*.

The sequence in Figure 10.7 illustrates relative changes in parts of the brain from fish to reptile to mammal. The *medulla* is

Figure 10.7 A comparison of the brains of a fish, a reptile, and a mammal.

— Cerebrum
— Cerebellum
— Medulla
— Olfactory bulb

— Cerebrum
— Medulla
— Cerebellum
— Olfactory bulb

— Cerebrum
— Cerebellum
— Olfactory bulb
— Medulla

simply the specialized anterior end of the spinal cord. The *cerebellum* is associated with various lower, or unconscious forms of behavior, and the *cerebrum* is the "gray matter," the thinking part of the brain. We will discuss all these structures, but there are a few preliminary points you might find interesting. First, not only is there an increase in general brain size (in relation to body size) as we go from fish to mammal, but there is also an increase in the size of the cerebrum in relation to other parts of the brain. In contrast,

Box 10.1 The Octopus   The octopus is a cephalopod, and like all cephalopods, is a vigorous predator. It can walk about on its "arms" or use jet propulsion by squirting water from its mantle cavity. It is soft-bodied and thus spends much of its time hidden safe in crevices of rocks. The octopus can quickly change its color or pattern to match its surroundings or blanch to a pale "white" (thus accenting its eyes) to escape predators. It may also project clouds of ink when threatened. The ink may temporarily blind an attacker and numb its smell receptors. If placed exposed into a tank it will immediately build a house out of any material available. The octopus is curious, active and perpetually hungry, so it is an excellent animal to use in behavior tests.

The octopus has large eyes and it sees about as well as man. It is primarily a visual hunter and at the sight of a crab, it turns dark, breaks out in blotches of color and rushes toward its prey. If the prey is behind glass the octopus will press against the glass in its excited efforts. If a tentacle should reach over the glass and contact the prey, the success doesn't seem to register with the octopus. It will continue to "pursue" the prey—almost as if the information from the tactile (touch) and visual receptors were not integrated.

The octopus can quickly learn to discriminate between horizontal and vertical visual images (possible because the neurons in its optic lobes are arranged in at right angles rather than randomly as in most animals). But the octopus is unable to distinguish between cylinders A and B which have both been cut away 30%. On the other hand, it can distinguish these from C, which has been cut away 20%. Apparently, discrimination is on the basis of what proportion of the object comes in contact with the suckers.

If an octopus learns to perform a pattern with one tentacle, there must be a time lag before that knowledge is transmitted to other tentacles. Such information has helped to formulate principles of long-term versus short-term learning.

the olfactory lobe doesn't follow this pattern. Olfaction has to do with the sense of smell, so which of these animals do you suppose would rely more on a sense of smell? (Remember to consider the olfactory lobe in relation to total brain size.) Other sensory lobes could also be singled out, such as the optic lobe, which has to do with vision. Which animal do you suppose would have a larger optic lobe with respect to its brain mass, an eagle or an elephant?

Note that there is no implication here that as one moves up the evolutionary family tree, each species is smarter than the ones below it. There may be such a trend within certain groups, but there are also many exceptions. For example, the octopus, which is a mollusk like the snail, is more intelligent than many species of vertebrates. The octopus brain, in fact, is structurally similar to the mammalian brain in terms of its complexity and organization.

The evolutionary development of the mammalian cerebrum, which culminates in the human brain, is undoubtedly one of the most crucial events in the history of life on earth. Such a statement is admittedly somewhat grandiose, but its validity becomes apparent when we consider the impact of the human species on the fragile life system which thinly covers the planet. It would be interesting, therefore, to know how such a brain came to be. What spurred its development? And, as usual, no one knows. One prevailing theory, however, is that the cerebrum developed as a tool to aid in the survival of one small, seemingly insignificant group of reptiles which lived in an exceedingly dangerous world dominated by the great dinosaurs. The line of reptiles which was to give rise to mammals branched off from other reptiles 180–200 million years ago.

Because of their size, these premammals were certainly no match for the speedy, incredibly powerful, and voracious dinosaurs. Since they couldn't outrun or outfight the "ruling reptiles," a premium was placed on their mental agility. In order to survive, the premammals were forced to *outthink* the dinosaurs. The dimmer members among them would rapidly have fallen to predators and thus would have failed to leave their "dim genes" in the next generation. So, with their very existence at stake, the premammals must have rapidly increased their reasoning ability, starting the cerebrum on its way toward dominating the brain of at least one line of animals.

The rapid development of the cerebrum and increasing reliance on intelligence must have been accompanied by other changes among the early mammals. Since the cerebrum is the learning center of the brain, and since learning occurs with experience, animals which were becoming highly cerebrated would

Vertebrates
(backboned animals)

Arachnida
(spiders)

Crustacea
(lobsters, crabs)

Myriapoda
(centipedes)

Insecta
(insects)

Platyhelminthes
(flatworms)

Mollusca
(oysters)

Echinodermata
(starfish, crinoids)

Brachiopoda
(lamp shells)

Arthropoda

Annelida
(segmented worms)

Bryozoa
(moss animalicules)

Coelenterata
(corals, jellyfish)

Porifera
(sponges)

Protozoa (amoeba)

Figure 10.9 Four neurons (right) found in human beings which show the diversity of these cells. A and B are cells with different kinds of cell bodies, axons and dendrites. C, a motor neuron with axons which run from the nervous system to the effector (in this case, a muscle). A sensory neuron, D, running from the receptor to the spine. Note that the sensory neuron has no true dendrites. The nodes in the myelin sheath are where one Schwann cell ends and another begins. A single nerve cell may be nine feet long such as those that run from the base of a giraffe's spine down its hind leg.

have had to develop a life pattern that gave them a chance to learn before they were exposed to the dangers of the world. Hence parental care developed. Even today the offspring of "learning animals" stay with their parents for extended periods. (How old were you when you left home?) During this time they are cared for and protected by their parents until they can gain enough experience to cope with the world. Also, by parental association, they usually learn from their parents or other members of the group. The term *group* is appropriate because most of the more cerebrated species are social animals, although there are notable exceptions.

Of course, in a high-risk world increased cerebration is not the only course open to natural selection. As an example of another evolutionary route, consider the "strategy" of birds. The noted psychologist Oscar Heinroth is said to have commented, "Birds are so stupid because they can fly." It is certainly true that if an animal can fly, it doesn't have to have great intelligence to escape most predators. A bird doesn't normally "outfox" its predators, since it has only to give a few flaps of the wings to escape terrestrial enemies. Even to escape airborne predators, such as hawks, most birds do not rely on their cunning. Instead, they respond to such threats by employing a few very stereotyped behavioral patterns, such as attacking in mass, aggregating to confuse predators, taking rather specific evasive maneuvers, or giving warning cries. Birds just have never had to develop great mental capacities, so don't be misled by the discerning frown on an eagle's face.

We will take a closer look at the complex and intriguing human brain, but let's first find out something about what nerves are, how they work, and how they can function in the regulation of bodily processes.

THE NEURON

A *neuron* is simply a nerve cell—although you may reach the conclusion that "simply" is hardly the word for it. Neurons come in a variety of sizes and shapes, and there are billions of them in the human body (Figure 10.9).

A neuron is composed of a *cell body*, which contains the nucleus of the cell; *dendrites*, which receive stimuli from other neurons and conduct impulses toward the cell body; and *axons*, which conduct impulses away from the cell body. Dendrites usually branch profusely, providing increased points of contact with other neurons. Most neurons have only one axon, and it is longer than the dendrites. The axon may also be branched, but not as much as the dendrites. For our purposes we can differentiate between dendrites and axons on the basis of their interaction with other

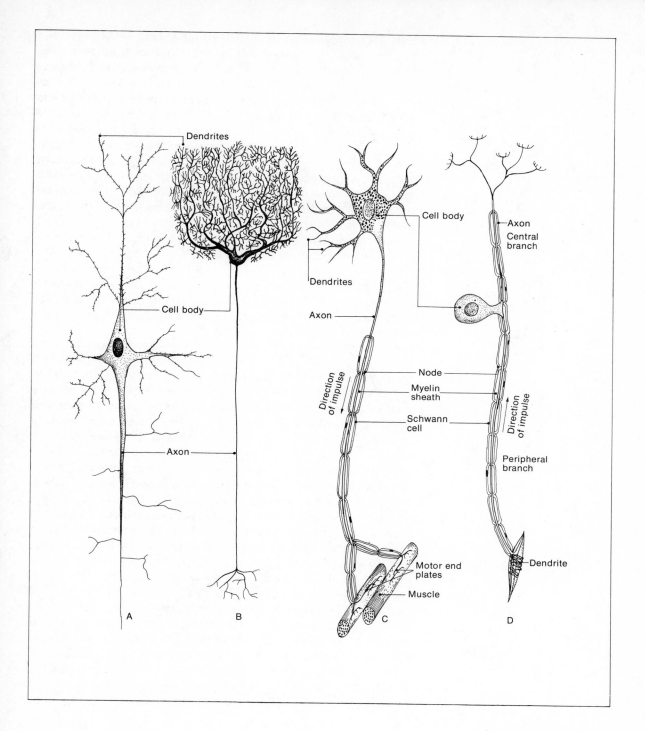

cells. Dendrites can be stimulated by other cells, while axons can't be stimulated. And axons, but not dendrites, can stimulate other cells. (You won't be surprised to learn that there are exceptions to these rules.)

Some vertebrate axons are enveloped in a fatty *myelin sheath*, which consists of the tightly rolled membranes of *Schwann cells* which surround the fiber. The myelin sheath serves to speed up the transmission of impulses along the axon. The presence of myelin gives tissue of the *central nervous system*, the brain and spinal cord, a whitish appearance. Nonmyelinated fibers are usually gray. The white myelinated fibers lie on the outside of the spinal cord, but retreat to the inside of the brain mass, leaving gray matter visible on the outside. Also, in most cases only myelinated fibers are capable of regeneration if they are severed; the gray areas of the spinal cord do not regenerate if damaged.

Nerves are bundles of whitish neurons, each wrapped in its myelin sheath. Some nerves are quite large, about the diameter of your finger. Although a nerve may contain many parallel neurons, each neuron can transmit an impulse independently of the others.

Another kind of cell in the vertebrate nervous system is the *glial cell.* No one is sure just what glial cells do, although they outnumber nerve cells by ten to one in the nervous system. Their branches often permeate the nerves, weaving between the neurons, and it is thought that perhaps they somehow support or nourish neurons. It is also suggested that they may store certain kinds of molecules necessary to the business of impulse transmission.

IMPULSE PATHWAYS

If you step on a tack, and are a normally sensitive and intelligent person, you may wish to take your foot off it. For you to be aware of your predicament, however, the condition of your foot has to register in your brain, which is some distance away. This message travels over neurons, as an *impulse.* But let's start at the bottom.

To begin with, the tack stimulates certain *receptors* in the sole of your foot. Receptors come in a variety of types, and each type specializes in a particular sensation, such as cold, pressure, or pain. In some cases bare nerve endings function as receptors. But in any case the stimulation of a receptor results in the transfer of that impulse to an *afferent neuron*, one which carries impulses *toward* the central nervous system. The impulse may be relayed from one neuron to the next until it enters the spinal cord, is transferred to an ascending neuron, and reaches the brain, where the message is registered: "Sharp pain in foot!" If the foot has not already been lifted by a reflex action (which we'll discuss later), you may say to yourself, "I really must remove my foot from the tack," whereupon an impulse is sent from the brain, down the spinal cord, out the spinal cord along the proper *efferent nerve*, one which carries the impulse toward the *effector*—in this case the muscle which raises the leg.

The junction of two neurons—the point at which the axon of one neuron transmits the impulse to the dendrite of another neuron—is called a *synapse*. Whereas theoretically an impulse can travel in either direction along a neuron, it can cross the synapse in only one direction; so in a sense the synapse acts as a one-way valve. We will look more closely at these fascinating junctures later.

THE MECHANISM OF THE IMPULSE

When a nerve fiber is at rest, certain chemicals are found in greater abundance within the cell, while others are more concentrated outside the cell. Furthermore, some of these chemicals are ionic, or electrically charged. For example, the resting neuron is literally bathed in sodium ions (Na^+), but these ions are scarce inside the cell. This difference results in a net positive charge outside the cell. In the cytoplasm of the neuron, however, the situation is reversed: the cytoplasm is negatively charged. It is true that there are a lot of positively charged ions in the cell fluid, such as potassium (K^+), but these are no match for the myriad of negatively charged ions found there, especially chlorine (Cl^-) and negatively charged proteins.

The electrical potential—the difference between the two levels of electrical charge—that is set up by this peculiar distribution of charged particles is called the *resting potential*. Typically, this potential difference is about 70 millivolts, or about 5 percent as much electrical energy as is found in a regular flashlight battery. The difference in the ion concentrations inside and outside the cell is maintained by considerable work on the part of the cell. As fast as positively charged Na^+ particles diffuse into the cell, the cell pumps them right back out again to maintain the resting potential. The energy for this pumping activity, as you might suspect, is provided by ATP.

Now, when a neuron is excited by either an electrical or a chemical stimulus, the cell becomes leaky to sodium ions, so that they rush in faster than they can be pumped out. The result of this shift in ions is that there is a change in the membrane potential of the cell. The change doesn't occur simultaneously along the entire length of the cell; beginning at the point of stimulation, a

Figure 10.10 The action potential occurs as the inside of the cell gains a positive charge. The charge is from about −70 to +20 millivolts as ion concentrations inside and outside the cell change. Changes in charges occur as an impulse passes. Note, the impulse travels from right to left. The absolute refractory period designates the period at which the nerve cannot respond to another stimulus. The relative refractory period designates the period when the cell can respond only to an unusually strong stimulus.

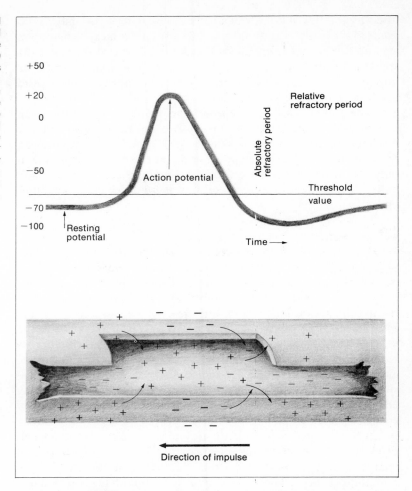

wave of inrushing sodium ions sweeps along the length of the cell, and as it passes, the cell fluid behind it immediately restores itself to its normal resting potential. Since the membrane is no longer separating positively and negatively charged particles, the cell at such points can be said to be *depolarized*.

This depolarization sweeps along the neuron at a regular rate of speed, depending on the neuron. The inrush of sodium ions at any point along the membrane results in a momentary net positive charge inside the cell at that point, so that the polarity is reversed. In other words, for an instant, the outside becomes slightly negatively charged and the inside slightly positively charged. This momentary positive charge inside the neuron is called the *action potential*. The impulse, then, is the passage of this action potential along the neuron. See Figure 10.10 for more explanation.

The *repolarization* of the cell begins as positively charged potassium ions rush out of the cell faster than the sodium can move in. This shift sets up a net positive charge outside again. As the membrane becomes less permeable to sodium, the "sodium pump" begins to move the sodium, which has rushed in, back out again. In the meantime, some ionic potassium begins to diffuse back to the neuron's interior. Thus the membrane potential of the cell is restored. Again, all this takes place in the briefest of periods so that most neurons can be depolarized and repolarized several times each second.

Two important qualities mark neural impulses. One is the fact that an impulse does not increase or decrease in strength as it moves along a neuron. Second, impulses operate on an "all-or-none" basis. This means that if a stimulus reaches the *threshold value* for a neuron, the lowest intensity at which a stimulus can be registered for that neuron, the cell fires, and it fires at only one intensity no matter what the strength of the stimulus.

If neurons operate in this all-or-none manner, how is it that we can detect various intensities of stimuli? If you put your hand on a stove, how do you know whether it is simply nice and warm or dangerously hot? Actually, we can make such discriminations in a number of ways. First, the more intense the stimulus (the hotter the stove), the higher the rate of impulses which pass along the neuron (although there are maximum limits). Also, different neurons in the same nerve may have different threshold levels. If some are easily stimulated and others require higher heat, then the stronger the stimulus, the more neurons will be stimulated. Hence the brain can interpret either the *frequency* of the impulses or the *number* of neurons stimulated to tell you whether you should remove your hand from the stove. Fortunately, the process doesn't take as long as the explanation.

THE SYNAPSE

Now let's consider how an impulse travels from one neuron to the next. Actually, what we are concerned with here is how one neuron activates another so that the impulse continues along a chain of nerve cells. How does the behavior of a neuron initiate a wave of depolarization along the length of another neuron? The story is an interesting one which centers on the secretion of certain remarkable chemicals.

When an impulse traveling along a neuron reaches the axon tip, it causes the buttonlike ending of the axon to secrete a chemical into the space between itself and the dendrites of the next neuron.

Figure 10.11 How synaptic transmission is achieved. Left, the relationship between the axon and the dendrite. Right, the transfer of acetylcholine across the synapse and the location of the enzyme acetlycholinesterase which breaks down this transmitter substance.

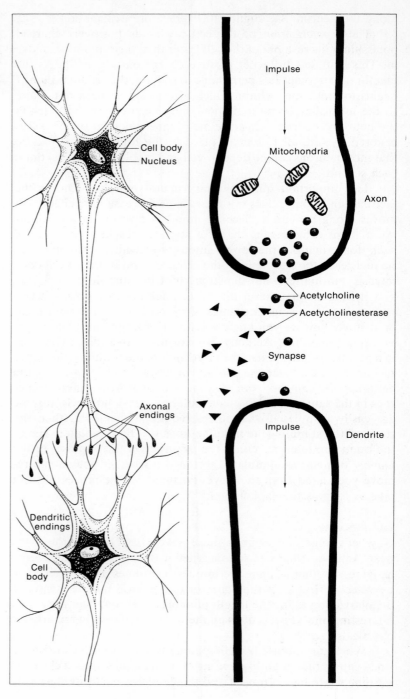

Box 10.2 Botulism　　The poison produced by the bacteria *Clostridium botulinum* is the most deadly poison known. Only 60 billionths of a gram will kill a man. The spores of the organism are found as a common inhabitant of the soil around the world. Fortunately, however, specific conditions must be met in order for these spores to grow. An anaerobic and nutrient-rich medium must be provided. Thus canned food which has not been properly sterilized is an ideal home for the microorganism. If canned food has been contaminated, however, it will produce the telltale clues of foul smell and buildup of gas (which may cause cans to swell). The signs of botulism are vomiting, constipation and paralysis of eyes and throat as breathing becomes more difficult. These begin one to three days after eating the poisoned food. The poison functions by interfering with the release of acetylcholine. This is the physiological basis of the paralysis.

The axon is enmeshed in the highly branched dendrites of the next neuron, so that there is a large total surface area over which the two neurons communicate. Any chemical secreted by the axon endings of one neuron, then, can be expected to immediately affect the next neuron. And this is just what happens.

In many of the neurons in our body the transmitting chemical is believed to be *acetylcholine*. The transmitting chemical is stored in small packets in the swollen tips, or "buttons," of the axon endings (Figure 10.11). When an impulse reaches the end of the axon, some of these packets move toward the membrane and rupture, thus releasing their transmitter substance into the synapse. The chemical quickly diffuses across the gap and contacts the dendrites of the next cell, where it sets up another action potential in that neuron.

After a few impulses have traveled down a neuron, causing release of the transmitter chemical into the synaptic space, it might seem that the chemical would build up in the synaptic area and cause the next neuron to continue firing in the absence of a real stimulus. But do not despair. Almost all types of dendrites secrete a substance into the synaptic space which quickly neutralizes any residual transmitter substance. When the transmitter substance is acetylcholine, the neutralizer is the enzyme *acetylcholinesterase*.

Humans sometimes accidentally come into contact with substances which inactivate these important neural enzymes, with horrible consequences. For example, certain insecticides, such as Malathion or Parathion, which are sprayed on crops across the country, have the advantage of being short-lived—unlike DDT. Within a few days after spraying the chemicals are believed to be relatively harmless. These chemicals, however, can block the effects

Figure 10.12 Insecticide being sprayed on crops to prevent crop destruction from insect pests. Many of these substances find their way into humans and can be dangerous. Also, it is inevitable that many harmless or beneficial insects which are part of the ecosystem may be exterminated.

of acetylcholinesterase. Keep in mind that the enzyme functions not only to keep chains of neurons from firing blindly in response to accumulating transmitter substance, but also to keep acetylcholine from diffusing over to nearby neurons and causing them to fire as well. What do you think might be the behavioral effect on a farmworker exposed to the insecticides? There have been several deaths, particularly among Chicano farmworkers, attributed to these chemicals. Conversely, other substances are known to block the release of acetylcholine.

In addition to excitatory transmitter substances such as acetylcholine, there are also numerous *inhibitory transmitter substances*, which are as yet largely unidentified. These substances, it seems, function at *inhibitory synapses* and counteract the effect of the excitatory synapses. Since each neuron manufactures only one type of transmitter substance, inhibitory substances are manufactured by inhibitory neurons, which converge with the axons of excitatory neurons at the dendrites of the next cell. This opposing system has been found at every central nervous system synapse which has thus far been extensively studied. Inhibitory synapses may function as a means of keeping neural excitation under control, so that a chain reaction of neural activity is not set up throughout the ner-

Figure 10.13 How a neuron is excited or inhibited. Here, two excitatory neurons and one inhibitory neuron synapse with the dendrites of another neuron. The darker axons are firing. In (1) the neuron is at rest with no axonal action; in (2) one excitatory neuron fires but it does not stimulate to threshold levels; in (3) both excitatory neurons fire and the second neuron is stimulated to fire; in (4) the inhibitory neuron is fired at the same time so that the threshold levels of the second neuron is raised above the effects of even both excitatory neurons; in (5) the inhibitory neuron firing alone will, of course, fail to excite the second neuron.

Labels within the figure: "Excitatory nerve fibers", "Inhibitory nerve fiber", "Cell body of second neuron".

vous system (Figure 10.13). It is believed that uninhibited neural activity of this kind may be responsible for the convulsions that occur in epilepsy.

THE HUMAN NERVOUS SYSTEM

Now that you know something about the way nerves work, let's see how they are organized in the human nervous system. The

Figure 10.14 A cross section of the spinal cord showing the distribution of white and grey matter and the pathway of a simple reflex arc. The arc involves three neurons (sensory, connector and motor) but simpler arcs may have only two. In such a case the sensory neuron would connect directly to a motor neuron. It will also be noticed that the motor neuron acts by stimulating an effector (muscle fiber). Thus muscle fibers also react to transmitter substances. The sensory nerve is stimulated by a receptor which secretes its own transmitter substance. In each case the transmitter substance may be the same.

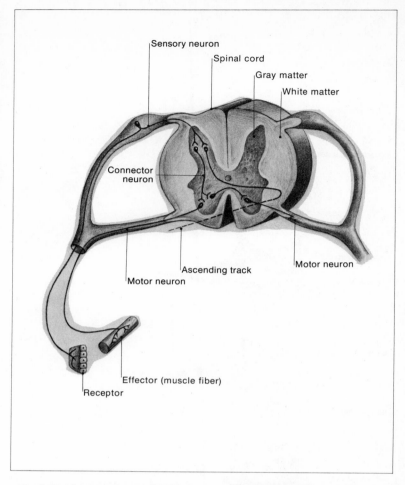

principles here are applicable to most mammalian systems since the neural organization is much the same throughout this diverse group. We will focus primarily on the central nervous system, which consists of the brain and spinal cord, and the fascinating peripheral nervous system.

THE SPINAL CORD AND THE REFLEX ARC

If you pride yourself on the fact that you are a "thinking animal," you may be a little disappointed to realize that your spinal cord can often receive information from the body's receptors, process it, and initiate the proper response before your brain even "knows" what has happened. Such a chain of events describes the *reflex arc* and illustrates something about the makeup of the spinal cord.

As illustrated in Figure 10.14, the impulse is generated in a stimulated receptor and transmitted to a sensory neuron, which enters the spinal column over the *dorsal nerve root*. The impulse is then transferred to a connector neuron, which in turn transfers it to the proper motor neuron. The motor neuron leaves the spinal cord over the *ventral nerve root* and travels outward to the effector (usually a muscle group).

Note that impulses can cross the spinal cord, so that effectors in the other half of the body can be stimulated. Also, the impulse can be transmitted to the dendrites of yet other neurons, which ascend the spinal cord to the brain. Thus even though the response has already been accomplished reflexively, your brain is informed of the change. Direct neural routing from the spinal cord in a reflex arc saves time because the distance the impulse has to travel from receptor to effector is shorter. And no time is spent in deliberating over the decision. You might come to think this is important if you should crawl out of your sleeping bag one night and step on a hot coal.

The reflex arc is structurally simple, but it may produce a highly coordinated response. For example, the reflex arc that takes place in the familiar knee jerk involves sensory and motor pathways which lie on the same side of the body and is thus an example of the simplest form of arc. However, the jerk of the knee entails a contraction of a *group* of muscles on the front of the leg, acting together. Each muscle must be activated separately. At the same time the motor neurons going to the back of the leg must be inhibited from firing. It wouldn't do to have both groups contracting at once. So even in this "simple" case, *several* two-neuron arcs enter into the reflex. When a reflex involves connecting neurons, it is apparent that the response can be increasingly complex and highly coordinated.

THE BRAIN

The brain itself is an exceedingly complex structure, and there are many things we just don't understand about it. However, the research goes on. We do know some things, and you're about to hear a few.

To begin with, the human brain is divided into three parts: the forebrain, the midbrain, and the hindbrain. The *forebrain* consists of the two cerebral hemispheres and certain internal structures, such as the thalamus and hypothalamus. The *midbrain*, logically enough, is the area between the forebrain and hindbrain. The *hindbrain* consists of the medulla, the cerebellum and the pons (Figure 10.15).

Paracentral lobe
Superior frontal gyrus area
Precuneus area
Gyrus cinguli area
Corpus callosum
Fornix
Cuneus area
Thalamus
Hypothalamus
Olfactory bulb
Pituitary
Cerebellum
Pons

Medulla

Spinal cord

Parietal lobe area
Precentral gyrus
Fissure of Rolando
Postcentral gyrus
Fissure of Sylvius
Frontal lobe
Prefrontal area
Occipital lobe

Temporal lobe

The Medulla

As a rough generality, the more subconscious, or mechanical, processes are directed by the more posterior parts of the brain. For example, the hindmost part, the *medulla*, is specialized as a control center for such visceral functions as breathing and heartbeat. As we have already seen, it also functions as a connecting area between the spinal cord and the more anterior parts of the brain.

The Cerebellum and Pons

Above the medulla and somewhat further toward the back of the head is the *cerebellum*, which is concerned with balance, equilibrium, and coordination. Do you suppose there might be differences in this part of the brain between athletes and nonathletes? Do you think this "lower" center of the brain is subject to modification through learning? The *pons*, which is the portion of the brainstem just above the medulla, connects the cerebellum and the cerebral cortex, accenting the relationship between the cerebellar part of the hindbrain and the more "conscious" centers of the forebrain.

The Thalamus and Reticular System

The *thalamus* and *hypothalamus* are located at the base of the forebrain. The thalamus is rather unpoetically called the "great relay station" of the brain. It consists of densely packed clusters of nerve cells, which presumably provide connections between the various parts of the brain—between the forebrain and the hindbrain, between different parts of the forebrain, and between parts of the sensory system and the cerebral cortex.

The thalamus contains a peculiar neural structure called the *reticular system*, an area of interconnected neurons which are almost feltlike in appearance. These neurons run throughout the thalamus and into the midbrain. The reticular system is still somewhat a mystery, but several interesting facts are known about it. For example, it bugs your brain. Every afferent and efferent pathway to and from the brain sends side branches to the reticular system as it passes through the thalamus. So all incoming and outgoing communications to the brain are tapped by this system. Also, these reticular neurons are rather unspecific. The same reticular neuron may respond to stimuli from the hand, the foot, the ear, or the eye. We don't know the functions of this system, but it is believed that the reticular apparatus serves to activate the appropriate parts of the brain upon receiving a stimulus. In other words, it acts as an arousal system for certain brain areas. The more messages it intercepts, the more the brain is aroused. Thus the reticular system may function importantly in sleep. You may have noticed it is much

Figure 10.16 The reticular system. The arrows illustrate a general arousal throughout the brain and a specific target being activated (shaded region).

easier to fall asleep when you are lying on a soft bed in a quiet room with the lights off. Under these conditions there are fewer incoming stimuli; as a result, the reticular system receives fewer messages, and the brain is lulled rather than aroused. On those nights when you suffer from the "big eye" and just can't sleep, the cause may be continued (possibly spontaneous) firing of reticular neurons (Figure 10.16).

The reticular system may also regulate which impulses are allowed to register in your brain and which will be filtered out or diminished. When you are engrossed in a television program you may not notice that someone has entered the room. But when you are engaged in even more absorbing activities, it might take a *general* stimulus on the order of an earthquake to distract you, whereas the *specific* sound of just a turning doorknob would immediately attract your attention. Such filtering and selective depression of stimuli apparently take place in the reticular system.

The Hypothalamus

The hypothalamus is a small body densely packed with cells which is extremely important in regulating the internal environment as well as certain general aspects of behavior. For example, the hypothalamus helps to control heart rate, blood pressure, and body temperature. It also plays a part in the regulation of the pituitary gland, as we learned earlier. And it controls such basic drives as hunger, thirst, sex, and rage. So now you know what's to blame for all your problems. Experimental stimulation of various centers in the hypothalamus by means of electrodes can cause a cat to act hungry, angry, cold, hot, benign, or horny.

The Cerebrum

For many people the word *brain* conjures up an image of two large, deeply convoluted gray lobes. What they have in mind, of course, is the outside layer of the two cerebral hemispheres, the dominant physical aspect of the human brain. The cerebrum is present in all vertebrates, but it assumes particular importance in humans. In some animals it may serve essentially as an elaborate refinement to implement behavior that could be performed to some degree without it. In more advanced animals it takes on a more functional importance.

For example, if the cerebrum of a frog is removed, the frog will show relatively little change in behavior. If it is turned upside down, it will right itself; if it is touched with an irritant, it will scratch; it will even catch a fly. Also, sexual behaviors in frogs can occur without the use of the brain—but we'll try not to extrapolate from that. Rats are more dependent on their cerebral cortex. A rat that is surgically deprived of its cerebrum can visually distinguish only light and dark, although its patterns of body movement seem unimpaired. A decorticated cat can meow and purr, swallow, and move to avoid pain, but its movements are sluggish and robotlike. A monkey whose cerebral cortex has been removed is severely paralyzed and can barely distinguish light and dark. The result in humans is total blindness and almost complete paralysis. Although such persons can breathe and swallow, they soon die.

It is apparent from this information that the cerebrum is more than just the center of "intelligence." It seems that, from an evolutionary standpoint, with increasing cerebration, more and more of the functions of the lower brain are transferred to the cerebrum. In this progression, even when the neural "center" of a certain function is not shifted to the cerebrum, a certain element of its control may be transferred there.

Generally, the degree to which neural control has been shifted to the cerebrum is reflected in the size of this structure. In other words, more "intelligent" animals have larger cerebral components in their brains. Increased dependence on the cerebrum may also be reflected by factors other than increased cerebral size. For example, the cerebrum may be more convoluted in more intelligent species. The deep convolutions of the human brain are lacking in the brain of the rat. Even the convoluted cerebrum of the highly touted and undoubtedly intelligent dolphin has fewer "layers" than the human cerebrum.

Hemispheres and Lobes

The human cerebrum consists of two hemispheres, the left and the right, each of these being divided into four lobes. At the back is the *occipital lobe*, which receives and analyzes visual information. If this lobe is injured, black "holes" appear in the part of the visual field which is registered in that area.

The *temporal lobe* is at the side of the brain. It roughly resembles the thumb on a boxing glove, and it is bounded anteriorly by the *fissure of Sylvius*. The temporal lobe shares in the processing of visual information, but its main function is auditory reception.

The *frontal lobe* is right where you would expect to find it—at the front of the cerebrum, just behind the forehead. This is the part that people hit with the heel of the palm when they suddenly remember what they forgot. One part of the frontal lobe is the center for the regulation of precise voluntary movement. Another part functions importantly in the use of language, so that damage to this part results in speech impairment.

The area at the very front of the frontal lobe is called the *prefrontal area*, if you follow that. Whereas it was once believed that this area was the seat of the intellect, it is now apparent that its principal function is sorting out information and ordering stimuli. In other words, it places information and stimuli into their proper context. The gentle touch of a mate and the sight of a hand protruding from the bathtub drain might both serve as stimuli, but the two stimuli would be sorted differently by the prefrontal area. A few years ago parts of the frontal lobe were surgically removed in efforts to bring the behavior of certain aberrant individuals more into line with the norm. Fortunately, this practice has been largely discontinued.

The *parietal lobe* lies directly behind the frontal lobe and is separated from it by the *fissure of Rolando*. This lobe contains the sensory areas for the skin receptors, and also the areas which detect bodily position. Even if you can't see your feet right now, you have some idea of where they are, thanks to receptors in the parietal lobe. Damage to the parietal lobe may cause numbness, but it can also cause a person to perceive his own body as wildly distorted. In addition, he may suffer from an inability to perceive spatial relationships in the environment.

By probing the brain with electrodes, it has been possible to determine exactly which area of the cerebrum is involved in the body's various sensory and motor activities. We can see the results of such mapping in the rather grotesque Figure 10.17. The pictures are distorted not out of any appreciation of the macabre, but to demonstrate that the area of the cerebrum devoted to each body part is dependent not on the size of that part, but on the importance it has come to have in the natural history of man. Thus we have the greatest number of *senses* in the face, hands, and genitals, but the greatest amount of control only in the face and hands (as you may already have found out).

Also note that the sensory and motor areas are not randomly scattered through the cortex, but that there is some order in their appearance. Thus the index-finger control area lies near the thumb control area, and the elbow area lies closer to the finger area than does the shoulder area.

One important fact about the cerebral hemispheres is that although they are roughly equal in potential, the two hemispheres are not identical. Moreover, any chemical or structural differences between them are accentuated by learning, because the control for certain learned patterns takes place primarily in only one hemisphere. One example of the difference between the hemispheres is seen in handedness. (It is interesting that less cerebrated animals such as rats and parrots also show handedness. Willy, my chinchilla, is a southpaw.) Other functions, such as speech, "I.Q.," and perception, are also more likely to be located in one hemisphere than the other. The primary speech center of righthanded people is located in the left hemisphere. In lefties, on the other hand (no pun), damage to the right temporal lobe may result in poor performance on I.Q. tests, but perceptual-test performance is relatively unaffected.

The primary route of communication between the right and left cerebral hemispheres is the *corpus callosum*. It seems that if one side of the brain learns something—for instance, by covering one eye while learning a visual task—the information will be transferred to the other hemisphere. However, if the corpus callosum has been cut, the material has to be taught anew to the other hemisphere by uncovering the other eye and repeating the training program (Figure 10.18).

It should be pointed out that a certain degree of compensation is possible between the two halves of the brain. For example, if one hand is injured, it is possible to learn to use the other hand. So, in spite of the natural handedness you were born with, retraining is possible as muscles controlling movement of the other hand develop, and neural events in the opposite side of the brain begin to

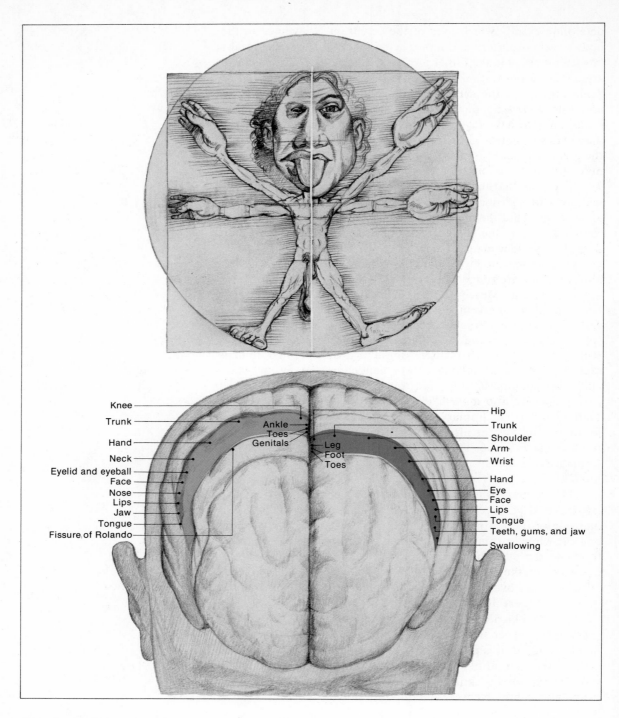

Knee
Trunk
Hand
Neck
Eyelid and eyeball
Face
Nose
Lips
Jaw
Tongue
Fissure of Rolando

Ankle
Toes
Genitals

Leg
Foot
Toes

Hip
Trunk
Shoulder
Arm
Wrist
Hand
Eye
Face
Lips
Tongue
Teeth, gums, and jaw
Swallowing

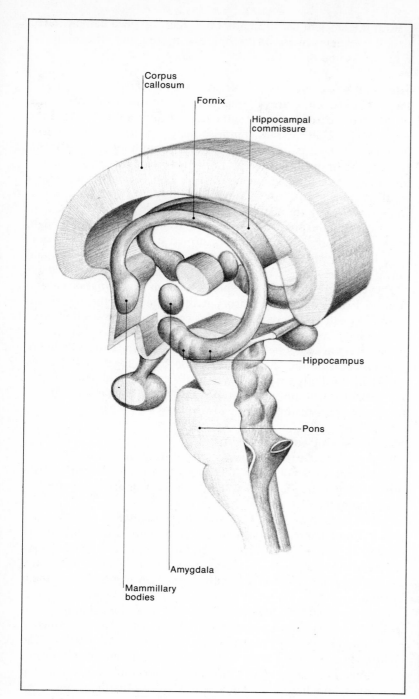

Corpus
callosum

Fornix

Hippocampal
commissure

Hippocampus

Pons

Amygdala

Mammillary
bodies

Figure 10.18 The corpus callosum of the human brain. Its function is to transfer information from one side of the brain to the other.

Figure 10.17 Function maps (opposite page) of the sensory area of the cerebral cortex (left) and the motor area (right). Note that the greater parts of the human brain are devoted to face, hands and genitalia. The sensory section is taken posterior to the fissure of Rolando, the motor section anterior to it.

permit the efficient use of that hand. It has also been demonstrated that if one hemisphere of the brain is injured, the other side can take over some of its tasks.

THE PERIPHERAL NERVOUS SYSTEM

Pairs of thick, white nerves emerge from the brain and either side of the spinal cord and innervate every receptor and effector in the entire body. These nerves comprise the *peripheral nervous system*. There are twelve pairs of *cranial nerves,* stemming from different parts of the brain, and these primarily innervate the senses, glands, and muscles of the head. These twelve pairs of nerves are found in all reptiles, birds, and mammals (Figure 10.19). Fish and amphibia have only the first ten pairs.

The human spinal cord gives rise to thirty-one pairs of tough, white spinal nerves. Each nerve is formed from the union of a dorsal and a ventral nerve root which emerge directly from the spinal cord. The dorsal nerve root is swollen by a huge ganglion at about the level where it enters the spinal column. The ganglion houses the cell bodies of all the sensory neurons entering the spinal column. The cell bodies of the neurons comprising the ventral nerve root (the motor neurons) lie imbedded in the spinal column, so these nerves need not be disfigured by ganglia. The two great nerves fuse just outside the spinal column and travel together for a way before increasingly smaller nerves begin to branch off from them—nerves which will ultimately branch into delicate neurons to innervate every part of the body.

THE AUTONOMIC NERVOUS SYSTEM

The *autonomic nervous system* is formed from a special set of peripheral nerves which serve the heart, lungs, digestive tract, and other internal organs. This system is largely under the control of the hypothalamus.

The autonomic nervous system may be divided into the *sympathetic* and *parasympathetic* nervous systems, which act antagonistically (Figure 10.20). Generally, the sympathetic system works to speed up certain body processes, with an increase in energy expenditure, and the parasympathetic system works to slow them down.

The sympathetic nervous system is activated in what has been called "fight-or-flight" reactions. It also functions in reproductive behavior, but perhaps we should avoid further alliteration. The fight-or-flight syndrome becomes apparent in certain emergency situations. For example, if a bear rushes into the room where you are quietly reading, your body will react "sympathetically." The

Optic II
Optic tract
Trochlear IV
Abducens VI
Stratoacoustic VIII
Vagus X
Hypoglossal XII

Accessory XI
Glossopharyngeal IX
Facial VII
Trigeminal V
Oculomotor III
Optic chiasm
Olfactory I

Figure 10.19 Human cranial nerves. The twelve large nerves shown originate in the brain and serve sensory and motor functions in the head and neck. An exception is the vagus nerve which passes through into the trunk serving the heart and parts of the digestive tract.

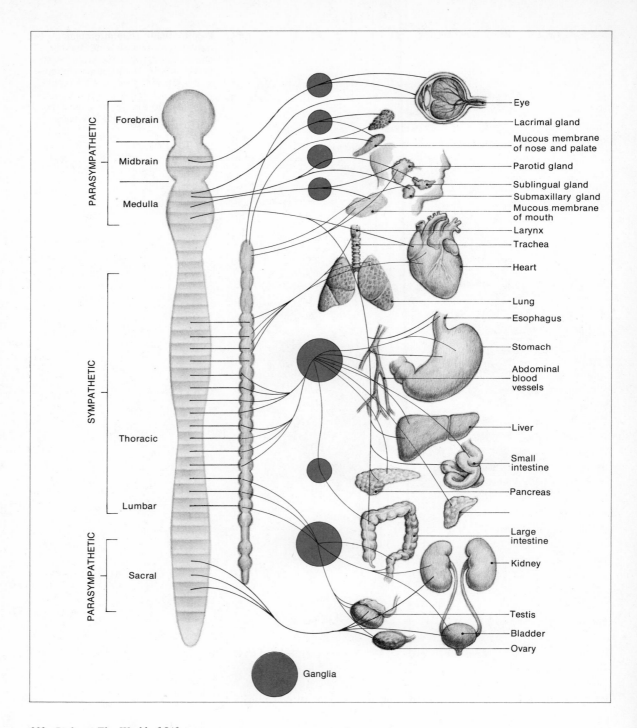

PARASYMPATHETIC

Forebrain

Midbrain

Medulla

SYMPATHETIC

Thoracic

Lumbar

PARASYMPATHETIC

Sacral

Eye

Lacrimal gland

Mucous membrane
of nose and palate

Parotid gland

Sublingual gland
Submaxillary gland
Mucous membrane
of mouth

Larynx

Trachea

Heart

Lung

Esophagus

Stomach

Abdominal
blood
vessels

Liver

Small
intestine

Pancreas

Large
intestine

Kidney

Testis

Bladder

Ovary

Ganglia

pupils of your eyes will dilate, the better to see the bear; peripheral blood vessels will decrease in diameter, so blood loss will be minimized in case you are scratched on your way out; heart and breathing rate will increase, bringing oxygen to your running muscles; and digestion will almost stop, since your visceral blood is needed elsewhere. After you, or the bear, have left, the parasympathetic system will take over and reverse all these responses.

It is interesting that, whereas the transmitter substance in the parasympathetic axon is acetylcholine, the sympathetic system's transmitter substance is *adrenaline.* You might wonder why adrenaline couldn't simply be released from the adrenal glands in an emergency situation, to travel in the blood and elicit these same changes throughout the body—why it was necessary to develop another, but similar, emergency system. The adrenal gland, in fact, may secrete adrenaline into the bloodstream in a stress situation, but the two emergency reactions illustrate an important difference between hormone and neural regulation. The sympathetic system releases its adrenaline *directly* into the proper effector. Only small amounts are released, so although the response is immediate, it is of short duration. Greater amounts of adrenaline may be secreted by the adrenal gland, and while these take longer to reach the effector, their effect is more long lasting. As a general rule, neural regulation is more immediate and short-term than hormonal regulation.

AUTONOMIC LEARNING

The autonomic nervous system is traditionally described as "involuntary," in that we do not normally exercise conscious control of our breathing rate, heartbeat, or urine formation. While it is true that the autonomic system can, and normally does, function in the absence of conscious control, there is evidence that some measure of conscious control is possible. In other words, we may be able to learn to control our autonomic reactions.

L. V. Dicara and N. E. Miller of Rockefeller University were able to teach rats to increase or decrease heart rate, blood pressure, intestinal contractions, blood vessel diameter, and even rate of urine formation. This was done by monitoring patterns and noting normal fluctuations over a period of time. As natural variations occurred, the ones in the desired direction were rewarded by electrical stimulation of the "pleasure center" of the brain and variations in the other direction were punished by unpleasant shock. If the rat was to slow its heart rate, when the heart slowed naturally, the animal would be rewarded; when it accelerated naturally, punishment would follow. Soon the heart began to beat more slowly.

Figure 10.21 Sherpa guide.
(Courtesy: Engelhard, Monkmeyer
Press Photo Service)

The researchers have produced similar results in humans and have taught at least one person to control his blood pressure. The amazing feats accomplished by some practitioners of Yoga are believed to be due to autonomic learning. Mountain tribesmen of the Himalayas are able to withstand extreme cold in their bare extremities. For example, climbers have reported their Sherpa guides sleeping barefoot under frigid conditions. It has been suggested that this is possible because they have "learned" to withdraw body fluids from these extremities so that cell membranes cannot be disrupted by the formation of ice crystals. The idea of autonomic learning has not yet received the attention it deserves in the Western world, but your imagination can give you some idea of its various possibilities.

PENFIELD'S MAPPING

Wilder Penfield, of the Montreal Neurological Institute, was one of a group of researchers who produced a fascinating body of evidence by stimulating the human brain with electrodes. Most of Penfield's work was done on epileptic patients in the course of surgical attempts to correct their condition. Epileptic seizures, which vary in intensity from a brief tingling or numbness to convulsions and unconsciousness, may result from injury to the cerebral cortex. Uncontrolled neural discharges from the injured area sporadically spread throughout the brain, causing these epileptic attacks. The surgery was an attempt to identify and remove the damaged area from which the discharges were emanating without removing other essential cortical matter. The patients were conscious throughout the operations (the brain feels no pain) and could respond to questions when various parts of their brains were stimulated. By carefully positioning the electrodes and recording the patient's response, Penfield was able, in effect, to "map" the brain, to determine which parts were correlated with what kinds of activities. Such mapping has resulted in the type of information we saw in Figure 10.17, as well as some important information about the nature of memory.

One of the most startling of Penfield's findings was that the human brain actually "remembers" virtually everything it has ever experienced, and that these experiences can be elicited anew. He found that stimulation of certain areas caused patients to "relive" some of their experiences. One patient *heard* an old familiar strain of music as if it were being played in the operating room when a certain point on the brain was touched. When the electrode was removed and reapplied to the same place, the music did not pick up where it had left off, but began again from the beginning.

Another patient saw himself sitting with relatives and friends in their house in South Africa. The event had actually occurred years before. Another patient "watched" a play she had seen years before, and another heard again Christmas carols being sung in her old church in Holland.

It was fascinating to learn that it is not only the important events which can be summoned from the past, but even everyday, humdrum occurrences which had been long forgotten. It seems that even extremely minute details of our lives are stored in the brain, literally millions of bits of information, which we may not be able to summon through normal attempts at remembering. The amount of detailed information which we store away unaware has also been demonstrated by hypnosis. For example, an elderly bricklayer under hypnosis described in detail the bumps on the bricks in a wall he had built when he was in his twenties. When the wall was checked, it was found that the bumps were there, just as he had said.

The events elicited by Penfield's probings are more than simple vivid recollections. The patients seemed actually to relive the experiences. Penfield himself has described the storage of such memories as being similar to recording by film or wire recorder. The recorder is able to run only forward, and it has only one speed, normal. In other words, electrode stimulation cannot elicit events in any sequence other than the one in which they happened, and when elicited, these remembered events are reexperienced at the same rate they occurred. As long as the electrode is held in place, the experience of a former day goes forward. It cannot be held still, turned back, or crossed with other periods.

As Penfield and Lamar Roberts state in their book, *Speech and Brain Mechanisms*:

The thread of time remains with us in the form of a succession of "abiding" facilitations. This thread travels through the ganglion cells and synaptic junctions. It runs through the waking hours of each man, from childhood to the grave. On the thread of time are strung, like pearls in unending succession, the "meaningful" patterns that can still recall the vanished content of a former awareness.

Penfield's work has provided many surprises and has given us new information on the vexing question of what memory really is. One theory is that memory stems from RNA changes in neurons, brought about by experience; it has also been suggested that memory involves the formation of new neural circuits developed in a learning situation. Moreover, there is some evidence that there is

Figure 10.22 Thread of time.

more than one kind of memory, and that each may involve different processes. In any case, the eventual explanation will have to account for the sometimes bizarre information provided by Penfield and his coworkers.

Now that we know something about the astonishing array of information stored in the brain, it is exciting to imagine its possible uses, if we could develop a precisely controlled and less drastic means of retrieval. On long train rides (remember trains?), or in boring company, we could reread a favorite book. There would be little need for testing pure "memory" as is so often done in classroom examinations, since each student would, in effect, be taking an open-book test. (In a hypnosis experiment one high school girl was able to "see" her notes on the blackboard during an examination after having read through them only once.) Other possibilities also come to mind. In cases of terminal disease, instead of facing death in apprehension and racked with pain, perhaps selective brain stimulation could enable a dying person to "relive" happier youthful days, to experience again long hikes with loved ones, robust optimism, and the companionship of old friends, now gone, as life dwindled away.

11
Behavior

(Courtesy: F. D. Schmidt, San Diego Zoo)

The field of study loosely called "behavior" can't exactly be called rowdy, but it certainly hasn't been as sedate or circumspect as some other areas of science. One reason is that early in its development, researchers branched off in several different directions. As each branch developed more or less independently and as researchers in each group became vaguely aware of what the others were up to, the air filled with criticism. Some considered certain behavior patterns to be genetically determined. Others said that all behavior could be accounted for through learning. Some believed that behavior should be studied under natural conditions. Others believed it should be studied in the laboratory where the environment could be carefully controlled. And no one had much use for anyone else's ideas. It would be nice if we could say that the dissension is past history, that we are now beyond all that, but many basic disagreements remain to this day. Let's see if we can discover what all the hollering is about.

ETHOLOGY AND COMPARATIVE PSYCHOLOGY

The study of animal behavior certainly isn't new. After all, the ancients had theories to account for such events as migration and the territorial behavior of certain animals. In the 1930s, however, new dimensions were added to the discipline with the development of a unified theory of *instinct*. The idea was born in Europe, although it is hard to trace to its precise origins. The development of the instinct idea is attributed primarily to the Austrian Konrad Lorenz and his coworkers, who described their theory in 1937 and again in 1950. The Dutch zoologist Niko Tinbergen, at Oxford University, amended the scheme in 1942 and 1951, and an idea resulted which, in spite of its imperfections, demanded serious consideration. And then the arguments began.

Lorenz and Tinbergen, who shared the 1973 Nobel prize in medicine and physiology with the German biologist Karl von Frisch, were the founders of an approach to behavior called *ethology*. Ethology, a discipline cultivated in Europe, traditionally con-

Box 11.1 History of the Instinct Idea Instinct was referred to in Greek literature at least 2,500 years ago, and, for some reason, it continued in the thinking of man with almost no supporting evidence, as very few other ideas have been able to do. In the first century, the Stoic philosophers developed the idea that animal behavior takes place without reflection. This was the beginning of the movement to separate the behavior of man from that of other animals. Man was officially removed from his place on the natural scale of animals (where Aristotle had placed him at the top, just above the Indian elephant) by Albertus Magnus (Big Albert) in the 13th Century. St. Thomas Aquinas, a student of Albertus, supported the separation of man and other animals based on the idea that man has a "soul," divinely placed in the embryo sometime before birth. This marked the beginning of a religious explanation of instinct. The reasoning was: There is a life after death and the part of man that continues to live is called the soul. The precise conditions, good or bad, under which the soul continues after death are determined by how the person behaved during life on earth. In order to make the "correct" decisions reason is necessary. Animals do not share in the hereafter and do not need reason. Instead, they are guided by blind instinct.

This philosophical notion of instinct was rejected by many people. For example, Erasmus Darwin, Charles's grandfather, believed that behavior has no innate quality at all, but that it is entirely learned. Charles disagreed, and, in fact, was willing to treat behavioral characters just as morphological ones, species-specific and subject to natural selection. The instinct idea was to fall into some disrepute in the last part of the 19th century, partly because of the tendency of some researchers to name an action as instinctive and to let the matter drop. Such an attitude obviously would not encourage the search for verifying data. So it came to be that in this century the idea of instinct was scoffed at, not considered valid subject matter for serious investigation.

Even by the end of the 19th century, however, some workers had not abandoned the instinct idea. C. O. Whitman, working on the behavior of pigeons at the University of Chicago, was accumulating evidence which in 1898 led him to declare, "Instincts and organs are to be studied from the common viewpoint of phyletic descent". Oskar Heinroth in 1910 described the behavioral similarities of related species and later noted that certain characters would appear in animals reared in isolation. An encompassing theory, however, was to be left to Heinroth's student, Konrad Lorenz.

cerns itself with innate behavioral patterns. Research is normally done in the field, under conditions as natural as possible, although ethologists have also relied in part on observations of the many tame or semitame animals which can usually be found sharing their house and yard.

The strongest disagreements with the ethologists' instinct theory came from American *comparative psychologists*. Their research has traditionally dealt with learning processes in laboratory animals, primarily the Norway rat. Although many of the arguments continue, there has been a closing of the gap between ethologists and comparative psychologists. Probably the final integration of the two approaches depends on further data from a third area of study, *neurophysiology*, which is concerned with how the nervous system works. Such information, however, is hard won and, at present, there just aren't enough researchers working in the nebulous gray areas. But let's take a look at the ethologist's concept of instinct and see what all the argument has been about.

Figure 11.1 Konrad Lorenz receiving his Nobel Prize from King Carl XVI Gustaf of Sweden. (Courtesy: Wide World Photos)

INSTINCT

It was once believed that, whereas human behavior is a result of learning associated with high intelligence, other animals respond only to unalterable "instincts" that are indelibly stamped into their nervous systems at birth. Since mounting evidence over the years did not bear out this sweeping generalization, the concept of instinct fell into some disrepute. In fact, many behavioral scientists declined to even use the word. The problem was compounded by the fact that while ethologists were seeking to clarify the concept as an explanatory tool, the word *instinct* found its way into the layman's vocabulary. Thus, as some were attempting to define the exact parameters of the instinct concept, others were using the term with wild abandon to explain everything from a baby's grip to swimming.

With all the confusion over the term itself, it is best to start with an explanation of what the instinct concept actually is all about. Let's first consider the components of an instinct.

Figure 11.2 In 1973, Niko Tinbergen shared the Nobel Prize with Konrad Lorenz and Karl von Frisch. (Courtesy: Wide World Photos)

FIXED ACTION PATTERNS AND ORIENTATION

It has long been known that animals are born with certain behavioral repertoires. The first time a tern chick is given a small fish, it manipulates the fish so that it will be swallowed head first, with the spines safely flattened. Dogs that have spent their entire lives indoors will attempt to bury a bone by making scratching motions on the carpet as if they were moving earth. The same dogs will turn around several times before lying down, although there is no

Figure 11.3 A fly-catching frog. The frog first orients itself with respect to the fly and then releases its fixed action pattern, the tongue flick. (Adapted from The Study of Instinct, N. Tinbergen, Clarendon Press, Oxford)

Figure 11.4 A graylag goose retrieving her egg. (Adapted from The Study of Instinct, N. Tinbergen, Clarendon Press, Oxford)

grass to trample nor spiders to chase out. Such patterns, which are innate and characteristic of a given species, are called *fixed action patterns*. When fixed action patterns are coupled with orienting movements, we can speak of an *instinctive pattern*. But this probably isn't too clear just yet.

To illustrate the relationship between a fixed action pattern and its orientation, consider the fly-catching movements of a frog (Figure 11.3). For the fixed action pattern to be effective, the frog must carefully orient itself with respect to the fly's position. The performance of the fixed action pattern, the tongue flick, results in the frog's gaining a fly. It is important to realize, however, that once the fixed action pattern is initiated, it cannot be altered. If the fly should move after the pattern is initiated, the frog misses since the sequence of movements will be completed whether the fly is there or not. When a frog catches a fly, one can easily distinguish the orientation movements from the fixed action pattern.

With other instinctive patterns the orientation may be indistinguishable from the fixed action itself. For example, if a goose sees an egg lying outside her nest, she will roll it back under her chin (Figure 11.4). She keeps the egg rolling in a straight line by lateral movements of the head. In an interesting experiment, it was found that the fixed and orienting components of the retrieval pattern could be separated. The orienting movements depend on the direction taken by the rolling egg. If the egg is removed during the retrieval the goose will continue to draw her head back (the

Figure 11.5 A falcon in flight. (Courtesy: G. Ronald Austing from National Audubon Society)

fixed component)—but in a straight line, without the orienting side-to-side movements. Thus we see that the fixed action pattern, once initiated, is independent of any environmental cue, so that once it starts it continues. The relationship between the fixed action pattern and its orientation component is similar to that between a car's engine and its steering wheel. Once the engine is started, it continues to run, but each change of direction requires a new movement of the steering wheel and thus depends on the specific environmental circumstance.

It has been found that an animal may not possess all its fixed action patterns when it is very young. Many of these patterns develop as the animal matures, as is the case with morphological characteristics. For example, wing flapping or other patterns associated with flight normally do not appear until about the time a bird is actually ready to begin flying, although there may be some incipient flapping just prior to this time. Also, the fixed action pattern and its orientation component do not necessarily appear simultaneously in the animal's behavioral repertoire. A baby mouse may scratch vigorously with its leg flailing thin air; later it will come to apply the scratch to the itch.

Now let's consider another drama, one which might put some of this together for us. After daylight a peregrine falcon begins its hunt. At first its behavior is highly variable; it may fly high or low, or bank to the left or the right. It is driven by hunger and would be equally pleased by the sight of a flitting sparrow or a

Figure 11.6 When a peregrine falcon spots its prey, it may show increasingly stereotyped behavior until finally it ends its instinctive sequence with the final consummatory pattern.

scampering mouse. Suddenly it sees a flock of teal flying below. Its behavior now becomes less random as it swoops toward them in what is the next stage of "teal-hunting behavior." First the random search, and then the swoop.

Almost all falcons perform this dive in just about the same way, although there are some variations. This is a sham pass to scatter the flock, and the falcon is likely to fall through any part of it. The teal, at the sight of the falcon, have closed ranks for mutual protection. (After all, it is hard to pick a single individual out of a tightly packed group. A predator is likely to shift its attention from one individual to another, and in failing to concentrate on a single prey, it may come up empty-handed.) The sham pass works though, and the terrified teal scatter. The falcon immediately beats its way upward, and after picking out an isolated target below, it begins a second dive.

If we were to watch this same falcon perform such a dive on several occasions, we might notice that the procedure is much the same every time. And if we were to watch several different falcons, we might notice that they all perform this action in much the same way. The greatest amount of variation is during the earliest part of such dives, since, at this stage, what the falcon does will be dictated in part by what the teal does. The falcon's options become fewer, and its behavior becomes more and more stereotyped, as it descends. Finally, there is a point at which the falcon's action is no longer variable. It now initiates the last part of the attack, the part which is no longer influenced by anything the teal does. At this instant the falcon's feet are clenched, and it may be traveling at over 150 miles an hour. Unless the teal makes a last split-second change in direction, it will be knocked from the sky, and the falcon's hunt will have been successful.

The falcon's energy will then be spent in following the teal to the ground, where, with rather unvarying and stereotyped movements, it will pluck the teal, tear away its flesh with specific movements of its head, and swallow the meat. The sequence of muscle actions in swallowing is very precise and always the same. The swallowing action itself seems to bring a measure of relief, so that the falcon does not feel like hunting again for a while. Theoretically, it is the desire to perform swallowing movements which brought the falcon out to hunt in the first place. In other words, the falcon was searching for a situation that would enable it to swallow. This probably seems a bit bizarre, but let's go on.

We can derive several points from this example. First, the state of the animal is significant. The falcon had not eaten for a time and was hungry. Hunger thus provided the impetus, or "motivation." In general, if an instinctive action pattern has not been performed for a time, there is an increasing likelihood of its appearance. Second, the earlier stages of this instinctive sequence were highly variable, but the behavior became more and more stereotyped until the final *fixed* action was accomplished. Third, the performance of this final action provided some relief. We have to assume this relief from the fact that once the falcon has performed this final action (swallowing), it does not immediately seek out conditions which will permit it to do so again. You will also note that the fixed action pattern provides the animal with a biologically important element—in this case, food.

Courting behavior in birds is also believed to be instinctive. In one experiment Daniel Lehrman of Rutgers University found that when a male blond ring dove was isolated from females, it soon began to bow and coo to a stuffed model of a female—a model which it had previously ignored. When the model was replaced by a rolled-up cloth, he began to court the cloth; and when this was removed the sex-crazed dove directed his attention to a corner of the cage, where it could at least focus its gaze. It seems that the threshold for release of the behavior pattern became increasingly lower as time went by without the sight of a live female dove. It is almost as though some specific "energy" for performing courting behavior were building up within the male ring dove. As the energy level increased, the *response threshold*, the minimum stimulus necessary to elicit a response, lowered to a point at which almost anything would stimulate the dove.

Invoking a mysterious "energy" seems less necessary to explain the hunting behavior of the falcon, since the reason the falcon hunts seems apparent: it is hungry. But then how do you explain hunting behavior in well-fed animals such as house cats? Perhaps the hunt is an instinctive action pattern in itself, so that its performance provides a measure of relief.

APPETITIVE AND CONSUMMATORY BEHAVIOR

There are labels, of course, for the things we have been talking about. First of all, instinctive behavior is usually divided into two parts, the *appetitive* and *consummatory* stages. Perhaps the clearest examples of appetitive behavior are seen in feeding patterns, from which the name is probably derived. Appetitive behavior is usually variable, or nonstereotyped, and involves searching—for example, for food. Some stages of appetitive behavior may be more variable than others, as we saw in the increasingly stereotyped hunting behavior of the falcon. The specific objective of this appetitive behavior is the *performance of consummatory behavior*. Consummatory behavior is highly stereotyped and involves the performance of a fixed action pattern. Swallowing is the consummatory pattern in feeding behavior and involves a complex and highly coordinated sequence of contractions of throat muscles (Figure 11.7). These contractions are the components of a fixed action pattern. As you have learned earlier, the performance of consummatory behavior is followed by a period of rest or relief, during which the appetitive patterns cannot be so easily initiated again.

Numerous experiments have demonstrated that the performance of consummatory behavior is the actual goal of appetitive behavior. This means, in effect, that a hungry animal is not looking for something to fill its stomach, but something to swallow. In an experimental demonstration the food swallowed by a hungry dog was shunted away from the stomach by tubes leading to a collecting pan on the floor. As soon as the dog had *swallowed* the proper amount of food, it stopped eating, apparently satisfied, although its stomach was still empty. The performance of the consummatory swallowing patterns had apparently provided relief.

RELEASERS

Instinctive actions are released by certain very specific signals from the environment. In the case of the falcon the diving sequence was released by the sight of the teal flying below. Courtship bowing is normally released in male blond ring doves by the sight of an adult female. Environmental factors which evoke, or release, instinctive patterns are called *releasers*. The releaser itself may be only a small part of any appropriate situation. For example, fighting behavior may be released in territorial male European robins, not only by the sight of another male, but even by the sight of a tuft of red feathers at a certain height within their territories. Of course such a response usually "works" because tufts of red feathers at that height are normally associated with the breast of a competitor. The point is that the instinctive act may be triggered by only certain parts of the total environmental situation.

The exact mechanism by which releasers work isn't known, but one theory is that there are certain neural centers, called *innate releasing mechanisms*—or IRMs. When these centers are stimulated by impulses set up by the perception of a releaser, the centers trigger a chain of neuromuscular events. It is these events which comprise instinctive behavior. Many such IRMs may be involved in any complex pattern, so that specific releasers must be presented in sequence. Or, as we have learned, the performance of one consummatory act (such as a falcon's knocking a teal to the ground) may act as the releaser for the next part of the pattern (such as plucking the teal before eating it).

ACTION-SPECIFIC ENERGY AND VACUUM BEHAVIOR

The energy which is believed to build up for each instinctive pattern is called *action-specific energy*. The continued buildup of this energy in the absence of a releaser may cause a progressive lowering of the threshold for release of a pattern, as we saw in the case of the

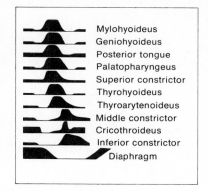

Figure 11.7 Diagrammatic summary of the sequence, timing and intensity of muscular contraction of swallowing in the dog. (From Doty and Bosma, 1956 *Journal of Neurophysiology,* 19:44–60)

male dove. Eventually, if no releaser is provided at all, the instinctive pattern may "go off" on its own. This is called *vacuum behavior*. Once, Konrad Lorenz noticed one of his tame starlings flying up from perches on the furniture and catching insects on the wing. He hadn't noticed any insects in his house, so he climbed up to see what the starling was catching. He found nothing. The well-fed starling had had no chance to release its insect-catching behavior, and so now the behavior was apparently being released in the absence of any stimulus.

THE BEHAVIORAL HIERARCHY

Niko Tinbergen proposed a *behavioral hierarchy* to define the relationships of the different levels of an instinctive behavioral pattern. According to this model, increasingly specific stimuli are needed to release each successive level of an instinctive pattern which lies "lower" on a hierarchical scale. As an example, he described the pattern of breeding behavior in the male stickleback fish—a common, but fascinating, denizen of European ditches. The reproductive behavior is activated by the lengthening days of spring, and at that time the visual stimulus of a suitable territory may elicit either aggression or nest building. Nest building will commence unless another male appears. In this case the sight of the competitor will elicit any of a number of aggressive responses, depending on what that male does. If the intruder bites, the defender will bite back; if the intruder runs, the defender will chase; and so on.

According to Tinbergen's model, the energy for an instinct passes from higher centers to lower ones (from left to right in Figure 11.9). This happens only when a block, or inhibitor, for a particular action is removed. Hence, the energy of the general reproductive instinct passes to a "fighting" center on sight of another male. The sight of that male fleeing then removes a block at the next level of behavior, so that energy flows to the centers for chasing behavior.

It is thus apparent, then, that the term *instinct* carries with it very specific implications. For a behavioral pattern to qualify as instinctive, it should meet the kinds of conditions we have talked about. For example, does the pattern show an appetitive phase? Does it ever appear as vacuum behavior? Is there a period immediately following its performance when the threshold for its release is raised?

The critical student will note that there are also certain problems with the instinct model. For example, a strictly hierarchical organization makes it hard to account for the fact that different appetitive actions can lead to the same consummatory

Figure 11.8 The sequence in courtship and mating in the three-spined stickleback, which is found in European ditches and streams. The male is attracted by the sight of a female swollen with eggs. He is torn between chasing her and leading her to the nest he has built and the result is a peculiar zig-zag dance. Once she is maneuvered into his "tunnel of love" he pokes her tail with his snout and thus induces her to lay eggs. He then promptly chases her off, returns to fertilize the eggs and then he spends days fanning oxygen-laden water over them until they hatch.

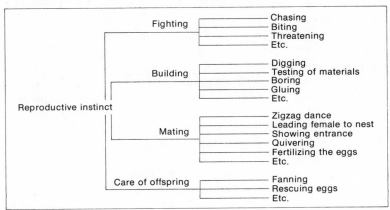

Figure 11.9 The hierarchical organization of the reproductive instinct of the male three-spined stickleback according to Tinbergen (1942).

act. Also, how is it that an animal can perform a complex chain of fixed action patterns without intervening appetitive patterns? The instinct model, as we will see, cannot be rigidly applied to all seemingly innate patterns. But generally, it works. It helps to explain what we see. What we have presented is the classical ethological interpretation of instinct. The scheme has been criticized by neurophysiologists who, for example, can find no neural basis for the notion of action-specific energy. Perhaps the model's usefulness lies in behavioral prediction and as a springboard for further research.

LEARNING

The importance of learning for an animal varies widely from one species to the next. For example, there are probably very few clever tapeworms. But then why should they be clever? They live in an environment that is soft, warm, moist, and filled with food. The matter of leaving offspring is also simplified; they merely lay thousands and thousands of eggs and leave the rest to chance—to sheer blind luck. In contrast, chimpanzees live in changeable and often dangerous environments, and they must learn to cope with a variety of complex conditions. They are long-lived and highly social, and the young mature slowly enough to give them time to accumulate information. We see, then, that learning is more important to some species than others. The *way* animals learn also differs from one species to the next, but we are only beginning to understand some of the differences.

Although learning may theoretically take place in a variety of ways, we will consider only three types of learning: *habituation*, *classical conditioning*, and *operant conditioning*.

HABITUATION

Habituation is, in a sense, learning *not* to respond to a stimulus. In some cases the first time a stimulus is presented, the response is immediate and vigorous. But if the stimulus is presented over and over again, the response to it gradually lessens and may disappear altogether. Habituation is not necessarily permanent, however. If the stimulus is withheld for a time after the animal has become habituated to it, the response may reappear when the stimulus is later presented again.

There are several ways in which habituation is important in the lives of animals. For example, a bird may learn not to waste energy by taking flight at the sight of every skittering leaf. A coral-inhabiting fish may come to accept the familiar sight of neighbors living within its own territory. A strange fish, though, which might be searching for a territory, would immediately be driven away.

Figure 11.10 Pavlov's apparatus with which he demonstrated classical conditioning in the dog. Note the graphs at the bottom.

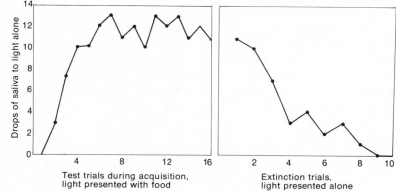

Test trials during acquisition, light presented with food

Extinction trials, light presented alone

Habituation is often ignored in discussions of learning, perhaps because it seems so simple, but it may well be one of the more important learning phenomena in nature. For example, it may function internally as a component of satiation. It may also help to explain why animals continue to react aversively to predators they see only rarely, while ignoring commonly seen but harmless species.

CLASSICAL CONDITIONING

Classical conditioning was first described through the well-known experiments of the Russian biologist Ivan Pavlov (Figure 11.10). In *classical conditioning* a behavior that is normally elicited by a given stimulus comes to be elicited by a substitute stimulus. Pavlov found that a dog would normally salivate at the sight of food. He then presented the dog with a signal light 5 seconds before food was dropped onto the feeding tray. After every few trials the light was presented without the food. On the basis of numerous experiments, Pavlov found that the number of drops of saliva elicited by the light alone was in direct proportion to the number of previous trials in which the light had been followed by food.

This conditioning process also worked in reverse. When the conditioned animal was presented over and over again with a light signal that was not followed by food, the salivary response to the light diminished until it finally was extinguished.

Pavlov's experiments demonstrated two important properties of classical conditioning: *generalization* and *discrimination*. It was found that if a dog had been conditioned to salivate in response to a green light, it would also respond to some degree to a blue or a red light. In other words, the dog was able to generalize regarding the qualities of lighted bulbs. However, it was found that, with careful conditioning, the dog could be taught to respond only to light with certain properties. Dogs do not see color well, so they are more likely to respond to intensity differences rather than color differences. If food was consistently presented only with a green light, and never with a red or a blue light, the dog would come to respond only to green lights, thus demonstrating clearly that it was able to discriminate differences between stimuli (discrimination) as well as the properties they had in common (generalization).

OPERANT CONDITIONING
Operant conditioning differs from classical conditioning in several important ways. Whereas in classical conditioning the reinforcers (or rewards, such as food) follow the stimulus, in *operant conditioning* reinforcers follow the behavioral response. Also, in classical conditioning the experimental animal has no control over the situation. In Pavlov's experiment all the dog could do was wait for lights to go on and food to appear. There was nothing he could do one way or the other to make it happen. In operant conditioning the animal's own behavior determines whether or not the reward appears.

In the 1930s B. F. Skinner developed an apparatus that made it possible to demonstrate operant conditioning. This device, now called a *Skinner box*, differed from earlier arrangements involving mazes and boxes from which the animal had to escape in order to reach the food.

Once inside the Skinner box, an animal had to press a small bar in order to receive a pellet of food from an automatic dispenser.

When the experimental animal (usually a rat or a hamster) was first placed in the box, it ordinarily responded to hunger with random investigation of its surroundings. When it accidentally pressed the bar, lo! a food pellet was delivered. The animal did not immediately show any signs of associating the two events, bar pressing and food, but in time its searching behavior became less random. It began to press the bar more frequently. Eventually, it came to spend most of its time just sitting and pressing the bar. This sort of learning Skinner called operant conditioning.

It is possible to build "stimulus control" into operant-conditioning situations. For example, a pigeon in a Skinner box may be required to peck a bar in order to receive food. If the apparatus is designed so that food is delivered only when the bar is pecked within a certain period after a sign lights up that says "peck," the pigeon will come to peck the bar mostly at that time. It will largely ignore the bar until the "peck" sign is lighted. If the pigeon pecks after a "don't peck" sign lights up, and that peck is not followed by food, the pigeon will come to ignore that sign through discrimination learning. It should also be pointed out that pigeons could be trained to peck in response to the "don't peck" sign. Pigeons can't really read.

The relative importance of each type of learning to animals in the wild isn't known at this point. It is likely that most adaptive, or beneficial, behavior patterns arise in nature as interactions of several types of learning.

The kind of learning of which an animal is capable is largely related to its evolutionary heritage. For example, M. E. Bitterman of Bryn Mawr demonstrated that there are strong species differences in "habit reversal" ability. In these experiments, once an animal had learned to discriminate between two choices for its reward, the reward was transferred to the opposite choice. Monkeys and rats showed the best performance, fish the worst, and turtles scored somewhere in the middle. The turtle scores were rather peculiar. They progressively improved in making spatial choices (for example, in differentiating high and low), but not in visual choices (such as in differentiating between a circle and a square). Thus it appears that habit reversal requires a specific kind of learning ability that is present in mammals, but not in fish, and exists in turtles only with respect to a certain class of problems.

HOW INSTINCT AND LEARNING CAN INTERACT

Much of the argument in the field of behavior has been over one simple question: is a certain behavior innate or learned? After years of bickering, the question itself is no longer regarded as valid. The

Box 11.2 Reward and Reinforcement Animals can be trained to perform rather complex acts by a system which involves rewarding those actions most closely approximating what the experimenter wants the animal to do. As time goes by, the animal is rewarded only when he does more and more specific acts. For example, in teaching a dog to open a door he may receive food for just walking to the general area of the door in his random strolls about the room as he looks for food. Then he may be only rewarded when he touches the knob, then only when he bites the knob and, finally, only when he turns the knob thus opening the door.

Food is often used as a reward because of its rather universal appeal. By using the word "appeal" we point out a problem with the word "reward." Reward implies some knowledge of how the food works in achieving conditioning. The psychologist B. F. Skinner prefers the term "reinforcement" which is defined as any char-

acter, such as food, which alters the probability of an animal performing a particular behavior. Thus, it is an operational term implying nothing about mechanism.

The success of any class of reward, of course, is dependent upon the nature of the animal. A bowl of water may serve as a stronger reward for a Norway rat than for a kangaroo rat, since the latter derives water from food sources.

Reward, or reinforcement, may be positive or negative. If you stay in school and obey the law you will be rewarded by approval of peers and parents, not to mention by money and possibly power. Thus the classroom where you sit is crowded with people who probably all obeyed traffic laws on the way to school. On the other hand if you walk into a bar and slap the largest man there across the face with a wet fish, you may get all your teeth knocked out. Such behavior is negatively reinforced and, partly for this reason, has not become popular in our society.

supposition that a particular behavior stems from one source or the other neglects the myriad ways in which behavioral components interact. In a sense, it's like asking whether the area of a triangle is due to its height or the length of its base. A better question is how do innate and learned patterns interact in the development of a particular behavior?

As an example of such interaction, consider the development of flight in birds. Flight is usually considered largely innate pattern. Obviously, a bird must be able to manage it pretty well on the first attempt, or the bird will crash to the ground as surely as would a launched mouse. It was once believed that the little fluttering hops of nestling songbirds are incipient flight movements, and that the birds are, in effect, learning to fly before they leave the nest. But then someone performed the inevitable experiments. In one set of experiments some nestlings were allowed to flutter and hop up and down, while others were reared in boxes which prevented any such movement. At the time in their lives when the young birds would have normally begun to fly, both groups were released. Surprise! The restricted birds flew just as well as the ones that had "practiced"!

Flight behavior, then, is obviously an innate, or unlearned, pattern. It is important to realize, though, that generally young birds do not fly as well as adults. The innate pattern can be improved upon by learning through practice.

In other instances the innate and learned components of a behavior can be more clearly differentiated. The process by which a squirrel opens a hazelnut has been shown to be a complex of learned and innate behavior. If you have ever watched a squirrel open a nut you have probably been impressed with its speed and efficiency. Red squirrels, for example, first cut a groove along the growth lines on one or both of the flat sides of the nut, where the shell is thinnest. Then they insert their incisors into the groove and break the shell open.

In one study the ethologist Irenaüs Eibl-Eibesfeldt reared baby squirrels without any solid food which they could practice handling. When they were finally presented with a hazelnut, they correctly performed the gnawing actions and even inserted their incisors into the grooves correctly. Thus these patterns are apparently innate. The problem was that the young squirrels gnawed all over the surface of the nut and started several grooves without really completing any of them. Eventually they were usually able to break open the shell, but only after a great expenditure of energy. In time, however, their performance improved. After the squirrels had handled a few nuts, they began to make their grooves at the thinnest part of the shell, where gnawing was easier, so that the shell then opened more easily. Most of the squirrels soon became proficient at opening hazelnuts, as their innate gnawing pattern was modified through learning to produce the complex adaptive result.

The relative importance of innate and learned components of behavior may vary widely. For example, whereas reproductive behavior in sticklebacks may be largely innate and nut opening in squirrels perhaps less so, the song in some songbirds may be largely learned.

Moreover, white-crowned sparrows, in order to learn their song properly, must hear the song of their species during a brief "sensitive period" early in life. During this period they are not even able to sing yet. If they are exposed to the song after this period, they will produce abnormal songs, lacking the finer details. Thus it may be said that learning plays an important part in the development of song in this species.

We see that some behavioral patterns are almost completely innate, such as the reproductive behavior of the stickleback fish or the sucking response in young mammals, whereas others are almost entirely learned, such as songs in certain birds or the backstroke in humans. Behaviors which entail both instinct and learning may depend more on one element than another. Such information suggests that it is better to ask how a certain behavior develops than to ask whether it is innate or learned.

Let's now take a look at two areas of animal behavior which have long been a puzzle to scientists and see if we can impart to you some of that same spirit of confusion.

RHYTHMS

Why does an apartment-dwelling cat become restless as evening draws on? How is it that some people are able to wake up at whatever time they want to, almost to the minute? Or better yet, how does a potato know what time it is? Questions about timing in living things have been asked for years, and in some cases for hundreds of years. Of course, scientists have come up with all sorts of data and theories, but the fundamental questions just haven't been answered yet.

The rhythmic nature of life should not be unexpected, in view of the cyclic nature of the world around us. The earth revolves around the sun with regularity, and it rotates at a constant rate on its tilted axis. And the moon has a predictable and recurring relationship with the earth. Our lives are segmented by daily periods of light and dark and seasonal periods of cold and warmth.

Figure 11.11 Squirrels are born with a general idea of how nuts are to be opened. Their first clumsy attempts are quickly improved by practice. (Courtesy: Sonja Bullaty from National Audubon Society)

Since the earth itself functions in rhythmic cycles, we should not be surprised that, through the eons of evolution, a certain rhythmicity or sensitivity to environmental rhythm has winnowed into the sensitive cytoplasm of living things.

About 300 years before Aristotle, men described plants which raise their leaves during the day and lower them at night (Figure 11.12). About 2400 years passed, however, before anyone noticed that the leaves would continue their "sleep movements" even when the plant was placed inside and apparently deprived of its environmental information concerning the time of day. The basic question hinges here. Some organisms continue to show rhythmic behavior after they are placed under "constant" conditions. Is the rhythm a response to environmental clues of which we are not aware, or do the organisms have their own *internal* cycles? If an organism takes its timing cues from the environment, it is said to be under *exogenous* control. If it has its own innate internal cycle which operates independently of the environment, its control is *endogenous*.

Part of the difficulty in determining the mechanisms of biological rhythms is the fact that there is such a profusion of these cycles. There are rhythms in the staccato firing of neurons. Hibernating animals such as bears build up layers of fat every autumn. Human births are most likely to occur in the early morning hours than in the early evening. Cytoplasmic substrates and enzymes ebb and surge with a certain regularity.

Most of the rhythms that have been studied are those which occur in a daily cycle, or approximately every twenty-four hours. It is undoubtedly not coincidental that this is about the length of the solar day. Interestingly, when organisms with a twenty-four-hour cycle are placed under constant laboratory conditions, their rhythms may drift to a twenty-three- or a twenty-five-hour cycle. These periods of about a day are called *circadian rhythms* (from *circa*, meaning "about," and *diem*, meaning "day").

Circadian rhythms have led most researchers to agree that timing is internal. It is suggested that *approximate* twenty-four-hour clocks are built into circadian organisms, but that environmental cues, such as day length, are needed to keep the clock precisely set. In other words, the rhythm is innate, but the fine tuning is environmentally dependent. The fact that some rhythms go through several cycles every twenty-four hours while others are monthly or annual does not upset the theory at all. It is suggested that these rhythms either are governed by some other internal clocks or are based on a circadian clock but have some longer or shorter cycle. A familiar example is the two hands of a clock which move at differ-

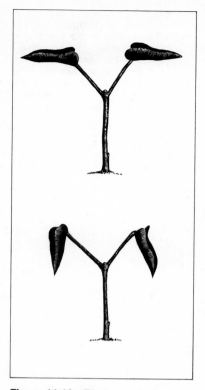

Figure 11.12 The normal sleep movements of plants. The rising and falling of leaves persist even when the plants are placed under constant conditions of light and temperature.

Figure 11.13 Diagram of the dual nature of biological rhythms according to the exogenous control theory. (From Brown, American Astronautical Society, Advances in the Astronautical Sciences, 17:29–39. 1964)

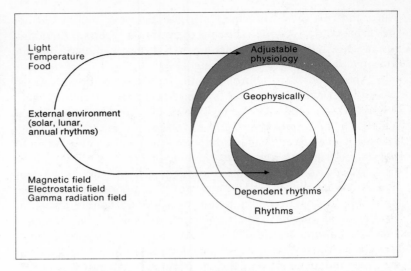

Light
Temperature
Food

Adjustable physiology

Geophysically

External environment
(solar, lunar,
annual rhythms)

Magnetic field
Electrostatic field
Gamma radiation field

Dependent rhythms

Rhythms

ent rates within the same twenty-four-hour period, but which are driven by the same mechanism. If you were to stick your finger in your ear every three minutes, the cycle of this behavior would be three minutes, but the timing would still be done with a twelve-hour clock. Another example is people who keep different hours. Some people are "day people" and others seem to be "night people," but they both use the same twelve-hour clock—and they usually marry each other.

Frank A. Brown, one of the more vigorous exponents of the exogenous control theory, conceptualizes a two-layered ring of rhythmic control (Figure 11.13). He visualizes a "core" of unchanging geophysical constants, which is overlain by a number of variable rhythms dependent on such environmental factors as light, temperature, or humidity. The geophysical influences, he believes, can penetrate all experimental situations so that the organism can receive changing environmental clues under what were thought to be constant conditions. In this scheme, only the outer conditions can be experimentally controlled. Among these subtle geophysical influences are atmospheric tides (measured as barometric highs and lows), which have secondary effects on such geophysical factors as gravity, geomagnetic field, electrostatic field, and background radiation.

Brown describes an experiment in which potatoes were kept under environmentally "constant" conditions and also shielded from the direct effects of barometric changes. The metabolism of these potatoes, it was found, varied with what was happening outside. In fact, one could look at the respiratory charts and tell

A

B

C

D

E

Animal Diversity

A. The Tamandua; its protrusible tongue, long snout and strong claws betray it as an anteater. (Courtesy: San Diego Zoo) B. African hunting dogs kill by disemboweling their prey even while it lives. (Courtesy: San Diego Zoo) C. Penguins, clumsy on land, are well adapted to fishing in frigid waters. They are however, a favorite food of the voracious killer whale. (Courtesy: Gail Weingart) D. A variety of species of foxes survive throughout the world, from arctic to desert climes. They are quick, nervous and live on small prey. (Courtesy: Hubbard Scientific Company. Copyright © 1970) E. The dingo is the wild dog of Australia. It is hunted relentlessly by sheepmen, but we know surprisingly little of its biology. (Courtesy: San Diego Zoo)

F. The deadly stone fish; instead of fleeing, the fish lies camouflaged on the bottom. At any sign of danger it raises its short spines. Penetration of the spines insures a quick death. (Courtesy: Sea World) G. The hornbill feeds on berries, small vertebrates and possibly the eggs of other birds. The male seals the nesting female into a hole in a tree with mud and feeds her through a narrow opening. (Courtesy: San Diego Zoo) H. The strong markings of the lion fish warn of its poisonous fins. (Courtesy: Sea World) I. Males of the lesser bird of paradise are highly conspicuous. They behave in a bizarre fashion to attract females. However, they may also attract predators. (Courtesy: San Diego Zoo) J. A sea anemone is a pouchlike animal which captures

F

G

H

I

all fish by stinging them with
ecial cells and drawing them into
body cavity. (Courtesy: Sea World)
Beetles comprise the largest
der of insects. They are stout
imals with heavy exoskeletons.
ourtesy: Hubbard Scientific

Company, Copyright © 1970) L. The
poisonous Arizona coral snake
advertises its dangerousness to
potential predators. Some harmless
snakes mimic their banded
coloration. (Courtesy: San Diego Zoo)
M. The dragonfly is a deadly hunter

of mosquitoes and a very fast flier
among the insects. Some species
take as long as five years to mature.
(Courtesy: Hubbard Scientific
Company, Copyright © 1970)

K

M

N. The newborn cheetah cub is helpless now but will soon be able to achieve speeds of about 70 miles per hour in brief sprints. The cheetah has a doglike appearance and cannot retract its claws. It can be tamed but in the wild it is an efficient hunter of African grassland antelope. It usually kills by holding its prey by the throat until it suffocates. (Courtesy: San Diego Zoo)

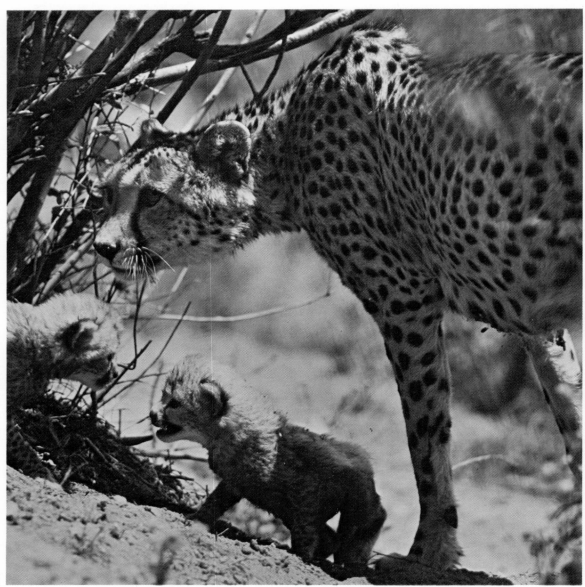

N

whether the outside temperature was 50 degrees above or 5 degrees below zero for any day, and whether the barometric pressure was high or low. Somehow, in spite of all the experimenters were able to do, clues were filtering through to the potatoes. An alternative explanation, of course, is that the potatoes were controlling the weather outside.

In another study Brown had fifteen oysters sent from New Haven, Connecticut, to his laboratory in Evanston, Illinois. He and his coworkers were investigating shell-opening activities of the oysters under constant light conditions. They expected the oysters to open, as if in feeding, in synchrony with the daily high tides in Connecticut. The oysters obliged for about two weeks, but then a strange thing happened. They began to shift their cycles until they were opening at what *would have been* high tide in landlocked Evanston. Somehow, the environmental information was getting through.

Figure 11.14 The question of how animals actually navigate remains largely unresolved.

ORIENTATION

There are undoubtedly people in South America who wonder where their robins go every spring. Those robins, of course, are "our" birds; they simply "vacation" in the south each winter. If you have been keeping a careful eye on the robins in your neighborhood, you may have noticed that a specific tree is often inhabited by the same bird or pair of birds each year. You may have wondered, since these birds winter thousands of miles away, how they find their way back to that very tree each spring. Such questions perplexed people for years, until in the 1950s a young German scientist named Gustav Kramer provided some answers—and, in the process, raised new questions.

Kramer found that caged migratory birds became very restless at about the time they would normally have begun migration in the wild. Furthermore, he noticed that as they fluttered around in the cage, they often launched themselves in the direction of their normal migratory route. Kramer set up experiments with caged starlings and found that their orientation was, in fact, in the proper migratory direction—except when the sky was overcast. At these times there was no clear direction to their restless movements. Kramer surmised, therefore, that they were orienting according to the position of the sun. To test this idea, he blocked their view of the sun and used mirrors to change its apparent position. He found that under these circumstances the birds oriented with respect to the position of the new "sun." This seemed preposterous. How could a stupid bird navigate by the sun when we often lose our way with road maps? Obviously, more testing was in order.

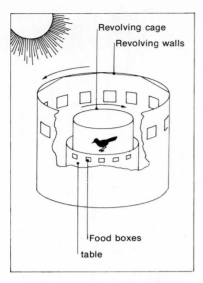

Figure 11.15 Kramer's orientation cage. The bird can see the sky through the glass roof. It is trained to expect food which lies in a certain direction according to the sun's position. The bird shifts direction when the apparent direction of the sun is changed by mirrors.

In another set of experiments Kramer put identical food boxes around the cage, with food in only one of the boxes. The boxes were stationary, so that the one containing food was always at the same point of the compass. However, its orientation could be changed with respect to the surroundings by revolving either the inner cage containing the birds or the outer walls which served as the background. As long as the birds could see the sun through any window, no matter how their surroundings were altered, they went directly to the correct food box. Whether the box appeared in front of the right wall or the left wall, they showed no signs of confusion. When the food was moved to a different box, they continued for a time to approach the old box until they discovered the new compass orientation of the new food box. On overcast days, however, the birds were completely disoriented and were unable to locate their food box very easily.

In experimenting with artificial suns Kramer made another interesting discovery. If the artificial sun remained stationary, the birds would shift their direction with respect to it at a rate of about 15 degrees per hour, the sun's rate of movement across the sky. Apparently, the birds were assuming that the sun they saw must be moving at that rate. When the real sun was visible, the birds maintained a constant direction as it moved across the sky. In other words, they were obviously able to compensate for the sun's movement. This meant that some sort of biological clock was operating—and a very precise clock at that (Figure 11.15).

Another study of starlings was conducted by K. Hoffman, who trained birds under natural light conditions to go to one particular food box, whose position was fixed. He then put the birds on an artificial light-dark schedule which was six hours behind local time. After about two weeks of this artificial regimen, when the birds were returned to normal light-dark conditions, they immediately approached a food box which lay at a 90-degree angle from the one they had been trained to choose. Their day had been altered by one-quarter (six hours), and their orientation was shifted correspondingly by a quarter-circle. After a week or two under natural light conditions the birds returned to their original directional behavior.

Such sun-compass orientation implies, of course, that starlings know the time of day and that they also know the normal course of the sun. In other words, incredible as it seems, they apparently know where the sun is supposed to be in the sky at any given time. It seems that part of this ability is innate. In another experiment, a starling reared without ever having seen the sun was able to orient itself fairly well, although not as well as birds that had been exposed to the sun earlier in their lives.

Sun-compass orientation exists in a variety of other animals including insects, fish, and reptiles. There is also some evidence of celestial navigation in nocturnal birds—navigation by the stars. More recent evidence indicates that some species of navigating birds may even be guided by the earth's magnetic field. It should be pointed out that this is an extremely vigorous field at present, and new data are appearing almost daily. Any synthesis at this point would be extremely tenuous. However, these examples provide some idea of the value of innate characteristics in orientation, as well as the surprising sensitivity of some species to environmental cues.

COMMUNICATION

If you should come upon a large dog eating a bone, walk right up to him and reach out as if to take the bone away. You may notice several changes in his appearance. The hair on his back may rise, his lips may curl back to expose his teeth, and he may utter peculiar guttural noises. He is communicating with you. To see what his message is, grab the bone.

It would be a remarkable understatement to say that communication is an important phenomenon in the animal world. It is accomplished in many ways and with innumerable adaptive variations. Its ultimate function, however, is to increase reproductive success in those animals able to send and receive signals. Communication may be directly involved with reproductive success as a component of mating behavior or precopulatory displays. It may also indirectly increase reproductive output by helping the offspring to avoid danger when parents give warning cries, or by simply helping the reproducing animal to live better or longer so that it may successfully mate again. Remember, the charge to all living things is "reproduce as much as possible or your genes will be lost." Communication helps animals to carry out that reproductive imperative.

Let's consider a few general methods of communication, and in so doing perhaps we can learn something about the ways in which animals influence the behavior (or the probability of behavior) of other animals. That, after all, is what communication is all about.

VISUAL COMMUNICATION

Visual communication is particularly important among certain fish, lizards, birds, and insects—and to some primates as well. Visual messages may be communicated by a variety of means, such as

Figure 11.16 The starling, in winter plumage. In its annual migration it apparently navigates by the sun. (Courtesy: A. W. Ambler from National Audubon Society)

Figure 11.17 An example of graded display. At the left the bird, *Fringilla coelebs,* is showing a low intensity threat. The posture at center is a medium intensity threat and at right is a high intensity threat. (From P. Marler, 1959 in Darwin's Biological Work. Ed. by P. R. Bell, Cambridge University Press)

through color, posture or shape, movement, or timing. As we have seen, color serves as a component of the releaser of territorial fighting in European robins. In grackles the female solicits the attention of the male by assuming a head up posture while fluttering her drooping wings. Some male butterflies are attracted to the female by her performance of a specific flight pattern. And as an example of communication by timing, fireflies are attracted to each other on the basis of their flash intervals, each species having its own frequency.

Because of the high information load possible with visual signals, subtle variations in the message can be conveyed by gradations in the display (Figure 11.17). Of course, graded displays are useful only to those species which are sensitive and intelligent enough to be able to utilize such subtleties. Another advantage of the high information load in visual signals is that the same message may be conveyed by more than one means. Such redundancy may be used either to modify the message or to emphasize it in order to reduce the chance of error in interpretation.

A visual signal may be a permanent part of the animal, as in the elaborate coloration of the male pheasant and the striking facial markings of the male baboon. These animals advertise their maleness at all times and are continually responded to as males by members of their own species. On appropriate occasions they may emphasize their "machismo" by behavioral signals, such as the strutting of the pheasant and the glare of the baboon. In other cases the visual signal may be of a more temporary nature, such as the reddened rump areas in female chimpanzees and baboons during estrus. As a more short-term signal, a male baboon may expose his long canine teeth as a threat (Figure 11.18).

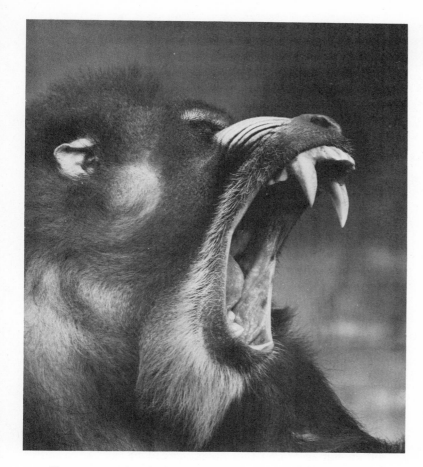

Figure 11.18 A male baboon displaying his large canine teeth as a threat signal. (Courtesy: San Diego Zoo)

Short-term visual signals have the advantage that they can be started or stopped immediately. If a displaying bird suddenly spots a hawk and "freezes," its position will not be given away by any lingering images in the environment. Also, in visual messages the recipient is notified of the exact location of the sender. The recipient can then respond to him in terms of his precise location, as well as his general presence and behavioral state (aggressive, romantic, or whatever).

Visual signals also have certain disadvantages. For example, if a sender can't be seen, it can't communicate. And all sorts of things can block vision, from mountains, trees, and fog to oversize hats. Visual signals are often useless at night or in dark places (except for light-producing species). Also, since visibility weakens with distance, long distances diminish the usefulness of such signals. As distance increases, the signal must become bolder and simpler;

Figure 11.19 Sound spectrograms of the mobbing calls of several species of British birds while attacking an owl. The sounds have qualities which make them easy to locate. (From Darwin's Biological Work. Ed. by P. R. Bell. P. Marler's "Developments in the Study of Animal Communication." 1959. Cambridge University Press. pp. 150–206)

hence it can carry less information. The only way to intensify a visual signal is to increase its visibility, either by decreasing the distance or by increasing its size or contrast. It isn't possible to pump more energy into the signal itself, as with certain other forms of communication.

SOUND COMMUNICATION

Sound plays such an important part in communication in our own species that it may surprise you to learn that it is limited, for the most part, to arthropods (joint-legged creatures such as insects) and vertebrates. Surprised, eh? We're all familiar with the song of the cricket and the cicada, so you are probably aware that insect sounds are usually produced by some sort of friction—such as by rubbing the legs together or the legs against the wings. These sounds are characterized by differences in cadence rather than in pitch or tone, as with most birds and mammals.

Vertebrates use a number of different forms of sound communication. Fish may produce sound by means of frictional devices in the head area or by manipulation of the air bladder. Land vertebrates, on the other hand, usually produce sounds by forcing air through vibrating membranes in the respiratory tract. Of course, they may communicate by sound in other ways. Rabbits thump the ground, gorillas pound their chests, and woodpeckers hammer your drainpipe early on Sunday mornings.

Most of the lower vertebrates rely on signals other than sound, but some species of salamanders can squeak and whistle, and there is one that barks! Frogs and toads advertise territoriality and choose mates at least partly by sound. Most reptiles do not communicate by sound, but territorial bull alligators can be heard roaring in more remote areas of the Southern swamps, where a few have escaped becoming status symbols as belts, shoes, and purses. Darwin described the roaring and bellowing of mating tortoises when he visited the Galápagos Islands in 1839.

Sound may vary in pitch (low or high, depending on the frequency of the sound waves), in volume, and in tonal quality. The last is demonstrated as two people hum the same note, yet their voices remain distinguishable. It is possible to graph the characteristics of sounds by use of a *sound spectrogram*. In effect, this is a translation of sound into visual markings, which makes possible a more precise analysis of the sound.

The function of the message may dictate the characteristics of the sound. For example, Figure 11.19 shows a sound spectrogram of the mobbing cries of several bird species. Such calls appear as brief, low-pitched *chuk* sounds, whose source is easy to locate.

When a bird hears the repetitious mobbing call of a member of its own species, it is able to locate the caller quickly and join it in driving away the object of concern—usually a hawk, crow, or other marauder. In contrast, when a hawk or other predator is spotted flying overhead, the warning cries of songbirds are usually a high-pitched, extended *tweeeeee*, which is difficult to locate. The response of a bird hearing this call is quite different from its response to the mobbing call. The warning cry sends the listener heading for cover, often in a deep dive into protective foliage, from where it may also take up the plaintive, hard-to-locate cry.

Sound signals have the advantage of a potentially high information load through subtle variations in frequency, volume, timing, and tonal quality. They are distinguishable at low levels, but they can also be made to carry over great distances by pumping more energy into them—in effect, becoming louder. In addition, sounds are transitory; they don't linger in the environment after they have been emitted. Thus an animal can cut off a sound signal should its situation suddenly change—for example, with the appearance of a predator. A further advantage is that an animal doesn't ordinarily have to stop what it is doing to produce a sound. As everyone knows (except perhaps builders of student housing), sounds, unlike visual images, can go around or through many kinds of environmental objects.

Sound communication also has its disadvantages. Sounds are rather useless in noisy environments. Thus some seabirds which live on pounding, wave-beaten shorelines rely primarily on visual signalling. Also, sounds attenuate with distance. And the source of sound is sometimes hard to locate, especially underwater.

CHEMICAL COMMUNICATION

You have probably seen ants rushing along single file as they sack your cupboard. You may also have taken perturbingly slow walks with dogs which stop to urinate on every bush. In both cases the animals are communicating by means of chemicals. The ants are following chemical trails laid down by their fellows, and the dog is advertising his presence.

Insects make extensive use of chemical signals, and certain chemicals elicit very rigid and stereotyped behavior. An "alarm chemical" is produced when an ant encounters some sort of threat, and as the chemical permeates the area, every ant of that species which detects the chemical reacts by rushing about in a very agitated manner. Female moths produce a "sex attractant" which attracts males from astonishing distances. The substance is called *bombykol*, and it is so potent that a male moth will be aroused by

Figure 11.20 Sound spectrograms of five species of British birds when a hawk flies over. Such calls are high-pitched, drawn out, and difficult to locate. (From Darwin's Biological Work. Ed. by P. R. Bell. P. Marler's "Developments in the Study of Animal Communication." 1959. Cambridge University Press. pp. 150–206)

Figure 11.21 In a beehive, the queen is constantly cared for and protected. She is the only hope the population has of carrying their own kinds of genes into the next generation. (Courtesy: Grant Heilman)

a glass rod dipped in a solution diluted to one part in a trillion!

The immediate and invariable response produced by certain chemical signals is comparable to the action produced by hormones. Whereas hormones work *within* an individual, however, these chemicals work *between* individuals and are called *pheromones* (from the Greek *pherein*, meaning "to carry," and *hormon*, meaning "to excite").

It was at first believed that pheromones are found only among the insects, species whose behavior is largely genetically programmed. But then it was found that pheromonal interactions also occur in mammals. For example, if pregnant rats of certain species smell the urine of a strange male, some component of that urine will cause them to abort their fetuses and become sexually receptive again. As we have already seen, the females of most mammal species also signal their sexual receptivity by some chemical means, with varying degrees of influence on the male, depending on the species.

Honeybees provide another interesting example of chemical communication in insects. The queen produces a "queen substance" which, when fed to developing females, causes them to become sterile workers. When the queen dies, the inhibiting substance is no longer present, and new queens begin to develop. At this time the undeveloped ovaries of some adult workers also suddenly begin to mature. Their eggs will produce drones—royal consorts, one of which will mate with the new queen on her nuptial flight.

The molecular makeup of several chemical signals in the animal world has been carefully analyzed. It was found that the backbone of the molecule usually consists of five to twenty carbon atoms. The molecular weight of the chains ranges between 80 and 300. It is reasoned that molecules with at least five carbons are needed to provide the variation and specificity for discrete messages. The maximum molecular size is more likely to be set by other factors. For example, such molecules are usually carried in the air, and they must therefore be small and light. Also, more energy is expended in the manufacture of larger molecules.

Chemical signals have the advantage of being extremely potent in very small amounts. Also, because of their persistence in the environment, the sender and receiver do not have to be precisely situated in time and space in order to communicate. Chemicals can also move around many sorts of environmental obstacles.

The specificity of chemicals, however, limits their information load. Moreover, because chemicals do linger in the environment they may advertise the presence of the signaler to arriving predators as well as to the intended recipient of the message.

Figure 11.22 The ranges of the golden-fronted woodpecker (left) and the red-bellied woodpecker (right) join in central Texas. In spite of their almost identical appearances, they generally occupy different kinds of habitats and do not interbreed. (Courtesy: Allan D. Cruickshank and Karl H. Maslowski from National Audubon Society)

WHY COMMUNICATE?

Now that we have some idea of how communication can be accomplished, we might ask, why is it necessary to communicate at all? You can probably sit right there and think of a host of reasons, but let's see if we can't organize the ideas a bit.

SPECIES RECOGNITION

First, communication may permit one animal to know that another animal is of the same species. This may not seem too important—unless one is interested in reproduction. After all, animals of two different species cannot normally produce healthy offspring. But a lot of species are very similar, and in the absence of some precise means of identification an animal might waste a lot of time and energy in trying to mate with a member of the wrong species. Therefore, species identification must somehow be quickly established.

For example, the golden-fronted woodpecker and the red-bellied woodpecker often share the same woods around Austin, Texas (Figure 11.22). The two species look very similar to the casual observer. The most conspicuous difference is that the golden-

Figure 11.23 A male salticid spider approaches a female only after inhibiting her predatory behavior by semaphore signalling. (From *Social Behavior from Fish to Man*, Etkin, University of Chicago Press, 1967)

fronted woodpecker has a small yellow band along the base of its red cap, which is absent on the red-bellied. Also, the red-bellied woodpecker has slightly more white in its tail. The habits of the species are much alike and their calls are very similar, at least to the human ear. Nevertheless, each of these birds is somehow able to recognize its own species. At least cross mating has never been observed, nor have any hybrids. No one knows exactly what cue the birds use, but to them, apparently, the signals are clear.

Interestingly enough, whereas the two species ignore each other sexually, they have come to treat each other as competitors and will eject a member of the other species from a territory as quickly as they would a member of their own species. Apparently, the sign stimuli which release sexual and territorial behavior are quite different in these species.

POPULATION RECOGNITION

Even when species identity has been established, it might also be to an animal's advantage to be able to recognize those individuals from its own particular breeding population. Consider a species of bird in which some subpopulations live in dark, moist lowlands and are therefore dark-colored; other subpopulations live on dry mountaintops and are light-colored, making them less conspicuous against the light-colored dry earth. If a lowland individual should mate with a highland individual, the offspring might be conspicuous in either type of terrain and might therefore quickly be taken by predators. Hence there would have evolved a tendency for members of each population to breed with their own type. Actually, there aren't many field data to show that population recognition is an important phenomenon in nature, but it seems that it would be. As support for the notion, it has been shown that population-specific signals exist in several kinds of animals.

INDIVIDUAL RECOGNITION

It may also be important for an individual to be able to recognize specific individuals within its own population. If you have ever watched gulls at the beach, you may have noticed that the adults all look alike. This is because you aren't a gull; all people probably look alike to them. If you should visit a gullery, the place where they nest, you might see thousands of "identical" gulls. Color banding and recording playback experiments, however, have shown that gulls can recognize their own mates on sight and are able to filter out the calls of their mates from the raucous din of the clamoring gulls wheeling above.

In species in which the sexes come together only for copulation, the mate-recognition problem becomes one of sexual identification. Male salticid spiders approach the female only for mating (Figure 11.23). This approach is performed with much waving of special appendages to inhibit predatory behavior on the part of the larger female. Mating is brief, and the male quickly clears out. Afterward, should the female encounter him, she will recognize him only as prey. We can see, then, that individual mate recognition is important only in species which establish pairs and that it increases in importance with the likelihood of interaction with other individuals.

Individual recognition is important for two primary reasons. First, one should be able to recognize one's own mate, especially when the rearing of offspring is done by both parents. Mate recognition ensures that pairs will attend to the nest which harbors their own offspring, thus increasing their chances of reproductive output. This is an essential factor in species which show a high level of coordination while rearing offspring. For example, some birds bring food to their mates which are incubating eggs. In the African hunting dog, the male is more likely to regurgitate food to his own mate when she is nursing pups, than to any other females.

The second reason individual recognition is important is that it functions in the maintenance of hierarchies. Once animals in a group know their rank with respect to the others, there will be less fighting over food or other commodities. In any group where rank is unknown, or not yet established, the incidence of fighting is likely to be high. Rank, however, can only be maintained when each animal can recognize the others individually so that it can respond to each according to its status.

Figure 11.24 An African cape dog regurgitating food to its pup. (Courtesy: Photo and Copyright © Baron Hugo van Lawick)

Dominance Hierarchy in a Flock of Hens

	2	3	4	5	6	7	8	9	10	11	12	13
Hen 1 pecks	2	3	4	5	6	7	8	9	10	11	12	13
Hen 2 pecks		3	4	5	6	7	8	9	10	11	12	13
Hen 3 pecks			4	5	6	7	8	9	10	11	12	13
Hen 4 pecks				5	6	7	8	9	10	11	12	13
Hen 5 pecks					6	7	8	9	10	11	12	13
Hen 6 pecks						7	8	9	10	11	12	13
Hen 7 pecks							8	9	10	11	12	13
Hen 8 pecks								9		11	12	13
Hen 9 pecks									10	11	12	13
Hen 10 pecks							8			11	12	13
Hen 11 pecks											12	13
Hen 12 pecks												13
Hen 13 pecks none												

Source: Adapted from W. C. Allee, *The Social Life of Animals* (New York: Norton, 1938).

Table 11.1 Dominance hierarchy in a flock of hens as determined by differences in access to food or water. The highest ranking bird may peck all others. The second highest may peck all but the higher one, and so on. A new bird introduced into such a group may have to fight frequently until its rank is established. In such a system not only the dominant bird benefits, but the subdominant as well. If the subdominant had challenged the dominant it would have gotten worked over and lost the commodity as well. So, rather than risk injury and lose time, energy and the commodity, it might do better to seek an unclaimed commodity elsewhere.

Box 11.3 Sexual Dimorphism You may have noticed that boys look different than girls . . . at least they used to. Even today, however, it's easy to tell a rooster from a hen. In some deep-sea fishes the sexes are even more dissimilar, the male may be an appendage fused to a knob on the head of a female. On the other hand, it may be difficult to tell a male mouse from a female and the sexes of seagulls are almost identical. Sexual dimorphism—or the different appearance of the sexes—varies widely among animals.

There are interesting evolutionary implications of sexual dimorphism. How did it start? With respect to vertebrates, one idea is that as they came under increasingly severe competition for commodities such as food, the males increased in size and strength in order to compete more successfully. The females were restrained from such changes by the demands of gestation (which placed a premium on a certain body size) and the fact that they were more closely associated with the young (which made it important that they be physically and behaviorally less conspicuous). Males which were best able to gain commodities might then have been disposed to share those with females, for reproductive purposes. This meant that females might then be more likely to choose the more vigorous males as mates. This tendency would have driven males toward increasing garish appearances and bold, aggressive and conspicuous behavior in order to display their vigor and to intimidate other males in disputes over commodities. Exceptions exist, of course, but generally, today males of most species, especially among birds and mammals, are larger, stronger, and more aggressive than females and dominant over them in social interactions. It provides some entertainment, and a lot of argument, to try to determine how this general condition could have arisen.

An interesting relationship between sexual dimorphism and domestic duties exists among some species. Consider an example from birds. The sexes of song sparrows look very much alike. The males have no conspicuous qualities which immediately serve to release reproductive behavior in females. Thus courtship in this species may be a rather extended process as pair-bonding (mating) is established. Once a pair has formed, both sexes enter into nest building, feeding and defense of the young. The male may only mate once in a season but he helps to maximize the number of young which reach adulthood carrying his genes. He is rather inconspicuous, so whereas he doesn't turn on females very easily, he also doesn't attract predators to the nest.

The peacock, on the other hand, is raucous and garish. When he displays to a drab peahen, he must present a veritable barrage of releasers to her reproductive IRMs. In any case, he displays madly and frequently and is successful indeed. Once having seduced an awed peahen, he doesn't stay to help with the mundane chores of child rearing, but instead disappears into the sunset looking for new conquests. The peahen he has just left may be an inferior bird, not likely to be able to rear young successfully. He could have discovered this by a more careful process of mate selection. But no matter, perhaps the next conquest will be a better bird. Alas, just as he attracts females, he also attracts tigers and his flurry of sexual activity may be rather quickly ended. So, instead of carefully rearing one brood each season for several seasons, he maximizes his reproductive success through another means.

Think of other species of birds and mammals. Which evolutionary route have they taken? What about humans?

(Courtesy: *Jack Dermid from National Audubon Society*)

AGGRESSION

The old image of "Nature, red in tooth and claw" has recently been superseded by the popular notion that other animals get along with their own kind and humans are the only animals that regularly kill members of their own species. It might be a good idea to take a look at the animal kingdom to see what is actually going on.

First of all, we should note that *aggression* is belligerent behavior which arises as a result of *competition*. A lioness may spring toward a group of wild pigs, single out one in the chase, catch it, and quickly smother it or break its neck. However, she is not acting out of aggression. She is being about as aggressive as you are toward a hamburger. However, if one of the big boars should turn and charge her, the lioness may momentarily show a component of aggression—fear. Fear is an important component of aggression; it is simply at one end of the behavioral spectrum of aggressive behavior. At the other end is sheer, unbridled belligerence.

FIGHTING

Let's consider the most obvious form of aggression, fighting. You can discount the old films you have seen of leopards and pythons battling to the death. Such fights simply aren't likely to happen. What is a python likely to have that a leopard needs badly enough to risk its life for, and vice versa? Although such fighting might occur in the unlikely event that one should try to eat the other, fighting is much more likely to occur between two animals that are competing for the same commodity. Obviously, the strongest competitor for any animal is going to be one that utilizes the same habitat in the same way. And what sort of animal is most likely to do this? Right! A member of the same species. Moreover, if there is strong competition for mates, fighting is even more likely between members of the same sex within that species. And this is in fact where most fighting occurs—between members of the same sex of a species.

Of course, fighting between species sometimes occurs. The golden-fronted and red-bellied woodpeckers exclude each other from their respective territories. Lions may attack and kill African cape dogs at the site of a kill. The lions don't eat the dogs; they just exclude them.

Animals may fight in a number of ways, but they usually manage to avoid injuring each other (Figure 11.25). There are several apparent benefits to such a system. First, no one is likely to get hurt. The competitor is permitted to continue its existence, it is true, but the possibility of having to compete with him again entails less risk than does serious fighting. Also, since animals are

Figure 11.25 In some cases fighting may occur between species which normally never even interact. Here, a raccoon approached a hare in the darkness and in the scuffle, the hare leaped over the raccoon and administered a severe drubbing to the raccoon's head with its powerful back legs. Some evidence is beginning to appear which suggests that animals of different species which share the same area may recognize each other individually and may react to each other on an individual basis. (Courtesy: Lynwood Chase from National Audubon Society)

Figure 11.26 Fighting male rattlesnakes, *Crotalus ruber*. The fighting usually involves each trying to push the head of the other to the ground, but the poisonous snakes refrain from biting each other. (From E. C. E. Shaw 1948, after Eibl-Eibesfeldt, Ethology. The Biology of Behavior, 1970)

most likely to breed with the individuals around them, the opponent may be a relative who is carrying some of the same genes; hence there is a reproductive advantage in sparing him. (Can you figure out why? We'll discuss it later.) In fact, if the population doesn't tend to scatter much, the competitor might even be one's own mature offspring; it could also be a prospective mate.

When fighting occurs between potentially dangerous combatants, the fights are usually stylized and relatively harmless. For example, horned antelope may gore an attacking lion, but when they fight each other, the horns are never directed toward the exposed flank of the opponent (Figure 11.27). Such stylized fighting does, however, enable the combatants to establish which is the stronger animal. Once dominance is established, the loser is usually permitted to retreat.

On the other hand, all-out fighting may occur between animals that are unequipped to injure each other seriously, such as hornless female antelope, or between animals that are so fast that the loser can escape before he is seriously injured, as in house cats.

We might ask ourselves, since it is useless to ask an antelope, why they don't gore each other. An incurable romantic might assume that it is because they don't want to hurt each other. In all likelihood, what the two antelope "want" has little do to with it. The fact is that they *can't* hurt each other. When the system works an antelope could no more gore an opponent than fly! In terms of the actual mechanism, it may be that the sight of an opponent's exposed flank acts as an inhibitor of butting behavior. Conversely, a facing view, under certain conditions, might serve as a release of very stereotyped fighting behavior. (One explanation of such behavior could be based on the instinct model. Reconstruct it yourself in terms of releasers, IRMs, and neuromuscular associations. The idea is further discussed later.)

Can we assume, then, that animals do not fight to the death? As a general rule, they don't, but there are exceptions. Accidents may occur in normally harmless fighting, but there are also some species which normally engage in dangerous fighting. If a strange rat is placed in a cage with a group of established rats, the group may sniff at the newcomer carefully for a long time, but eventually they will attack it repeatedly until they kill it. Guinea pigs and mice often fight to the death. The males of a pride of lions may kill any strange male they find within their hunting area, and a pack of hyenas may kill any member of another pack that they can catch.

IS AGGRESSION INSTINCTIVE?

Our discussion of aggression so far suggests that aggressive behav-

Figure 11.27 Male Thompson's gazelles engaged in ritualized fighting. Only head-on sparring occurs. An attack to a vulnerable flank almost never occurs. (Courtesy: Stanley N. Botwinik from National Audubon Society)

Figure 11.28 Hornless females of the Nilgau antelope often butt the flank of an opponent. Horned males of the same species almost never attack in such a manner, nor do horned females of other antelope species. (After Walther Jahrbuch der Gesellshaft von Opel-Rreigehege Tierforshung, 1961)

Figure 11.29 Firemouth cichlid fish. (Courtesy: Gene Wolsheimer from National Audubon Society)

ior in many animals is largely innate. But there are many who contend that aggressive behavior, especially among mammals, is directly attributable to learning. Others, of course, stress that aggressive behavior has both learned and innate components. The argument has extreme significance for our species. If aggression is socially disruptive in an increasingly crowded world, then we must ask how it can be controlled. As a practical example, does television violence serve to *release* aggression, and hence dissipate it, or does it serve as a behavioral model for aggression? Since there is a tendency in most academic disciplines to focus on aggression as a cultural, or learned, phenomenon, let's consider other evidence here which suggests an instinctive component.

In certain species of highly aggressive cichlid fish, the males *must* fight before they are able to mate. If a male's reproductive state is appropriate and a female is available, the male will frantically seek an opponent upon which to release his fighting behavior. Finding none, he will often attack and kill the female. He is then ready to mate, but of course it is too late. The behavior can be described in terms of Tinbergen's hierarchical model. In removing the inhibitory block for mating (by fighting), he can move to the next level of reproductive behavior, but he has destroyed the releaser of that behavioral level, the female.

Among mammals, rats will learn mazes in order to be able to kill mice, and it is thus assumed that the killing is a consummatory act which reduces tension. Squirrel monkeys, after being angered by an electric shock, will learn a task in order to be able to attack a ball. The opportunity to attack serves as the reward.

Does an animal learn aggression when it is shielded from all opportunities to learn such behavior? Rats and mice have been reared in isolation, so that there was no opportunity for them to learn aggression. When a member of the same species was introduced into their cages, however, they attacked, showing all the normal threat and fighting patterns.

HUMAN AGGRESSION

Since aggressive behavior does appear to have innate components in some species, what about man? Is there any evidence that members of our own species are genetically driven to behave shabbily toward each other? It is difficult, of course, to isolate behavioral components in a species as complex as our own. This is partly because we have the capacity to override both learned and innate behavioral tendencies by conscious decision. For example, we rarely see couples copulating wildly on the tables of romantic little restaurants, no matter what their mood. This ability to override our "urges" has led some observers to assume that those urges—romantic, aggressive, etc.—must be culturally based, not innate. Hence to correct any disruptive aggressive tendencies in our group, we need only to change our culture and/or our institutions.

However, if the behavior in question does have an innate basis, cultural stifling of it can only be cosmetic. Do you think the people in the restaurant have forgotten their urge, even though they are intensely talking about something else? An aggressive person may have learned that smiling and speaking in low tones reaps him greater rewards, but if his aggression is innate, then we can expect him to be aggressive in small ways, or to direct it into harmless channels, or perhaps to show sporadic outbursts of violence when his guard is down. And, of course, smiles and soft voices can in themselves be the vehicle for extraordinary violence. Some of the most horribly bloodthirsty of programs have been proposed in smiling, congenial, reassuring, and even kindly tones by seemingly gentle and eminently reasonable souls. The point is that if aggression is an innate tendency in our species, we can expect it to permeate our social patterns. It may be temporarily suppressed, and can perhaps be rechannelled or disguised, but it will exert a subtle and abiding influence in our daily lives.

In an interesting experiment it was shown that aggression could be provoked in people which, when released, caused some relief. First, an experimenter provoked a roomful of students to anger (that was the easy part), causing their blood pressures to rise. The students were then divided into two groups. Each group watched as the experimenter attempted to perform a series of simple tasks. One group was instructed to administer painful electrical shocks (or so they thought) to the experimenter whenever he made an error. The other group could only inform the experimenter of his errors by harmlessly flashing a light. It was found that the group that believed they were actually inflicting pain to the experimenter (who was also an actor) showed a rapid decrease in blood pressure, a sign of relief. The group that was not given an opportunity to release aggression maintained their high blood pressures for a much longer time. In other experiments blood pressure was found to drop quickly if the subjects were allowed to hurl verbal insults at the experimenter. In both cases the release of tension was short-term, as is the case with other types of instinctive patterns.

The arguments on both sides are much more complex than this brief summary might suggest. The important point, however, is to recognize that these alternative positions exist, and to attempt to structure our social programs as broadly as possible.

Figure 11.30 An injured porpoise may be carried to the surface for air by its comrades. (After Siebenaler and Caldwell, Cooperation Among Dolphins, Journal of Mammology, 1956)

COOPERATION

Cooperation seems much nicer than aggression, and most of the animal stories of our youth (our pre–Jack London youth, that is) involved animals which helped each other in some way. Certainly cooperation is highly developed in some species, and its complexity and coordination in certain instances surpasses imagination. But again we have the problem of sifting fact from fiction—a difficult, if not impossible, task for those with a casual interest in the subject.

Cooperative behavior occurs both within species and between species. As examples of *interspecific* (between species) cooperation you will recall the relationship of the rhinoceros and the tickbird. The little bird gets free food while the rhinoceros rids itself of ticks and harbors a wary little lookout. The highest levels of cooperation, however, are most likely to be found between members of the same species.

Let's consider a few examples of *intraspecific* (within species) cooperative behavior. Porpoises are air-breathing mammals, much vaunted in the popular press for their intelligence. In fact, certain of their actions support the claim. Groups of porpoises will swim around a female in the throes of birth and will drive away any predatory sharks that might be attracted by the blood. They will also carry a wounded comrade to the surface so it can breathe. Their behavior in such cases is highly flexible, in that it is often adjusted to the subtleties of the environmental circumstances, rather than being a stereotyped behavior pattern. Such flexibility indicates that their behavior is not a blind response to innate genetic influences.

Group cooperation among mammals is probably most common in defensive and hunting behavior. For example, yaks of the Himalayas form a defensive circle around the young at the approach of danger, standing shoulder to shoulder with their massive horns directed outward. This defense is effective against all predators except humans, since the yaks often pitifully maintain this stance while they are felled one by one by rifles. Wolves, African cape dogs, jackals, and hyenas often hunt in packs and sometimes cooperate in bringing down their prey. In addition, they may bring food to members of the group which were unable to participate in the hunt.

Mammals such as the ones we have considered so far might be expected to show the highest levels of cooperative behavior as a result of their intelligence. It is a bit surprising, therefore, that among all animals social behavior and cooperation are most highly developed in the lowly insects (Figure 11.31). The complex and highly coordinated behavior patterns of insects are generally considered to be genetically programmed, highly stereotyped, and not much influenced by learning—except perhaps for very restricted classes of learning. But the arguments for this are beyond our present scope.

In honeybee colonies the queen lays the eggs and all other duties are performed by the workers, sterile females. Each worker has a specific job, but that job may change with time. For example, newly emerged workers prepare cells in the hive to receive eggs and food. After a day or so their "brood glands" develop, and they begin to feed larvae. Later they begin to accept nectar from field workers and pack pollen loads into cells. At about this time their wax glands develop, and they begin to build combs. Some of these "house bees" may become guards, which patrol the area around the hive. Eventually each bee becomes a field worker, or forager. She flies afield and collects nectar, pollen, or water, according to the needs of the hive. These needs are indicated by the eagerness with which the field bees' different loads are accepted by the house bees.

If a large number of bees with a particular duty are removed from the hive, the normal sequence of duties can be altered. Young bees may shorten or omit certain duties and begin to fill in where they are needed. Other bees may revert to a previous job where they are now needed again.

The watchword in a beehive is *efficiency*. In the more "feminist" species, the drones (males) exist only as sex objects. Once the

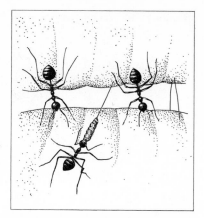

Figure 11.31 Weaver ants sew leaves together. As one group pulls the leaves together, another group passes silk-spinning larvae back and forth, pressing them to the edges of the leaves. (After F. Dolflein 1905, from Eibl-Eibesfeldt, Ethology, The Biology of Behavior, 1970)

Figure 11.32 This field worker is performing only one of the many jobs she has held in her lifetime, and one of the last. Unless she is needed to work in the hive again, she will probably continue to gather nectar until the end of her days. (Courtesy: Alvin E. Staffan from National Audubon Society)

queen has been inseminated, the rest of the drones are quickly killed off by the workers. They are of no further use. The females themselves live only to work. They tend the queen, rear the young, and maintain and defend the hive. When their wings are so torn and battered that they can no longer fly, they either die or are killed by their sisters. But the hive goes on.

ALTRUISM

We don't wish to shatter anyone, but, well . . . most of the Lassie stories aren't really true. Consider what would happen to the genes of any dog that was given to rushing in front of speeding trains to save baby chickens. The reproductive advantages would be considerable to chickens, but dogs with those tendencies might be selected out of the population by the action of fast trains. Their reproductive activities would therefore be cut short. In contrast, the genes of a "chicken" dog that spent his energy, not in chivalrous deeds, but in seeking out estrous females, would be expected to increase in the population. So what kinds of dogs are likely to predominate in the next generation?

If an animal is going to engage in "unselfish" deeds, its best reproductive bet lies in those deeds which advance the genes of other members of its own species, or perhaps the genes of those species which can help its own species. And as we will see, although selfless behavior does occur both within and between species, it is most likely to occur within species.

Altruistic behavior may at first seem to have obvious advantages, but the question is a prickly one, especially when we consider how such altruism might have evolved. *Altruism* may be defined in a biological sense as an activity which benefits another organism but at the individual's own expense.

It is easy to see how certain forms of altruism are maintained in any population. For example, pregnancy, in a sense, is altruistic. The prospective mother is swollen and slowed. Much of her energy goes to the maintenance of the developing fetus. At birth she is almost completely incapacitated and is in marked danger at this time. Pregnancy is clearly detrimental to her. So why do females so willingly take the risk? It may help to understand the enigma if we remember that the population at any time is composed of the offspring of individuals who have made such a sacrifice. Thus the females in the population may, to one degree or another, be genetically predisposed to make such a sacrifice themselves.

However, altruism on this basis doesn't explain why a bird may feed the young of *another* pair, or why an African hunting dog will regurgitate food to almost *any* puppy in the group. Why, also, would a bird which may have no offspring of its own give a warning cry at the approach of a hawk, alerting other birds at the risk of attracting the hawk's attention to itself?

To answer such questions we must look past the answers that first come to mind. It may seem cynical, but we must start with the premise that birds don't give a hoot about each other. A bird that issues a warning call isn't thinking, "I must save the others." At least, there is a simpler explanation of its behavior.

Keep in mind that the biologically "successful" individual is the one that maximizes its reproductive output. One way of accomplishing this is for the organism to behave in such a way as to leave as many individuals carrying its own type of genes as possible in the next generation. This would explain parental care. However—and this may not be so readily apparent—an individual can also leave its type of genes in the next generation by helping a relative's offspring to survive. An individual shares genes in common with a cousin, although fewer, of course, than with a son or a daughter. Hence there is theoretically a point at which an individual could increase its reproductive output by saving its nieces and nephews (provided there were enough of them), rather than its own offspring. From the standpoint of reproductive output, the organism would be better off leaving 100 nephews than one son.

To illustrate, suppose a gene for altruism appears in a population (notice that this sets up the mechanism for the continuance of the behavior). As you can see from Figure 11.33, altruistic behavior would most likely be maintained only in groups in which the individuals are related (or have some kinds of genes in common). It might be expected, then, in "nontransient" populations, those which don't move around much, and those in which there is little mixing with outside groups—in other words, where there is a high probability that proximity indicates kinship.

Keep in mind that *no conscious decision* on the part of the altruist is necessary. It simply works out that when conditions are right, those individuals which behave altruistically increase their kinds of genes in the population, including the "altruism gene." Nonrelatives would be benefited by the behavior of altruists, of course, but there is an increased likelihood of individuals nearer an altruist to be related to it.

It has been determined mathematically that the probability of altruism increasing in a population depends on how closely the altruist and "benefitee" are related. In other words, the advantage to the benefitee must increase as the kinship becomes more remote, or the behavior will disappear from the population. For instance, altruism toward brothers and sisters which results in the

Figure 11.33 In this population, A and A' are non-altruists. They behave in such a way as to maximize their own reproductive success, but do nothing to benefit the offspring of other individuals. In another segment of the population (B and B') a gene for altruism has appeared which results in individuals benefitting the offspring of others in some way. It can be seen that, assuming the altruistic behavior is only minimally disadvantageous to the altruist, generations springing from B and B' are likely to increase in the population over those from A and A'. It should be apparent that the altruistic behavior is likely to be greatest where B and B' are most strongly related, so that B shares a maximum number of genes in common with the offspring of B' and vice versa. The idea is that B, for example, can increase its own reproductive success by caring for the offspring of a relative with whom it has some genes in common. After all, reproduction is simply a way of continuing one's own kinds of genes.

death of the altruist will be selected for if the net genetic gain is *more* than twice the loss; for half-brothers, four times the loss; and so on. To put it another way, an altruistic animal would gain reproductively if it sacrificed its life for more than two brothers, but not for fewer, or for more than four half-brothers, but not for fewer, and so on. Therefore we can deduce that in highly related groups, such as a small troop of baboons, a male might fight a leopard to the death in defense of the troop. By the same token, a bird will give a warning cry when the chance of attracting a predator to itself is not too great, and when the average neighbor is not too distantly related.

This model, developed by J. B. S. Haldane in 1932 and expanded by W. D. Hamilton in 1963, helps to explain the extreme altruism shown by social insects, such as honeybees. Since workers are sterile, their only hope of propagating their own genotype is to maximize the egg-laying output of the queen. In some species the queen is inseminated only once, so all the workers in a hive are sisters and have half their genes in common. (Why only half?) In such a system, then, almost *any* sacrifice is worth *any* net gain to the hive and to the queen.

The discussion of the development of altruistic behavior is admittedly somewhat esoteric. However, it seems important to consider even our most cherished and our most despised behaviors in the context of man as an evolutionary product.

We now come to one of those disconcerting changes in topics. The following material is included because you may find the information useful in your own decision-making processes.

MINDBENDING

Let's consider a few biological implications of what are loosely called "drugs." The term is not very precise, and it probably evokes images of banal lectures during NFL commercials. Maybe "mindbenders" is a better word. In any case, what we're referring to are those substances which can produce a temporary change in neural events, attitudes, thoughts, or behavior. Since this covers a lot of things, from cigarettes and aspirin to speed and heroin, let's take them by classes and consider a few characteristics of each.

TOBACCO

Tobacco is the dried leaf of the plant *Nicotiana tabacum*. When it is burned, its by-products are able to cross the thin-walled alveoli of the lungs and enter the bloodstream. The spread of the smoking habit is curious, in view of the negative reinforcement involved in the first attempts. Recent medical evidence also provides some

Figure 11.34 Tobacco has long been used by man. Only recently has medical evidence pointed out that it is harmful to one's health.

discouragement; however, smoking was a well-established habit in Western culture long before its medical effects were known. So far cigarette smoking has been associated with higher incidences of lung cancer, emphysema, heart and circulatory ailments, and certain birth defects. Smoking may also permanently paralyze the tiny cilia which sweep the breathing passages clean, and cause the lining of the respiratory tract to thicken irregularly. Often an accompanying deep, hacking cough appears as the body attempts to rid itself of nicotines, "tars," formaldehyde, hydrogen sulfide, resins, and other goodies. To make matters worse, it seems that one often finds himself seated in a restaurant next to some chain smoker who is trying spasmodically to clear his respiratory passages.

The word *habit* is applicable here because a smoker may

actually become physiologically dependent on the products in the smoke. The smoker may suffer various withdrawal symptoms when he attempts to break the habit. In many cases even the development of serious circulatory and respiratory ailments is not a sufficient deterrent to enable a smoker to give up cigarettes. It is interesting that, in this country, millions of tax dollars are spent on research to control the effects of smoking, and on the other hand, large federal subsidies are provided to tobacco growers to ensure a continued supply.

CAFFEINE

Caffeine, which is one of the components found in coffee and tea, acts as a stimulant on the central nervous system. It is widely used in many cultures throughout the world. At one time in the Near East, coffee drinkers were put to death—perhaps no worse a fate to some than having to start the day without coffee.

The behavioral effects of caffeine include increased alertness and decreased fatigue and boredom. In small to moderate amounts, it may improve performance in boring or repetitive tasks but it doesn't help in more complex intellectual tasks such as reading or doing long division. It may help keep you awake so you can perform those tasks, however, since it can inhibit sleep. High doses can cause nervousness, irritability and a "jangled" feeling.

It is possible to develop *tolerance* for the drug so that increasingly higher doses are needed to produce the same effect. Withdrawal symptoms include headaches and irritability. But the effects are minimal, so withdrawing coffee-drinkers are generally not considered dangerous. There are no known effects of the long-term use of the drug.

MARIJUANA AND HASHISH

Marijuana (also known as dope, pot, grass or Mary Jane) is a preparation made from the Indian hemp (*Cannabis sativa*). It was cultivated throughout the United States in World War II to produce fibers for ropes when the supply of hemp from the Philippines was cut off. The wild progeny of those plants has distressed law officials no end in their effort to shield American youths from what was once called "The Killer Weed." Its active ingredient is a group of chemicals called tetrahydrocannabinols (THC). THC is highest in a preparation called *hashish* or *hash* which is made from certain types of marijuana. The effect of marijuana on most users is usually a mild euphoria, sometimes expressed as a happy or giggly mood.

The arguments which have emanated over marijuana have often been ill-conceived and irrational, as you well know, and it

Figure 11.35 Marijuana and hash with a variety of hash pipes. (Courtesy: Gail Weingart)

would be satisfying to say that the data are now in and that we have at last arrived at some valid conclusion regarding its effects. Alas! Such is not the case. Many questions remain unanswered.

Much of the information which we do have, however, indicates that the threat from marijuana may have been overemphasized. It is known that marijuana may alter functions in the cerebral cortex and other parts of the brain, but its short-term and long-term effects seem, at present, less disastrous than those of alcohol (see below). It is probably naive to assume any drug is harmless, however. Hopefully, unbiased medical research will soon provide us with information regarding the actual physiological effects of using marijuana.

It is known that the effects of marijuana are highly variable from one individual to the next and may be partly dependent upon the setting and mood in which the drug is taken. Marijuana seems to affect short-term memory so that a person may not be able to complete a sentence because he forgot how it began. It may also lower the attention span and the ability to change quickly from one topic to another. In certain mental tasks such as information processing, marijuana may slow performance just as alcohol does. On a group of other sorts of mental tasks, marijuana has no appreciable effect. One reason for the lack of such effect may be the ability of some users to "compensate" for the effects of marijuana.

Compensation refers to the ability to consciously override any effect of the drug.

With exceptionally high doses of THC the perception of the individual may change so that the world seems distorted and sensory input becomes exaggerated. The individual may not be able to cogently direct his thought patterns and he may experience paranoia (possibly brought on by the knowledge that he could go to jail).

Marijuana is generally not considered addictive, and there seems to be minimal development of "tolerance" so increasing doses are not needed. There may be a psychological "dependence" on the drug—which means that it has produced "good" effects in the past and the user may desire it again because of its rewarding effect.

Cannabis sativa has been cultivated in the Near Eastern countries since the time of the Crusades, where its THC was extracted to produce the resinous concentrate hashish. This substance was used as a drug by the *hāshshāshīn*, a Moslem terrorist group whose notorious violence was thought to be a result of addiction to the drug. More detailed accounts of the hāshshāshīn, whose name is the source of the word assassin, indicate that hashish was actually given as a reward for their murderous deeds—the "visions of glory" promised by their leader, who, by the way, was once a classmate of Omar Khayyam's.

Hashish has been widely used throughout India and Asia for centuries. Although it is highly potent and may produce marked hallucinogenic effects, there are no indications that it actually is addictive, as once supposed. It is common practice in India for nursing mothers to apply a minuscule amount of the substance to a breast to lull fretful infants as they nurse.

ALCOHOL

Ethyl alcohol is the active ingredient of wines, beer and whiskies. Basically it is a central nervous system depressant. It produces its well-known stimulatory effects by depressing the nervous system's inhibitory controls. One of its major effects is an increase in the level of aggressiveness in many individuals. It produces a marked decrease in motor performance and certain mental abilities (for instance, cognitive skills). Tolerance quickly develops in users. Paradoxically, those who can "hold their liquor" may be suffering the greatest damage from the drug because various other organs may not develop the same tolerance the brain does.

Alcohol is addictive and disuse may produce severe withdrawal symptoms, the most drastic of which are *delerium tremens* or DTs.

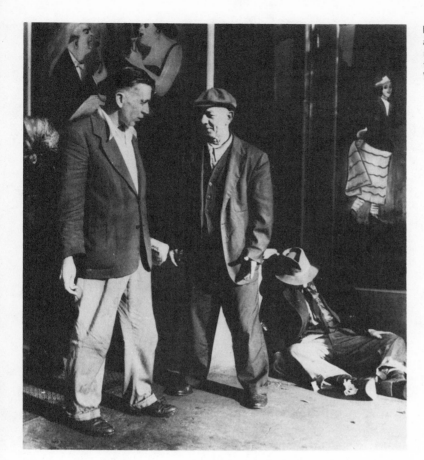

Figure 11.36 Most of the down and out inhabitants of the country's many "Bowerys" are addicted to the drug alcohol. (Courtesy: Wide World Photos)

Long-term use of alcohol may cause changes in the central and peripheral nervous systems, the liver, the stomach and the intestines, but the effects vary greatly from one individual to the next. One alcoholic's mind may go, while his drinking companion may only suffer cirrhosis of the liver, a potentially fatal disease. Treatment of alcoholism is difficult and may largely involve psychotherapy such as is provided by Alcoholics Anonymous. Interestingly, a recent news article reported the relief of many parents in a large West-coast city at learning that their children were shifting from pot to booze.

OPIATES

Opiates are generally what are referred to as "hard drugs." They are *narcotics* in that they relieve pain and produce sleep or stupor. Drugs in this group include *heroin*, *opium* and *morphine*. Users

Figure 11.37 Common paraphernalia used for injecting heroin. The small balloons are filled with the white powder and are often sold this way. (Courtesy: Gail Weingart)

rapidly develop marked tolerance, and in nearly all cases continued use results in addiction. Consequently, the user not only needs the drug continually, but he needs increasingly higher doses. Since such drugs are expensive, in many cases addicts resort to crime to raise the necessary money.

Injection of opiates causes a "rush" of a pleasurable sensation, or a violently negative reaction in beginning users. The rush is followed by a great sense of well-being, which is accompanied by a marked decrease in physical drive. The accompanying drowsiness produces the nodding behavior you can see in almost any New York subway. It isn't known how opiates behave in the nervous system. However, the withdrawal symptoms produced by abrupt discontinuation of the drug are usually violent. In cases of heavy addiction, withdrawal may produce such a shock to the system that death results.

The crime associated with heroin addiction has caused great consternation, especially in our larger cities. One partial solution has been the initiation of programs to maintain addicts on another kind of opiate called *methadone*. Methadone diminishes the craving for heroin, but it is also addictive. Individuals maintained on methadone through supervised medical programs, however, have been able to live relatively normal lives and to maintain satisfactory relationships with other people, including their families. Some people on methadone maintenance have held steady jobs for years while raising families and participating in community life.

COCAINE

Cocaine, or "coke," is legally classified as a narcotic, but biologically it only partly fits the description. It relieves pain, but it does not induce sleep or stupor. It acts through stimulation of the nervous system and produces a feeling of euphoria.

The use of cocaine originated with the Indians of the Andes, who chewed the leaves of the coca bush, often as they carried large loads up steep mountain trails. The practice in this country was to pulverize the bitter alkaline extract of the leaves and to sniff the powder. However, this habit causes destruction of the mucous membranes and may even dissolve the partition between the nostrils; so now heavy users usually inject the cocaine.

Cocaine has recently become a fashionable drug in certain segments of the American middle class, partly because it is not considered addictive. However, it can result in highly aggressive behavior and feelings of paranoia.

AMPHETAMINES

Amphetamines, often loosely referred to as "speed" or "uppers," include a number of commercial drugs such as Benzedrine, Dexedrine, and Methedrine. Their chemical properties are similar to adrenaline, and they cause great bursts of energy which may successfully overcome any feelings of fatigue. Students and truck drivers often take low doses in order to stay awake during midnight cramming sessions and long hauls, respectively. The drugs are also used to decrease appetite in weight-watchers. Because amphetamines are effective in improving performance on rigorous physical tasks, they are sometimes taken by athletes. Low doses do not impair skills or judgment.

The greatest abuse of amphetamines occurs with injection of the drug. Injection may produce an initial rush, followed by a feeling of vigor and euphoria which may last several hours. However, this is followed by general aching, discontent, and irritability. To delay this letdown, the user may boost himself with another injection. The high may thus last for days, during which time the user usually fails to eat or sleep. The end of this period may be followed by exhaustion, severe depression, paranoia, aggressiveness, extreme irritability and emotional overreaction.

Users develop a tolerance for amphetamines and after prolonged use may show withdrawal symptoms. These are less severe than with opiate withdrawal.

Figure 11.38 Barbiturates. (top)
Dexidrine in two forms. Bennies
(amphetamine sulphate) and
methaquolone (bottom). (Courtesy:
Gail Weingart)

BARBITURATES

Barbiturates, or "downers," are sold under a variety of trade names,
including Nembutal, Seconal, Tuinal, and Amytal. They are all

sedative-hypnotics which act on the cerebral cortex, midbrain, and brainstem areas to reduce anxiety and induce drowsiness and sleep. These effects are accompanied by loss of muscular coordination and slurring of speech—effects similar to those induced by alcohol.

Barbiturates are highly addictive and rapidly produce tolerance. Withdrawal symptoms may be as severe as with opiate or alcohol withdrawal. The heavy barbiturate user is likely to be confused, obnoxious, stubborn, and irritable. In contrast to the placid disinterest of the opiate user, barbiturate users may be particularly aggressive and violent.

Barbiturates and alcohol acting together may cause death by suppressing the breathing centers, and because both drugs cause mental confusion such accidental deaths are frequent.

PSYCHEDELICS

Psychedelics are a group of drugs which produce hallucinations and various other phenomena that mimic the experiences produced by certain mental disorders. These drugs include lysergic acid diethylamide (LSD), mescaline, peyote, psilocybin, and various commercial preparations such as Sernyl and Ditran.

LSD is probably the best known of these substances, although its use has probably diminished somewhat since its heyday in the late 1960s. LSD usually produces responses which follow a particular sequence. The initial reaction may be a sensation of weakness, dizziness, and nausea. These symptoms are followed by a distortion of time and space. The senses may become heightened and strangely intertwined—sounds can be "seen" and colors "heard." Finally, there may be changes in mood, a feeling of separation of the self from the framework of time and space, and changes in the perception of the self. The sensations experienced under the influence of psychedelics are unlike anything encountered within the normal range of experiences. The descriptions therefore can only be puzzling to nonusers.

The effects vary widely from one individual to the next, and from one "trip" to the next. The frame of mind of the user and his setting may play a large part in determining the nature of the experience. Users have often claimed new insights, but there is no evidence of psychological improvement or the ability to relate these experiences in any meaningful way even to subjective "reality." LSD, however, has been used medically to treat conditions such as alcoholism and severe paranoia. On the other hand, some users experience bad trips or "bummers" which may or may not have long-term effects. Bad trips may be terrifying experiences and can occur in experienced users for no apparent reason. There are

352 *Biology: The World of Life*

some instances in which the mind has not recovered at all, and others in which the individual has done himself irreparable damage in the interim.

There has been a lot of discussion about possible chromosome damage due to LSD use. A nagging suspicion among users has been that this fear was generated largely by biased reports by those who are merely opposed to use of a drug which they don't use or understand. Actually, there is no unequivocal evidence that pure LSD causes chromosome aberrations. On the other hand, "street" LSD is likely to be impure, or laced with small amounts of nerve poisons such as strychnine, and the detrimental effects that have been observed could be due to such impurities. Among users of illicit LSD there is, in fact, a higher incidence of chromosome damage and a higher rate of spontaneous abortion, or miscarriage.

Figure 11.39 Artist's conception of a drug trip.

12
Communities and Competition

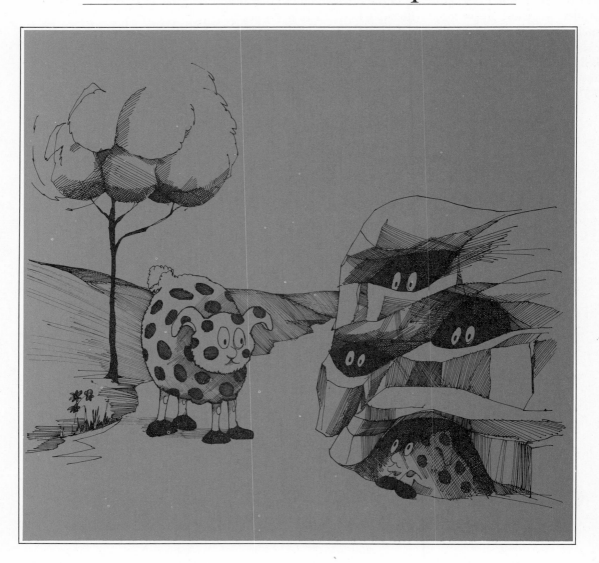

And NUH is the letter I use to spell NUTCHES
Who live in small caves known as NITCHES for NUTCHES
These NUTCHES have troubles, the biggest of which is
The fact that there are many more NUTCHES than NITCHES
Each NUTCH in a NITCH knows that some other NUTCH
Would like to move into his NITCH very much
So each NUTCH in a NITCH has to watch that small NITCH
Or NUTCHES who haven't got NITCHES will SNITCH

As any idea is sifted through the public consciousness, finally to be encompassed in a "household word," often many of the valuable nuggets of the concept are left behind. Such is the case with certain terms associated with ecology. The term *niche* has been bandied about rather freely in the past few years and has come to refer to anything from a writer's specialty, to a particular socioeconomic group, to a habitat. Even some biologists have been guilty of confusing niche and habitat. Let's see what some of these terms mean.

HABITAT AND NICHE
It has been said that if the habitat is an organism's address, the niche is its profession. The *habitat* may be defined as the place where the organism is found, and it can be described in several ways. For example, an animal may live in a desert habitat, or more specifically, in a briny desert pool. Furthermore, it may live in a certain part of that pool, its *microhabitat*. Wherever an organism lives, however, it interacts with its surroundings in myriad ways. It is affected by its environment, and may, in turn, influence this environment. The organism may also interact with other living things around it. The sum of all such interactions, along with the organism's own requirements, describe its *niche*. Actually, as you might imagine if you are beginning to gain some insight into the

"scientific mind," there is quite a bit of haggling over how these terms should be precisely defined, but let's not become involved.

It has become axiomatic that two species cannot occupy the same niche indefinitely. If two species were to find themselves in such a situation, it is generally predicted that they would be unequally adapted to that niche, so that in time as one species decreased in number, the other would eventually come to replace it. In effect, this means that when you walk through the woods and see various species of small seed-eating birds, each species is probably interacting with the environment in a different way. This principle was nicely demonstrated by the ecologist Robert H. MacArthur, who showed that five species of American warblers that feed in spruce forests actually divide the trees up into different feeding zones.

According to the population ecologist L. C. Birch, competition occurs where organisms utilize resources which are in short supply, or when they harm each other while seeking resources which are not in short supply. (Probably the latter cases are much rarer in nature.) Where species that share a habitat utilize its resources differently, it is assumed that they are reducing competition among themselves. For example, in an area inhabited by five species of titmice, each has a different foraging pattern—each species prefers a particular kind of tree and particular parts of trees. However, the degree of differences in their behavior changes with the season. In fact, it is found that they forage in almost identical ways during periods when food is abundant, a time when there is presumably a low level of competition.

Animals may also divide up a habitat in other ways. For example, they may utilize the same resources at different times of the day, or at different times of the year, or they may utilize different commodities within the same part of the habitat.

The real role of competition in the affairs of living things is not very well understood at present. In fact, many of our ideas are based largely on supposition or are supported by rather tenuous examples. However, never deterred, we plunge ahead.

THE LAND ENVIRONMENT

It seems safe to say that no two organisms interact with their environments in identical ways. Two tapeworms living in a raccoon's intestine, if they were asked, would probably describe their world differently, depending on their precise point of attachment and their individual perceptive and reactive tendencies. Two wildebeest living on an African plain would certainly have different experiences, and might be expected to perceive their environment

Figure 12.1 The feeding zones of five species of North American warblers in spruce trees. The darker areas indicate where each species spends at least half its feeding time. By exploiting different parts of the tree, the species reduce their competition for food and thus they can occupy the same habitat. (From "Population Ecology of Some Warblers of N. E. Coniferous Forests", MacAurther, *Ecology* 39:599–618)

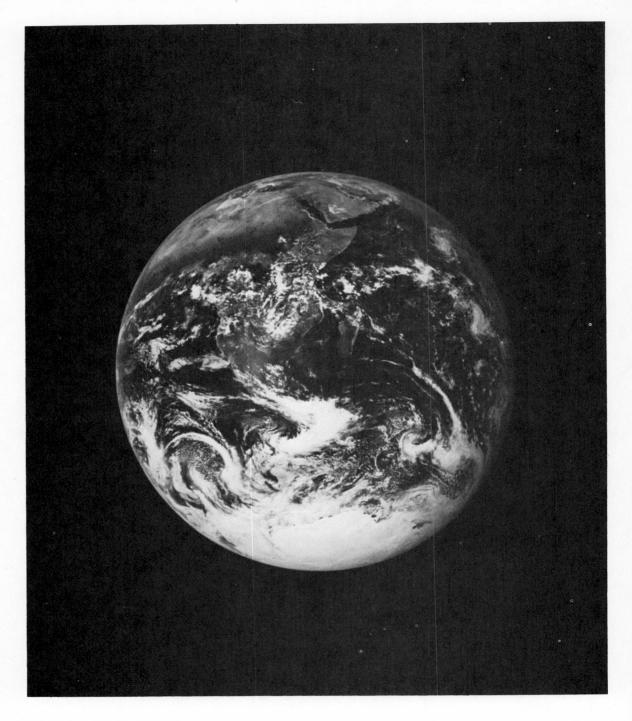

quite differently. Even identical twins see the world from opposite sides of a baby carriage. The point is that it is hard to generalize about the environment of any species. So now, of course, we will generalize.

We tend to think of our earth as a ponderous place which unfailingly provides its denizens with those things necessary for life. We often fail to realize, however, that life exists only in an exceedingly thin film which veils the surface of this immense ball—a delicate shell wherein the wondrous forces of sunlight and water interact. This fragile film is responsive to a number of influences, and hence is highly variable from one place to another. Furthermore, each place is in itself unstable, so that it changes with time. The different kinds of "places" in which life exists over the earth at present can be categorized roughly according to their physical and biological properties. Keep in mind, however, that these are merely arbitrary divisions of the great, complex, and intergrading areas of the earth. But it makes discussion easier.

BIOMES

By using some imagination we can divide the earth's land into several kinds of regions called *biomes* (Figure 12.3). Biomes are defined on the basis of the plants they support. Of course, the makeup of the plant community is dependent on other factors, such as soil conditions, weather, day length, competition, and the nature and abundance of the resident plant eaters. A moment of reflection, if all is going well, will show that the nature of any animal community is, in turn, largely determined by the nature of the plant life there, since plants are needed to trap the sun's energy and to provide shelter, building materials, and hiding places for the animals.

Let's briefly consider the seven generally recognized types of biomes.

Temperate deciduous forests are the type of forest which at one time covered most of the eastern United States and all of Central Europe. The dominant trees in these forests are hardwoods, such as oak, maple, beech, poplar, and hickory. These areas are subject to harsh winters, times when the trees shed their leaves, and warm summers, periods of rapid growth and rejuvenation. Before the new leaves begin to shade the forest floor in the spring, a variety of herbaceous *annual plants* may appear, plants whose life cycle consists of only one season; they bloom anew each spring, only to die in the fall. Rainfall may average 30 to 50 inches or more each year in these forests. The seasonal changes are moving and fascinating. People who live in such biomes describe a certain joy which swells

Figure 12.2 The earth is actually a very delicate sphere. When one small area is disturbed, it is likely that the result will be felt in other areas. (Courtesy: NASA)

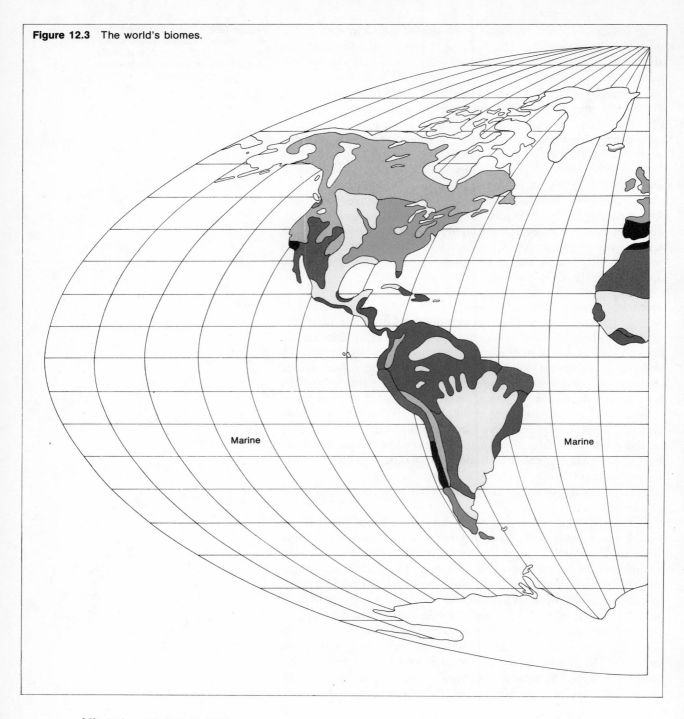

Figure 12.3 The world's biomes.

Marine

Marine

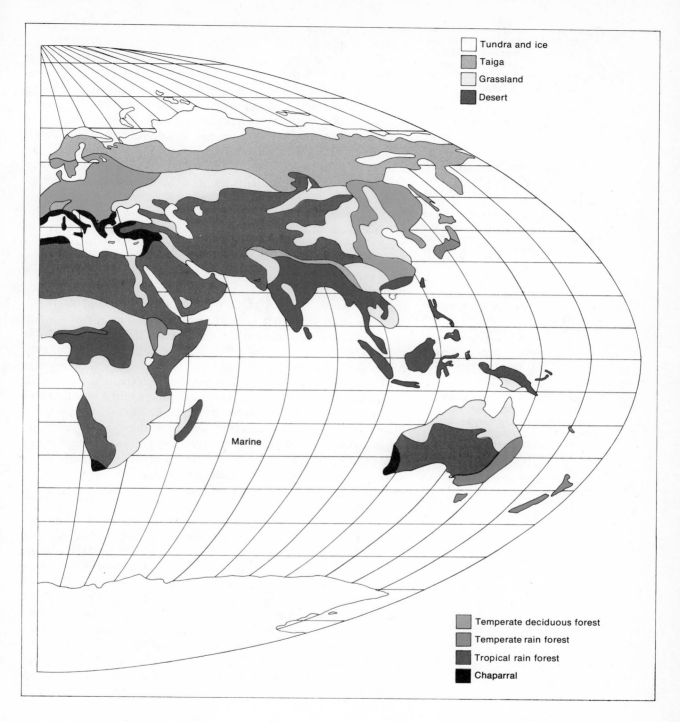

Tundra and ice
Taiga
Grassland
Desert

Marine

Temperate deciduous forest
Temperate rain forest
Tropical rain forest
Chaparral

Figure 12.4 Temperate deciduous forests (top left) once covered the eastern United States. Before shedding their leaves for the harsh winter season, these forests burst into a dazzling display of color, only to quietly renew their growth in the cool and temperamental spring. (Courtesy: U.S. Forest Service)

Figure 12.5 Temperate rain forests (top right) are dark and damp places which have a sense of brooding and somberness. One may walk quietly over the mossy carpet and see species which are found in no other areas throughout the world. (Courtesy: U.S. Forest Service)

Figure 12.6 Grasslands (bottom left) are found in many places throughout the world. This area is in Idaho. Such areas support an incredible array of animal life from tiny insects to large mammals. (Courtesy: U.S. Forest Service)

Figure 12.7 An American desert, (bottom right) a difficult place for most organisms to survive. Those which have adapted to such areas have morphological, physiological and behavioral characteristics which permit them to survive with little water to save water, and to quickly take advantage of brief rainfall. (Courtesy: U.S. Forest Service)

within them each spring, and the secret pensiveness which overcomes them in the fall as the days darken and the forests become more silent. Perhaps we are exceeding the technical description of the biome, but these are my favorite places.

Temperate rain forests have a rather mild climate with heavy rainfall throughout much of the year. Unlike temperate deciduous forests, plant growth in these regions generally continues throughout the year. Also, whereas temperate deciduous forests are characterized by a variety of trees in roughly equal numbers, the temperate rain forests tend to be dominated by a single species.

Grasslands occur in many parts of the world and are exemplified in the United States by the American prairie. They are characterized by an abundance of grasses, small bushes and shrubs, and in some parts of the world, thickets of bamboo, which is a type of grass. The soil in such areas is usually porous. Trees may appear along streams or rivers. Rainfall averages between 10 and 40 inches each year, but may be erratic in its timing. Grasslands are present in both tropic and temperate zones. You may be surprised to learn that grasslands support more species of animals, and a greater biomass, than any other kind of terrestrial habitat.

Deserts are characteristically hot in the daytime and cold at night. The 10 inches or less of rain which falls each year usually comes in sudden downpours, so that much of it runs off, sometimes causing flash floods and marked erosion. Directly after a rainfall, the annual plants of the desert take advantage of the moisture and explode in an orgy of growth and seed production. Other desert plants meet the water-shortage problem in other ways, either by storing water in their tissues or by reaching far underground to tap water tables with their, of course, tap roots. Deserts harbor few kinds of animals other than reptiles and insects. The preponderance of these species is chiefly because they are cold-blooded. (Can you figure out why?)

Tropical rain forests, or jungles, are found in the Amazon and Congo Basins and in Southeast Asia. The temperature in such biomes doesn't vary much throughout the year. Instead the seasons are marked by the torrential rains which fall almost daily during the summer. In some areas there may also be pronounced winter rainy seasons. These regions support many species of plants. Trees grow throughout the year and reach tremendous heights, with their branches forming a massive canopy overhead. The jungle floor may be dark and steamy, sparsely settled by smaller plants. Jungles literally swarm with insects and birds. Animals may breed throughout the year as a result of the continual availability of food. Since there is no harsh winter season, some species are coming into

Figure 12.8 Tropical rain forests are productive throughout most of the year and support a variety of animal life. When the overhead canopy is heavy, the floor may be continuously shaded and relatively free of small plants. (Courtesy: U.S. Forest Service)

Figure 12.9 A tundra area near Mt. McKinley. Tundra is especially prominent in the far north and in higher areas of the Rockies. In summer the area is almost startling in its impact on a lowland visitor because of its myriad ponds, riverlets and profusion of strange and fascinating plants. (Courtesy: Charlie Ott from National Audubon Society)

Figure 12.10 The taiga of Alaska's interior. Here, three species dominate the area: white spruce, aspen and willow. In other areas, a single cone-bearing species may dominate. (Charlie Ott from National Audubon Society)

production at all times. Competition is generally considered to be very keen in such areas because of the abundance of species.

Tundra regions are covered throughout most of the year by ice and snow. This biome is most prevalent in the far north, but in the United States it may be seen in the high Rocky Mountain regions. Summers usually last two to four months, just long enough to thaw a few feet of the soil above the *permafrost*, or permanently frozen soil. Thaw brings soggy ground, and many ponds and bogs appear in the depressions. The plant life consists mostly of lichens, herbs, mosses, and low-lying shrubs and grasses. Such plants must be incredibly hardy to be able to live in such a place. Their hardiness disguises their fragility. Once disturbed, these areas take very long periods to restore themselves. The marks of wagons which passed over the tundra a hundred years ago are clearly visible to this day.

Taiga refers to the northern coniferous forests which cover much of Canada and Siberia. The taiga has a longer growing season than the tundra. Annual rainfall is low, usually less than 20 inches. The forest may be dominated by a single kind of tree such as pine, spruce, or fir, and the forest floor is usually covered by a carpet of needles, with only a few shrubs or other plants pushing through. One may move silently on the muffling needles through the

Canadian taiga, observing a host of mammals from porcupines to moose.

ECOSYSTEMS AND COMMUNITIES

Biomes can be broken down further into *ecosystems*, specific areas within the biome. Ecosystems are theoretical entities in which the component species interact as a self-sustaining unit. An ecosystem, then, is considered an independent unit (although its independence would be hard to prove). An ecosystem might be a pond or a wooded area or a field. The organisms which live within an ecosystem comprise a *community*, and of course, they interact with each other as well as with their physical environment. Now, within a community each species has a particular role; it interacts with its environment in a certain, theoretically circumscribed, way. The sum of those interactions is its *niche*. So there you are.

THE WATER ENVIRONMENT

The earth's water may be classified roughly as fresh water or salt water, although not all bodies of water fall neatly into one category or the other. For example, Lake Pontchartrain, near New Orleans, is *brackish*, or a mixture of salt and fresh water.

Fresh water has about 0.1 percent salt, whereas seawater has about 3.5 percent salt. Each has its own importance in the earth's drama. Fresh water, for example, is drinkable; seawater is not (at least for us). Also, only fresh water can be used to irrigate crops. Now let's consider fresh water and salt water separately and note the unique importance of each.

FRESH-WATER BODIES

In general fresh water provides a less stable habitat than does seawater. Bodies of fresh water tend to evaporate to a fraction of their former size, and also to become muddied. Because of their relatively small volumes, they become polluted rather easily, and, in some cases, they can be made to age prematurely. Because of their peculiar vulnerability, human activity has, in many cases, drastically altered their character as ecosystems. In the Great Lakes, for example, the parasitic lamprey and the destructive alewife have begun to flourish, whereas these habitats will no longer support the edible lake trout and whitefish (Figure 12.11). One reason for this change is that the use of detergents and chemical fertilizers has increased the amounts of phosphate and nitrate which are washed into the water. Bacteria can break these down, but only with the use of great amounts of oxygen. Algae and other plants flourish under such conditions, but as they die, they return their nitrates and phos-

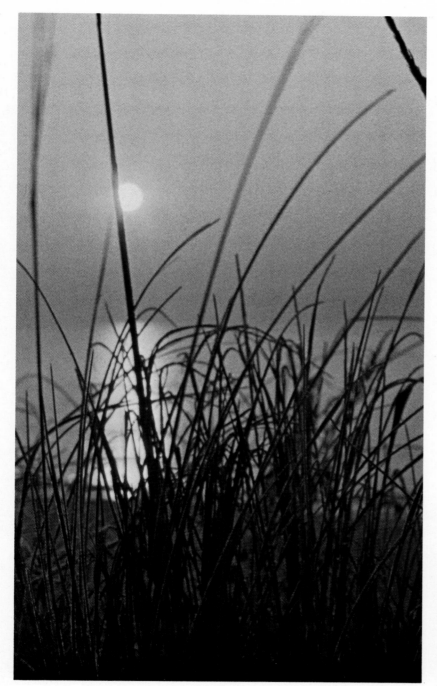

A

Plants

A. The sun rises on the dunes of Amagansett, Long Island. The dunes are constantly shifting in response to the whims of the wind. Such dunes are held in place, buffering land and sea, by a variety of grasses. In many areas, beach buggies and other recreational vehicles are threatening to destroy these fragile grasses, leaving the dunes unprotected from violent winter storms. (Courtesy: Tom Mahnken)

B. In the fall, many leaves turn bright colors before falling to the ground. (Courtesy: Tom Mahnken) C. The mushroom is not a plant, and it cannot make its own food. Instead, it utilizes the breakdown products of dead matter which have fallen to the forest floor.

B

urtesy: Harry Crosby/Photophile)
Broad-leaved aquatic plants, each
ching as much sunlight as
sible, may cover entire lake
faces and drastically alter the
ure of the life below. (Courtesy:
Weingart) E. Beavertail cacti have adapted to dry conditions by storing precious water in their cells, lowering the rate of water loss, and being able to quickly absorb brief desert rains through a shallow root system. (Courtesy: M.S. Breitenback/ Photophile) F. The majestic giant Sequoia once covered much of the northern hemisphere. In recent years the forests have fallen to lumbering interests until less than fifty small groves remain. This is the Sierra Redwood. (Courtesy: Roger Holden/Photophile)

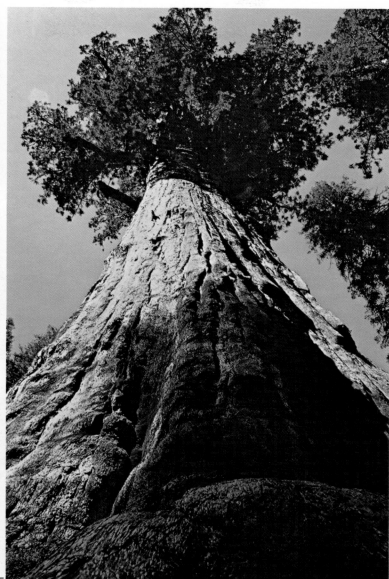

F

G. Lichens are a mutually advantageous combination of algae (which can photosynthesize) and fungi (which form strong attachments to surfaces). (Courtesy: Gail Weingart) H. The white brittle bush reduces competition for water by poisoning any plants which try to grow near it. I. Spanish moss, an epiphyte, derives life's necessities from the air. It is actually not a moss, but related to pineapples. J. Mangroves grow in shallow salt water. They are braced by arching "air roots." K. Another epiphyte. This one grows on trees, but does not parasitize them. L. The Joshua tree is actually a desert lily. (H—L Courtesy: Hubbard Scientific Company, Copyright © 1973)

G

H

I

J

K

L

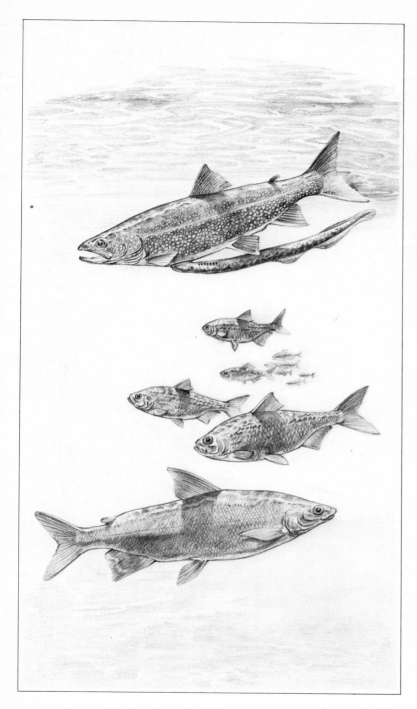

Figure 12.11 As a result of human activities the edible lake trout (top) and whitefish (bottom) are now almost non-existent in the Great Lakes. On the other hand, the parasitic lamprey and the destructive alewife have become common. One reason for this change is that the use of detergents and chemical fertilizers have increased the amount of phosphate and nitrate which is washed into the water. Bacteria can break these down but only with the use of great amounts of oxygen. Algae and other plants flourish under such conditions, but as they die, they return their nitrates and phosphates to the water and cause bacteria to use ever more oxygen. Finally, the lake cannot support animals which require a lot of oxygen, such as trout, but the new conditions are fine for alewives.

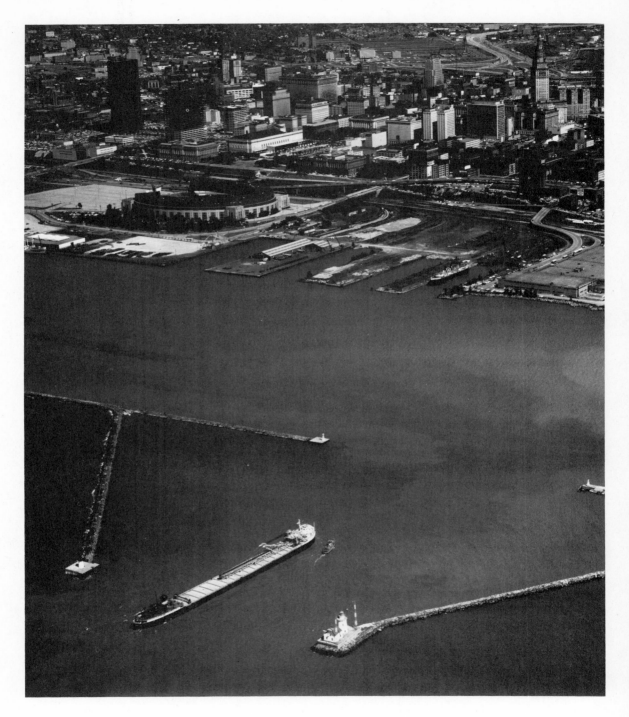

Box 12.1   A Dead Lake   Lake Erie was once one of the world's most beautiful lakes. Its waters were pure and abounded with life. It provided food, and lent its beauty, to man for literally thousands of years. What man, camped beside it in earlier times, spending leisurely days swimming and fishing and reveling in its quiet sunrises, could have guessed its fate? The growing human population, however, was demanding new goods. The production of those goods required great amounts of water. So factories were set up along the banks of the immense lake—factories which would use the water as a coolant, and the lake itself as a dump. Thus millions of tons of poisons of all sorts were poured directly into the water. After all, this was the cheapest means of disposal. Cheap production increased profits. But the question is, did the industrialists have the *right* to pour their garbage into a lake which they didn't own? As their profits increased, the lake was changed to the extent that it became dangerous to swim in it or drink it. The industrialists, of course, when threatened with belated lawsuits in recent years, argued that their dumping of poisons into the lake resulted in cheaper goods for people, and *that*, after all, is what the people *really* wanted. You can immediately see that the question can be extended to apply to industries which vomit their poisonous fumes into the air. Whose air is it, after all?

   Because of new demands being placed on the earth's fresh water, it is very likely that you will come to change your feelings toward the uses of fresh water before long. Whereas you may have, somewhere around the second grade, learned that water is something people drink and plants need, you may soon come to think of it in terms of its mining and allocation. Already, the Supreme Court has determined that only agriculturalists in one area of California may *vote* regarding water use; other citizens have no voice in the matter.

Pollution in Lake Erie (left). The Cuyahoga River adds its polluted waters to Lake Erie at Cleveland. The darker areas outside the breakwater are less polluted. (Courtesy: Laurence Lowry from National Audubon Society) phates to the water, causing the bacteria to use increasing amounts of oxygen. Consequently, the species which require a lot of oxygen, such as trout, are now almost nonexistent, whereas those which thrive on the algae and plankton are reproducing rapidly.

THE OCEANS

Oceans cover about three-fourths of the earth's surface. In fact, if the earth's surface were smoothed out, it would be covered entirely by water. The average depth of the oceans is about 3 miles, but there are places where the water is seven miles deep.

It has been suggested that the oceans will one day provide us with most of our food. However, we presently take only 3 to 5 percent of our food from the seas, and even if we were to double that in the next decade, our food problem would not be resolved. You may be surprised, and disturbed, to learn that most of the deep ocean waters are, like the deserts, almost completely devoid of life.

The oceans are not simply huge, still bodies of water, broken only at the surface by waves. In fact, the deeper waters are continuously in motion as great silent currents hold sway beneath the surface. Actually, these currents are primarily the result of a surface phenomenon, the *trade winds*. The "trades," once so important to commercial sailors, are prevailing winds—winds which hold rather steadily from the same direction—which are caused by the difference in air temperature between the poles and the equator, coupled with the effects of the earth's rotation. These great ocean currents caused by these winds spin in wide circles, clockwise in the northern hemisphere, counterclockwise below the equator. One such ocean river is the famed Gulf Stream, which follows the Atlantic coast as far out as Bermuda and then swings eastward to northern Europe.

Where the currents are deflected upward by the mountains of the ocean floor, cold water from the ocean depths wells up to the warmer surface. This cold water carries with it the ages of accumulated sediment from the floor, sediment rich in nutrients which forms the basis for the bloom of life at the ocean's surface. The nutrients carried by upwellings are utilized by tiny chlorophyll-bearing organisms called *phytoplankton*, which in turn serve as a food source for *zooplankton*. These minute plants and animals, barely visible to the naked eye, form the basis of the ocean's food pyramid (Figure 12.12). One cubic foot of seawater may contain over 12 million phytoplankton. These food producers, however, can live only near the ocean's surface, since the sunlight necessary for them to carry on photosynthesis cannot penetrate below about 600 feet.

Figure 12.12 The food pyramid of the ocean. The tiny phytoplankton at the base capture the energy of the sun. These are eaten by animals larger than themselves, which are, in turn, eaten by larger animals. At the top are the largest carnivores of the sea. (There are no large herbivores in the open oceans as there is on land.) What is the position of man in the food chain of the sea? Note that each level is comprised of far fewer organisms than the one below it. The reason for this will be discussed later.

Figure 12.13 A deep-water fish. (Courtesy: Sea Library/Mick Church)

Not only are phytoplankton important as primary food producers in the oceans, but they also manufacture much of the earth's oxygen. It has been estimated that they produce well over half the oxygen we breathe. Thus the importance of maintaining viable oceans is apparent. Such knowledge can only be disturbing as we continue to see the world's oceans used as a dump. The sight of the oil companies' new supertankers, which will carry dangerous, partly refined petroleum across the fragile oceans, may also cause frowns of consternation on the faces of the informed. There is slim consolation in the knowledge that even among smaller and more maneuverable oil-bearing vessels there were an estimated 5,000 collisions in one recent ten-year period. But let's not get into all that now.

Men diving to the ocean's depths in bathyscaphs have reported that they can see light at depths as great as 2000 feet. The light that reaches these depths is pale blue, since the reds and oranges of the light spectrum have been filtered out by the water above. As a result, many of the fish which live there are reddish in color when they are viewed in the light at the surface. At their natural depths they appear dark and shadowy, since there is no red light for their pigment to reflect and their red color absorbs what blue light there is.

It was once believed that nothing could live below about 1,800 feet because of the tremendous pressure of the water at such depths. At about the time Darwin wrote on the origin of species, however, a submarine cable broke in the Mediterranean and was hauled up for repairs from a depth of about 6,000 feet. To the astonishment of biologists, it was found to be covered with all sorts of living things. So notions concerning the effects of pressure had to be revised. Now it is believed that living things can exist at any depth, as long as they are able to develop pressures inside their bodies equal to the pressure outside. Fish that have been brought up alive from depths below 2,000 feet have literally exploded when they reached the surface.

The deepest reaches of the ocean must be peculiar environments indeed. We are just beginning to learn something about them, and the incoming information is fascinating. It seems that the deep ocean bottoms, untouched by the currents above them, are still and unchanging places. So far we have discovered only few kinds of animals there, and most of these are weird creatures indeed (Figure 12.13). Plant life is represented only by a few kinds of fungi, species which do not need light.

The ocean bottom itself is an ooze composed of the chalky and glassy corpses of tiny plants and animals which have rained down steadily onto its surface throughout the ages. Animals which have been seen on the bottom often have long appendages to keep them from sinking into the primal ooze. Because of the utter stillness of the deepest ocean bottom, very delicate and fragile creatures can live there, but we are probably aware of only a small part of the life that exists in these abysmal regions. Some animals escape our attention by burrowing into the ooze at the bottom. Divers have seen many strange holes and burrows but no one has ever seen the animals themselves.

COASTAL AREAS

The land-ocean interface is called by romanticists and builders alike "the seashore." The romanticists would write new eulogies if they knew the full impact of the seashore in their lives. The builders might (and then again, might not) become a bit more cautious.

Life is much more abundant in coastal waters than in any other part of the ocean. The edges of the continents extend beyond the shore for anywhere from 10 to 150 miles as *continental shelves.* At these relatively shallow depths, usually less than 200 feet, sunlight easily penetrates the waters, and a wide variety of plants can grow there. The plants, of course, provide many species of animals with food, shelter, and surfaces for attachment. Coastal areas are generally composed primarily of either rock, sand, or mud. And each type of area has its own special qualities.

Rocky coasts are perhaps the most dramatic, and to coastal navigators the most awesome, with jagged, surf-bathed rock formations that only suggest the immense reefs below. These coastal areas boast a remarkable array of plant and animal life. Much of this life is distinctly zoned along the rocky surfaces. For example, a zone dominated by the periwinkle snail is clearly delineated from the barnacle zone, which is, in turn, strongly demarcated from the seaweed zone. The zones are not occupied exclusively by a single species; they are simply identified according to the most prevalent species occupying them. In each zone a number of species of plants and animals are intermixed with the predominant life form, thus enriching the life of the area.

Sandy seashores have fewer forms of life, which is perhaps a good thing for timorous bathers. Maybe you have felt some relief as you waded out into the surf and found that you didn't step on some mysterious and toothy bottom dweller. The reason they are so scarce is that wave action causes a constant shifting of the sandy bottom, depriving the less mobile species of a fixed surface for attachment. Any animal which lives in such shifting areas must be able to get around quickly, as crabs do, or be able to dig and with-

Figure 12.14 The rocky coast boasts a variety of life, as divers who have risked buffeting against the jagged surfaces by the irresistable surging seas have found. (Courtesy: U.S. Department of Interior, National Park Service Photo)

Figure 12.15 The sandy seashore provides good swimming and wading areas for humans because the shifting sands beneath the water's surfaces make it difficult for bottom dwellers to live there. (Courtesy: Gordon S. Smith from National Audubon Society)

stand burial, like clams. Animals which live in tidal areas and are not mobile probably encounter exceedingly severe conditions at low tide, when they are subject to the parching sun and to terrestrial foragers such as shorebirds.

From an aesthetic standpoint *mud flats* are probably the least appealing part of the seashore, and this perhaps explains the general lack of interest in protecting them. They aren't especially spectacular as scenery, and they are poor places to spread a towel. Also, they tend to smell peculiar—to say the least. Mud flats do not harbor as many life forms as rocky coasts, it is true, but they support much more life than do other kinds of coastal areas.

Mud flats are submerged at high tide and exposed at low tide. Bottom dwellers, snails, insects, shrimp, crabs, fish, and birds abound in these areas. Each square yard of the mud flat can support thousands of such individuals. Many sea animals begin their lives in mud flats, only later moving out to take their place in the mysterious pageant of the open ocean. Hence the life of the ocean itself is, in a very real sense, dependent on the preservation of the unsightly mud flats. Unfortunately, mud flats and salt-water marshes have attracted the eye of commercial developers, who have not overlooked the opportunities provided by the public's lack of interest in them. Up and down the coasts our bayous, marshes, and bays are being filled in for use as industrial sites, "waterfront" housing tracts, and high-rise apartment complexes and condominiums. San Francisco Bay alone is now only two-thirds of its original size, much of it having been replaced by long stretches of paved, chromed and neon-lit tourist areas.

In order to argue effectively—or in the absence of opportunities to act, to feel properly incensed—regarding the abuse of our environment we must know something about how species interact with their environment and with each other. Is any species able to affect the well-being of another? In what ways do species interact with their environment? Does the passage of any species from the earth affect us in any real way? Often there aren't specific answers to such questions, but there is enough information available for us to make some reasonably intelligent decisions about how we should regard the life around us. Some of the fundamental relationships between living things may be easier to visualize if we compare the behavior of species on islands and continents.

ISLAND LIFE

Because islands differ from continents in important ways, we have gained some important insights from observations of how species behave in the two types of environments.

Figure 12.16 Torrey Pines State Park, in Southern California, is the only place that these particular trees exist. The condominium development borders on the State Park. Note the mud flats. (Courtesy: Debra McQuiston)

Figure 12.17 Small, treeless islands do not support a great variety of life. Larger islands which have more altitudinal variation and plants of a variety of heights are likely to support a wide variety of animals. (Courtesy: Photophile)

Islands are, of course, much smaller than continents, but size is not the only geographic difference. To begin with, the continents are generally considered to be much older than their offshore islands. Also, islands are surrounded by water which acts as a natural barrier to the passage of living things to and from islands and thus impedes gene flow. The climate on islands may be less variable than that on continents because of the stabilizing effect of the surrounding water.

Continents generally boast a wide variety of habitat types. (Think of the range of habitat types available in the United States.) Islands, on the other hand, may only provide a few habitat types—partly as a function of their smaller size. Some islands are no more than small stretches of sand or rock, barely rising above sea level. Others, on the other hand, may be impressive bodies with tall mountains and lush valleys. So, you can see, it is difficult to generalize regarding just what sorts of habitats will be available on islands. However, in general, there are likely to be fewer habitat types available on islands than on their nearby continents.

The distance of an island from the nearest continent is another important factor. First of all, this will affect what kinds of species will reach the island, since water is a greater obstacle to some animals than to others. No one is surprised to find birds on islands. Birds are generally good travelers, so any animal reaching an island is likely to confront birds. That animal doesn't necessarily have to put up with snakes, however, since most reptiles are poor travelers. A walk through a Puerto Rican forest is enhanced by the sight of so many beautiful birds and the knowledge that no poisonous snakes are about.

Islands which lie close to a continent may be colonized by the poor travelers as well as the good ones, whereas the more distant islands would probably ordinarily be accessible only to the good travelers. Thus, fewer species normally reach the more distant islands. It should also be apparent that more species are likely to be found on older islands since the longer they have been there, the greater the probability of poorer travelers somehow managing to reach them.

We can expect that, generally, islands will have fewer species than their nearest mainland areas, since some species have particular problems in reaching islands. We can also see that there may be other reasons for the paucity of species on islands. For example, when we recall that islands may have fewer habitat types than mainlands, it is easy to see how an animal might reach an island, only to find that there was no suitable habitat. What would happen

Box 12.2 Grass on the North Slope One of the oil companies involved with developing the oil of Alaska's North Slope placed an ad in certain magazines not long ago. The ad was part of the oil companies' program to sell their image to the public rather than their product. In the ad they boasted that they were reseeding all the disrupted areas with a newly developed grass which grows five times faster than the grass which normally occurs there. I wrote them a letter pointing out that the faster-growing grass might easily displace the naturally occurring grasses through processes of competition. Also, I added, if the newer grass were to become dominant, would it be able to withstand the periodic aberrations in weather conditions that the natural grasses had survived for countless generations? If not, then segments of the ecosystem there, which are dependent upon grass, would be placed in a precarious position.

They sent me a brochure.

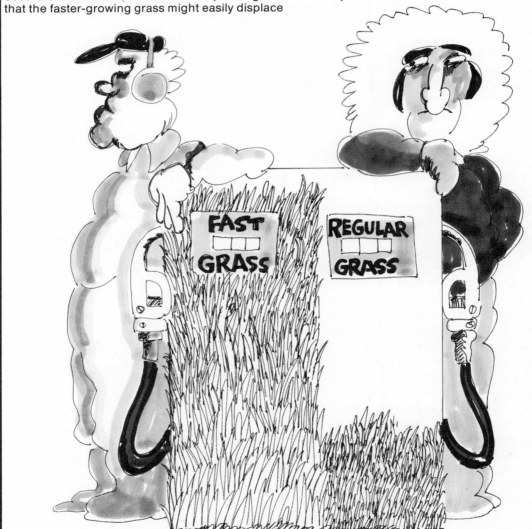

if a woodpecker landed on an island that had no trees?

If the island is low and flat, as smaller and more distant islands frequently are, the colonizer may have particular problems. As you know, the character of land changes with elevation. Hence an island that rises from sea level to lofty heights will provide more environmental variation, and hence more kinds of habitats, than a flat island with the same land area. As a result, islands with a range in elevation can be expected to support more kinds of species.

BIOTIC FACTORS

Even those island-colonizing species which are able to find suitable habitats are not likely to be as well adapted to the island habitat as they were to their mainland home. After all, the island will undoubtedly be different in some ways. A colonizing species, then, may survive rather tenuously for a short time (perhaps a few thousand years) until it becomes better adapted to the island situation.

And what about food? Any animal reaching an island is likely to confront a reduced food supply, whether it is a plant-eater or predator. The reasons for this are that probably few food species have been able to colonize the island. And those that have managed to become established may have changed during their long period of isolation so that their original predators are no longer particularly well suited for exploiting them.

Under the circumstances, then, any animal that reaches an island may have to either change its feeding habits or become extinct. As an illustration, suppose that a woodland bird species requires an acre of woodland per bird to sustain itself. When the birds reach an island, they may find that only a half-acre is available. However, if no grassland species has already colonized the island, the woodland birds may be able to enlarge their niche by feeding part of the time in the grassland. Even if they are not particularly well suited to this habitat, in the absence of grassland competitors they may be able to survive. And in time they may become well adapted to the new grassland-woodland habitat.

If a continental grassland species should appear on the island before the woodland birds become adapted, however, the latter may not be able to compete successfully with the newcomer. If it is forced from the grassland, it may become extinct. However, if the grassland species should arrive after the woodland species has adapted to grassland, the newcomer itself may be unsuccessful in colonizing. So you can see that on islands, as in romance, there appears to be an advantage in getting there first. As you can also see, competition is another reason for the paucity of species on islands.

It may seem that we are saying that a colonizing species moves into "unoccupied niches" on an island or that it adds other "niche units" to its own. However, niches are not preexisting ecological slots which can be filled or vacated; a species might more accurately be said to have *broadened* its niche if, on an island, it begins to interact with its environment in more ways than it had previously on the mainland.

The evolutionist Ernst Mayr, as a result of his long studies of island species, has concluded that the most successful island colonizers among birds are generally those species which eat plants (often more abundant and accessible than other food), are social (solitary species are most easily stopped by geographic barriers), are good travelers, and are exploratory and behaviorally flexible. Flexibility is important, since any island-colonizing species is likely to be confronted by novel conditions.

LESSONS FROM WOODPECKERS

Woodpecker watchers in the United States might notice that the red-bellied woodpecker is highly successful at getting food from the branches of trees, but that this species hardly ever forages among the leaves on the forest floor. In contrast, another species of woodpecker which occupies the same forests, the red-shafted flicker, is a proficient ground forager. If the red-bellied woodpecker were to attempt to forage in the leaf litter in those areas where flickers, thrushes, and other ground specialists live, it would probably end up with less food than if it continued to search among the branches.

Now, in each generation there are undoubtedly some red-bellied woodpeckers which show a tendency toward ground foraging. However, the number of eggs laid and the number of young reared depend partly on how much food the parents can find. So those ground-foraging red-bellied woodpeckers are likely to rear fewer young than their tree-searching brothers. Thus the tendency toward ground searching is continually selected against as long as the ground specialists are around.

The red-bellied woodpecker also doesn't spend much time trying to catch insects on the wing. The woods are full of flycatchers and phoebes, species which are quite proficient at catching food in this way, and if the woodpecker were to try to compete with them, it would probably come up short again. The red-bellied woodpeckers of any generation, then, are primarily the offspring of those woodpeckers which had a tendency, possibly reinforced individually through success, to search the bark and limbs of trees for food.

On Jamaica there is only one species of woodpecker which lives there year round. This may be because not many species of woodpeckers have reached the island in sufficient numbers to become established. And those that reach it now may be excluded by a well-established woodpecker—one which now occupies an expanded niche because it didn't have to compete with similar species during its period of establishment.

We might ask why natural selection would push the Jamaican woodpecker into new adaptive zones. Why didn't the woodpecker that originally colonized Jamaica simply become better at feeding in the old way? Why branch out (no pun)? Remember that because of the problems of insect colonization the Jamaican woodpecker probably has only a few species of insects available to it. It is true that each insect species may be well represented, but populations tend to fluctuate. Many species undergo cyclical rises and falls in number, especially in areas with seasonal variations in climate. For example, on Jamaica the winter months are dry and somewhat harsh. This means that the numbers of some insect prey are likely to be lower at this time.

If the Jamaican woodpecker restricted its foraging habits to only a few bark-dwelling species, the seasonal reductions of any of these prey species would have a marked effect on the foraging success of the birds. However, if the woodpeckers expand their niche to include more flying insects, ground-dwelling insects, and plants in their regular diet, their food supply is likely to be more stable, since probably not all of these diverse species will become scarce at the same time. In other words, if a woodpecker becomes a foraging "generalist" (despecialized), then as one type of prey decreases it can switch to another, so that it can rely primarily on different foods at different times of the year. Thus it is to the woodpecker's advantage to expand its niche, and niche expansion is possible because of the absence of close competitors.

These, then, are a few of the theoretical advantages of niche expansion for the Jamaican woodpecker. But what sort of pressures could cause the woodpecker to make these changes? After all, birds aren't very bright and aren't likely to think, "I really should see about expanding my niche, since that seems to be a good move for us island colonizers." One probable impetus relates to competition. Let's see if we can put it together.

COMPETITION

As a species invades an island, its numbers are likely to be low at first. They may remain low as the population adapts to the specific ecological situation on that island. Colonizing animals are not likely to be prepared to exploit most efficiently what the island has to offer. Their low number, however, does not mean that there is little competition between them. The level of competition would be determined by the abundance of the resources. If members of a group of island colonizers all tended to forage in the old ways, they might rapidly fall into competition for the same scarce food.

In each new generation, however, there will be those individuals which are predisposed to forage in ways a bit different from their parents. These birds would have been selected against on the species-packed mainland, but out here on the island they may be able to utilize food sources which had been previously largely ignored. They may even be able to find more food than their brothers who forage in the conventional way. Thus, while competition for food keeps the numbers of the strongest competitors low, the aberrant foragers may be at an advantage. By finding more food, they are able to produce more eggs and to feed more young. If the aberrant foraging pattern were to any degree genetically determined, some of the offspring of these birds might tend to forage in the new way.

Of course, as their numbers grew, there would also be increasing competition over the new foraging area, so advantages might continue to be bestowed on the more variant individuals among *their* offspring. Thus the niche of the species as a whole would continually broaden. As any other commodity, such as nest sites, fell into short supply there might be the same sort of expansion into new ecological areas. In time, then, any animal which colonizes an island may come to show a much broader range of behavior than it arrived with.

Now let us consider other possibilities which may be open to island colonizers. It probably isn't particularly important that you know about what's happening on islands as such, but by considering the possible events involved in island colonization, perhaps we can unveil certain important lessons of ecological interaction.

ADAPTIVE RADIATION

The Galápagos Islands, as you may recall, are an archipelago about 580 miles off the coast of Ecuador which is inhabited by a wide assortment of animals. The finches Darwin observed on his historic visit are perhaps not so spectacular as the large lizards which clamber from the surf-pounded rocks to graze on seaweed beneath the surging seas. However, Darwin's keen eye noticed something even more remarkable about these thirteen species of small birds. They are all 4 to 8 inches long, and both sexes are drab-colored browns and grays. There are six species of ground finches on the islands. Of the four species found on most of the islands, three live on the ground and take seeds of different sizes, as evidenced by their different-sized bills (Figure 12.18). The fourth species has a long bill and eats a lot of prickly pears.

There are also six species of tree finches in the Galápagos. One of these has a parrotlike beak and eats seeds and fruit. Four have bills well suited for eating insects, each species specializing on insects of certain types. But the strangest of all is the woodpecker finch. This little bird has a bill like a woodpecker, but it lacks the long protrusible tongue of the woodpecker with which to probe deep into crevices. So instead it pries grubs out of holes by using a long twig or cactus spine.

The thirteenth species of finch has diverged markedly from the other finches. In fact, by external appearance and behavior it should not be considered a finch at all, but a warbler. Only the examination of its internal structure unmasks it as an aberrant finch.

SPECIATION

So why have these little birds become so significant in the biological

Figure 12.18 Three species of "Darwin's finches" from Indefatigable Island. *Geospiza fortis* (top left) and *Geospiza fuliginosa* (bottom) are ground feeders but their bill-size differences indicate they take seeds of different sizes. *Geospiza scanders* (top right) has a long pointed bill and lives on the fruit and blooms of cactus. (After Eibl-Eibesfeldt 1970, *Ethology, the biology of behavior*, Holt, Rinehart and Winston, New York)

Figure 12.19 The woodpecker finch, *Camarhynchus pallidus*, is a tool user. It uses twigs of cactus spines as probes and levers to extract prey from holes or crevices. (After Eibl-Eibesfeldt 1970, *Ethology, the biology of behavior*, Holt, Rinehart and Winston, New York)

world? Their fame is due to Darwin's suspicion, supported by subsequent years of careful research by others, that the birds are all descended from the same stock. The prevailing idea is that long ago the islands were colonized by a species of ground-dwelling finch from the mainland. The colonizers were probably blown out to sea by some weather turbulence, and in time they became established on one or more of the islands. As their numbers increased, they could easily have spread throughout the islands. However, the rate of gene exchange among the birds of the various islands was low enough to ensure virtual isolation of each population. In time, then, as each small population was subjected to the specific and rigorous conditions of its particular island, changes would have appeared in their genetic constitution through the process of natural selection. These changes, of course, would have been reflected in the appearance and behavior of the species.

Box 12.4 Gaining an Island Toehold Ten thousand years is a rather brief period of time in geological history. In some cases, such as on islands, speciation might occur at much faster rates. The time element may be briefer on islands for two reasons: (1) the colonizers may show higher variability than the parent population; and (2) conditions may be stringent in the new habitat. The colonizers will, of course, not contain all the variations in the parent population, but because of their low numbers, any variations which they do carry will not be "swamped" by the genes from others, as they would in a large population. Thus differences may arise quickly in the progeny of small colonizing populations. Since colonizers are not likely to be well-adapted to their island, there may be strong selection operating on the variation so that changes in the population may take place rather rapidly.

It should be obvious, then, that the odds are against successful speciation, especially on islands. A small population attempting to colonize an area with new, and therefore somewhat inhospitable, environmental conditions is almost sure to meet with disaster. Perhaps many species of birds were blown out to the Galapagos by offshore winds, and perhaps most failed. But, finally, one didn't.

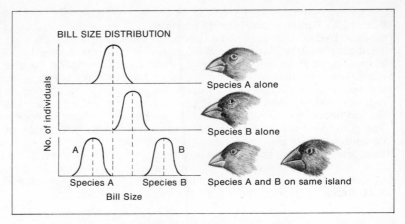

Figure 12.20 Character displacement in bill size in two hypothetical island species of finches which have come to share an island. The shift in bill sizes where the species come in contact is believed to be a means of reducing competition for food.

Under such conditions the population on each island would ultimately have become so different that they could no longer interbreed. The result would be *speciation*, or the formation of new species. It has been roughly calculated that each population would have to have been isolated for about 10,000 years in order to form a new species. We don't know how precise the time measurements are, but we do know that nature is in no hurry. It wouldn't matter whether the colonizers became established or not. The fact that they did was merely a fortunate interaction of characteristics between the colonizers and their new habitat.

Some of the differences between species may become more pronounced if the species occupy the same habitat. This phenomenon, called *character displacement*, may occur in response to high levels of competition between closely related species.

It is generally believed that speciation normally occurs only when populations are genetically isolated. In other words, the likelihood that the same stock will give rise to two species is greater when that stock is split into two separate populations which are prevented from interbreeding. A very few individuals could probably be exchanged between populations without interfering too much with the accumulation of genetic differences. If the exchange of genes between separate populations is too high, however, any modifications which might have appeared in either population would be mitigated or "swamped" by the genes from the other population, thus reducing the level of differences between them.

SPECIATION ON CONTINENTS
Of course, genetic isolation can occur on continents as well as islands. Otherwise this discussion wouldn't maintain whatever level of biological interest it has managed to muster. The Grand

Canyon, for example, has divided a population of Arizona squirrels into two subgroups. The south side of the canyon is inhabited by the Abert squirrel, while the Kaibab Plateau on the north side is inhabited by the Kaibab squirrel (Figure 12.21). In the period in which they have been separated, differences between them have accumulated to the point that they are now considered separate species.

From a biological standpoint, insular conditions may exist on continents. Each pond, for example, is isolated from other ponds to some degree. The plants and animals which inhabit ponds may or may not be good travelers, so their progeny may or may not be likely to reach other ponds. Hence if a pond remains isolated for a long period of time and doesn't dry up, its endemic species may eventually generate new life forms, perhaps not particularly different in appearance from other forms, but nonetheless unable to interbreed with them. Such genetic isolation may be found in species in lakes or mountaintops or caves or any other area surrounded by conditions markedly different from its own.

OCEANS AS ISLANDS

In a sense the Atlantic and Pacific Oceans are "islands," isolated from each other by the Americas, and each has developed its own life forms. For example, the deadly sea snake ranges throughout much of the Pacific and is common along Central American coasts, but it is absent in the Atlantic. However, a sea-level canal is now being considered to replace the lock system of the Panama Canal. The new canal would allow species which have been separated for eons to suddenly come together. The sea snake, of course, would have access to the Atlantic. Imagine the surprise of bathers! The snake would also pose other problems. Experiments have shown that Atlantic fish species do not recognize the banded sea snake as dangerous, as do the Pacific species which evolved with it. As a result, the Atlantic species try to eat the snake, and of course, these attempts are fatal.

Think of the ecological chaos in terms of predation and competition which will result from the mixing of thousands of species which are now complete strangers to each other. However, there are those with political and financial clout who argue that a sea-level canal is commercially cheaper than one which utilizes a lock system.

THE WEB OF LIFE

We have already considered a number of factors that might help to explain the high extinction rate of species on islands. Another

Figure 12.21 The Kaibab squirrel (top) and the Abert squirrel (bottom) are two species which have arisen from subpopulations created by the formation of the Grand Canyon. (Courtesy: John J. Staphlet, Sonja Bullaty from National Audubon Society)

Figure 12.22 A banded sea snake. (Courtesy: San Diego Zoo)

Figure 12.23 A hypothetical food web. Can you tell which organisms serve as food sources to the various species? This diagram is greatly simplified—for example, there are also omnivorous birds, such as crows, and mice may sometimes eat bird food. The predatory fox may also have to share its food with lynx and wolves.

reason for this phenomenon is simplicity itself. Because islands have fewer species than nearby continents, the island ecosystem is theoretically "simpler." Ecologically, simplicity means instability, and in an unstable ecosystem, extinction rates will be higher. Let's examine this point more closely, since it has important implications for our own species. Again we must deal in speculation to some degree because much of the essential data just aren't in.

Figure 12.23 shows a hypothetical and greatly simplified food web. However, these food relationships illustrate a point. Suppose the only carnivorous (meat-eating) mammal in this food web were the fox. Then, if some fox disease were to sweep the area, what would be the repercussions on the rest of the life there? We can see how the numbers of herbivores, such as rabbits, squirrels, mice, and some birds, might rise. These in turn might put new pressures on their plant food, and in so doing destroy the habitat of many other species while depleting their own food supply. The resulting initial increase in the numbers of herbivores would provide more food for owls and hawks, which could in turn be expected to increase their own numbers. The resulting abundance of such predators might then reduce the numbers of other animals, such as insect-eating birds and toads. What might then happen to the numbers of snakes? And the numbers of insects?

Now then, suppose the foxes had to share their food with lynx and wolves, as well as with hawks and owls. In this case a rampant fox disease would have markedly less effect on the ecosystem. Of course, elimination of the foxes would result in some changes in the system, since the fox has a somewhat different niche from the wolf or the hawk. But perhaps the system would continue largely as before until the fox population was restored. The point here is that a simpler system—for example, one with a single "top" predator—is inherently easier to upset.

Let's draw on a more abstract example to show the effect of the extinction of a single species in an interdependent system. Figure 12.24 shows five species which depend on each other. Dependency may be of a variety of types. For example, fleas depend on dogs for living quarters and transportation, as well as for food. Some starlings depend on old woodpecker holes for nests. Hawks depend on mice and starlings and woodpeckers for food, but not much on fleas.

Now, suppose in this system C somehow becomes extinct. Since A, E, and D depend on C, we might reasonably expect their numbers to be reduced. B, however, depends on A, E, and D. Even a slight reduction in the numbers of all three might have such an impact on B, that it might also follow C into oblivion. This would

Box 12.5 The Redwoods   It stood 390 feet tall. Nothing on earth could match it. It had stood as a slender sapling in the cool coastal air, perhaps moving slightly with the light breezes, on the very day Caesar finally decided to move against Britain. But all that happened a long way from the area which would be named California. The tree lived through it all. And some of it must have been pretty strange, but not as strange, perhaps, as what went· on in its last few years.

As the great ages passed, the sapling continued to grow. No one marked the time when its crown reached above all the others because it was only one of a vast forest of such trees which ran from north of San Francisco deep into Oregon. The straight, tall and insect-resistant trees naturally caught the eyes of lumbermen, even before the turn of this century. Then, as the population and industry of California grew, such demand was placed on the trees that some citizens tried to establish a national park to save some of the trees. But the lumbermen blocked it. In what has been called "one of the great swindles of all time," they arranged to change nearly all the redwood lands from public to private ownership.

After World War II, California experienced a population surge of unprecedented dimensions. Factories and homes were being built feverishly and lumber was needed. Redwood was ideal for lawn furniture too. The conservationists, led by the Sierra Club, pushed again for parks to be set aside. By 1960 there were two areas left which were relatively unscarred and of park caliber. While conservationists dickered among themselves over which area was the best, the lumbermen were busy in Sacramento and Washington. They also effectively muddied the waters by cynically proposing a number of other sites. The easily confused public became confused. The lumbermen were confident. One boasted that it takes fives years to get a national park bill through Congress and in five years there wouldn't be anything worth fighting for. As the arguments continued, the trees kept falling.

Then, in the summer of 1965 the 390 foot giant was discovered—by the lumber companies. No announcement was made to the public but the raucous chain saws and great log movers were quickly brought in. Without fanfare, the great tree was felled, cut into twenty-foot lengths and hauled away. In fact, every tree in the redwood stand at the junction of Bond and Redwood Creeks was brought down—right in the very heart of the Sierra Club's proposed park site. The crash of timber was drowned out only by the cheerful ring of the cash register. Trees were still falling when Hollywood actor Ronald Reagan, candidate for governor, spoke before a convention of the Western Wood Products Association. His words and wisdom were memorable: "A tree's a tree. How many do you need to look at? See one, you've seen 'em all."

It had been the tallest living thing in the world.

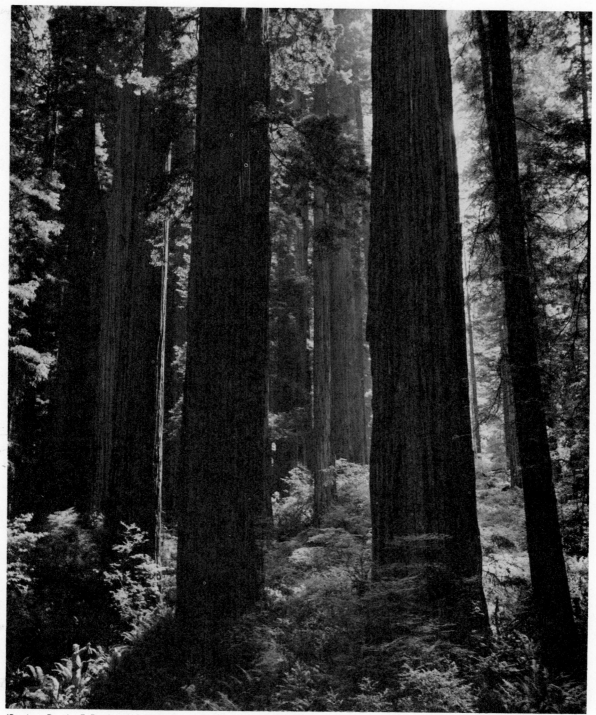

(Courtesy: Douglas F. Roy from U.S. Department of Agriculture, Forest Service)

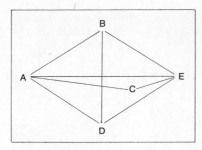

Figure 12.24 Interdependency of five theoretical species.

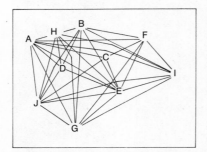

Figure 12-25 Interdependency of ten theoretical species.

mean that the animals on which A, E, and D were dependent would be reduced by *half*. How might this affect their fate? If A is more susceptible to the loss than E or D, its numbers might then drop quickly. What effect might this have on the two remaining species?

Now, suppose that instead of a five-species system, we have a ten-species system, as in Figure 12.25. Here the system is much more complex, although it is still far simpler than any actual biotic system would be. It is easy to see, however, that the loss of any single component might be more easily absorbed in this system than in the simpler one. Thus, the more complex a system is, the more inherent stability it has. One reason islands have fewer species than continents, then, is that they have a higher extinction rate, as components of their relatively simple and interdependent eco-systems drop out.

EXTINCTION AND US

Are there any lessons here for us? Should we care if over 150 species of animals have disappeared from the earth in the last fifty years? Should we be concerned that there are literally thousands of species whose very existence is presently endangered—largely because of our activities? Extinction, after all, is the natural end of populations. Species are born, and they mature, and they die. Some live a long time, perhaps as long as 2 million years; some die more quickly. We have hastened the extinction of many species we know about, and we have undoubtedly contributed to the demise of many more. In fact, there are undoubtedly many unobtrusive species which have lived among us during our time on earth—but which have disappeared as a result of our activities without our ever having known they even existed.

It is hard to explain the rationale of many of us who are concerned about such matters. I have never seen a sei whale, yet I don't want them to become extinct. Moreover, I felt this way long before I understood anything about how they might be an important part of an ecosystem. Possibly such feelings merely reflect the cultural attitude that it is "nice" to wish other living things well; thus the attitude is rewarded. I feel nice.

There are, of course, more cogent reasons for abhoring the extermination of any species. For one thing, the kind of attitude that encourages or sanctions the destruction of other species constitutes an intrinsic threat to our own well-being. This may be similar to the sort of reasoning which causes us to seek to weed out violence from our society—the fear that we, ourselves, might one day fall victim to it. The extinction of other species could also threaten us indirectly by simplifying the system of which we are a part or, more

Figure 12.26 Irresponsible whale hunting continues, endangering the existence of these magnificent mammals. This sperm whale has been killed and is on the way to the slaughter house. (Courtesy: The Sea Library/Peter Lake)

directly, by destroying parts of the ecosystem upon which we directly rely. For example, if we continue to poison the oceans because we are willing to believe only a few bottom dwellers are directly affected, we might eventually overstep some critical threshold and trigger the wholesale death of plankton, thus finding ourselves without a major source of the world's food and with our oxygen supplies dwindling.

This is all very rational, but none of this precipitated my gut response when I heard the Japanese were going to continue to hunt the whales. I simply found it sad. For some reason, I wanted those whales to continue to share the planet. If you had the same reaction, and you don't ever expect to see or eat a whale or use its oil, you might try to analyze the roots of your own response.

13
Populations

"At last! One nation, indivisible."

(Courtesy: Koren; © 1970 The New Yorker Magazine, Inc.)

The tuntong is almost extinct! Almost no one has noticed and even fewer care. Upon hearing the news, a sensitive and intelligent person might say, "That's too bad," and then, "What the heck is a tuntong?" It is true enough that we can't be informed on the status of every species on the earth. Most of us, in fact, know of only a few of the hundreds of species on the "endangered" list. (How many do you know of offhand?) It probably isn't necessary for anyone to have such information at his fingertips, even if it were available. However, at this critical time in our history we should know certain things about how numbers of any species behave, how they change, and what causes them to flourish or dwindle.

There are a number of reasons this is important. If you have ever stood alone amid the browns, yellows, and reds of a thicketed field on a cold autumn afternoon and watched a flock of geese beating its way southward against a gray and heavy sky, the presence of those geese will be important to you. You may not be able to persuade a land developer that their breeding grounds are important on the basis of the strange stirrings which swelled within you that day, but you will have placed a new value on a remote species, one with which you have shared an important moment in your life.

The numbers of living things are also of concern to us for reasons other than aesthetic, reasons which may even appeal to the crassest profiteer. As we have seen, the more species there are, the greater the stability of the biological system they comprise. In addition, we may be dependent on other species, both directly and indirectly, for our own food, oxygen and shelter. Finally, learning more about the population dynamics of other species may give us some insight into the status of our own. Providing information which might lead to that insight is basically the goal of this chapter.

POPULATION CHANGES
First, we should remind ourselves of the obvious: the numbers of living things change. Some of these changes are short-term or

Figure 13.1 The numbers of some species, such as these gannets, seem overwhelming. But whereas some species seem to maintain their populations size, some sporadically swell and fall in number. Yet others prove to be highly vulnerable and subject to rapid extinction, especially when they compete with us or take some commercial value. (Courtesy: Eric Hosking from National Audubon Society)

periodic and easily explainable. Grasshoppers, plentiful in summer, disappear in winter. Some birds disappear from American forests in the winter as they move to more hospitable areas. These are familiar and rather predictable phenomena. We all expect grasshoppers and birds to be back in spring—except, perhaps, for those permanently lost as a result of our use of DDT and other insecticides and our destruction of their habitats.

Apart from these short-term fluctuations, you may have noticed that the numbers of many species don't seem to vary much from year to year. For example, the number of birds around your house has probably been about the same from one summer to the next. However, if you think back, you may also notice that some species you remember as having been fairly common just aren't around anymore. Roger Tory Peterson, the noted ornithologist, has reported a decline in many species of songbirds around his home in Connecticut over the past several years. Have they simply moved to another habitat? Or are these species quietly slipping into oblivion, into a very final extinction?

And what about those sudden increases in numbers of certain animals? Where do all the locusts which suddenly appear to ravage African crops come from? Certain populations seem to blossom for some mysterious reason, and we hear of "invasions" or "plagues." Yet, in the face of an environment which allows such changes, most species seem to maintain rather stable numbers. How? Let's begin by considering how populations normally behave.

POPULATION GROWTH

If we were to place a few bacteria in a suitable medium, or a few sheep on an uninhabited but ecologically hospitable island, the growth of their populations might be expected to show similar patterns, at least at first. Since we are assuming that their new environment provides them with the necessities of life, their growth will be limited primarily by the rate at which they can reproduce, rather than by environmental factors. Figure 13.2 shows the rate of increase for a theoretical population.

Note that the population size becomes greater with time because increasing numbers of individuals produce offspring. If each individual in the original group leaves two offspring, the population will remain fairly small for several generations (one bacteria divides to leave two, these leave four, which leave eight—and these are all low numbers). But once the population increases a bit, the numbers begin to skyrocket. (What would be the result after only ten generations? Twenty generations?) The reproductive capacity of any population, when it is unrestricted, is called its *biotic potential.*

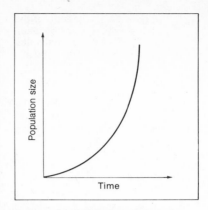

Figure 13.2 Theoretical growth. Time is plotted against population size. In such a case there is nothing operating to restrict growth, so the population is achieving its reproductive (or "biotic") potential.

CARRYING CAPACITY

The full biotic potential of any species is not likely to be reached because of the inherent limitation imposed by the environment. *Environmental resistance* is encountered when some essential factor or factors in the environment come into short supply and thus reduce the population's rate of increase. The limit of biotic potential set by environmental resistance defines the *carrying capacity* of the environment. As a population approaches the carrying capacity of its environment, then, its numbers begin to level off (Figure 13.3).

In some cases a population may overshoot the carrying capacity of its environment. When this happens its growth rate may fall off abruptly only to pick up again when the numbers fall below the carrying capacity. In more stable populations the numbers may oscillate rather closely around the carrying capacity (Figure 13.4).

If a population drastically overshoots the environment's carrying capacity, the result may be a severe depletion of numbers called a *crash* (Figure 13.5). In some species such crashes are a normal part of the population cycle. The numbers of annual plants and of many insects crash each autumn, for example, but their progeny appear each spring to reproduce madly in the few months they have available.

Some crashes are more permanent. A bacterial population may rise rapidly, but if the population lives in a petri dish on a laboratory shelf, it must eventually run out of food even as it

Figure 13.3 Diagram of the theoretical relationships between biotic potential, environmental resistance and carrying capacity.

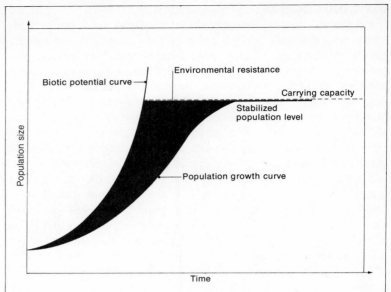

Figure 13.4 A theoretical population reaching the carrying capacity of the environment and then oscillating at that level.

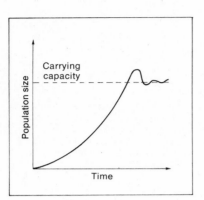

Figure 13.5 Diagram of a theoretical population which has drastically overshot the environment's carrying capacity, and then crashed.

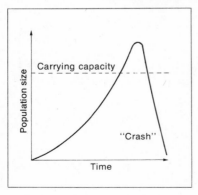

befouls its environment with its own waste products. After it crashes, it cannot recover. As another example, if environmental resistance in the form of predators (including hunters), disease, or food and water limitations is somehow reduced for a group of animals, such as elephants, their numbers may increase to the point that they eventually devastate their habitat. Oversized herds of elephants may eat all the young plants in their habitat which, if left alone, would eventually have provided much more food—and would have continued to do so for generations afterward. The result of such devastation is a lowering of the carrying capacity of the habitat for long periods afterward. In fact, it is possible to permanently lower the habitat's carrying capacity. If stripping an area of its foliage should increase erosion, the habitat may *never* recover. Fortunately, humans are a lot smarter than elephants.

REPRODUCTION RATE

There is an argument among biologists regarding whether mechanisms are operating which maintain population numbers within certain limits—mechanisms which only occasionally break down. It seems, however, that the relative stability of so many species would indicate that some sort of stabilizing influence is operating. Any such influence, if it exists, could theoretically operate at two levels by affecting: (1) how many offspring are produced, and (2) the longevity or mortality of organisms in the population.

HOW MANY OFFSPRING ARE PRODUCED?

The reproductive potential of most species, even the slowest breeding ones, is incredibly high. Charles Darwin estimated that a single pair of elephants could leave over 19,000,000 descendants in only 750 years. The fact that the entire world was not teeming with elephants indicated to him that some elephants were not reproducing, and that the reproducers were somehow "selected" by the environment.

Whereas elephants are very slow breeders, other species are more prolific. For example, the reproductive potential, for a *single year*, of the housefly is shown in Table 13.1.

Table 13.1 Projected populations of the housefly, *Musca domestica*, for one year, a period which encompasses about seven generations. Do all these offspring survive? The numbers are based on each female laying 120 eggs per generation and half of these developing into females.

Table 13.1—Reproductive Potential of the Housefly

Generation	If All Survive But 1 Generation	If All Survive 1 Year But Produce Only Once	If All Survive 1 Year and All Females Produce Each Generation
1	120	120	120
2	7,200	7,320	7,320
3	432,000	439,320	446,520
4	25,920,000	26,359,320	27,237,720
5	1,555,200,000	1,581,559,320	1,661,500,920
6	93,312,000,000	94,893,559,320	101,351,520,120
7	5,598,720,000,000	5,693,613,559,320	6,182,442,727,320

SOURCE: Adapted *from* E. V. Kormondy, *Concepts of Ecology* (New Jersey: Prentice-Hall, Inc., 1969), p. 63.

But what determines the rate at which an organism reproduces? It should be apparent that simple laws of energy would not allow an elephant to become pregnant several times a year. There is just no way that a female could find enough food to produce that many offspring. And what would happen to the helpless baby elephants already born as new brothers and sisters appeared on the scene?

It may seem puzzling, then, to learn that each animal theoretically reproduces to its biological limit. Can it be true that animals leave as many offspring as possible? You might intuitively agree when you recall that any generation is primarily made up of the offspring of the most successful reproducers. The next question we might ask regards method. How does an organism maximize its reproductive success?

TAPEWORMS

The tapeworm living so comfortably in the idyllic environs of your

Figure 13.6 Life cycle of a tapeworm.

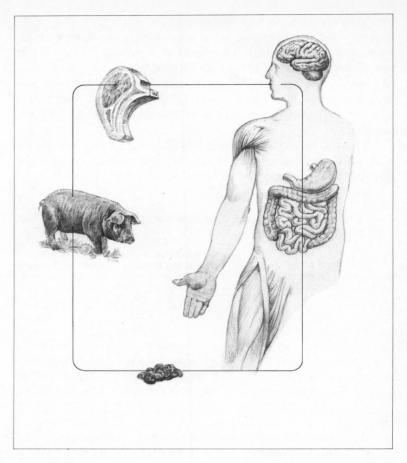

intestine is lucky to be there—although its luck is not yours. It is lucky because the life of a tapeworm is fraught with risk. The odds are against its finding a host, you. In order to reach a human host, the egg must pass out with the feces of the previous human host. Then it lies around in the sun and rain until it is eaten by a pig. Not only must it survive the elements, but it must also be found by a pig, so already the odds are against it. Once the egg has been eaten by the pig, it changes its appearance, moves through the intestinal wall of its host, and on through the bloodstream until it comes to lie in the pig's muscle. Here it forms a protective capsule around itself and waits. It waits for a human to eat the pig's flesh without cooking it too well. When this happens, the worm attaches to the intestine of its human host, grows into an adult, and begins to lay eggs.

There is obviously a strong probability that not all these conditions will be met, and thus most eggs will not develop into adult tapeworms. So what is the tapeworm's answer to such a demanding life cycle? It has become capable of self-fertilization (perhaps at the risk of going blind) and it lays thousands and thousands of eggs. It's whole life is devoted to egg production. It exists only to lay eggs, eggs, EGGS! A few of these, hopefully, will wend their way, by chance, through the complex maze which is the tapeworm's life cycle.

CHIMPANZEES

So that's the tapeworm's response to its particular situation. But what about other animals? Are they *all* in the numbers game? Let's consider the chimpanzee, an animal which has developed a different evolutionary strategy. While the female chimpanzee is sexually receptive, she may copulate with almost any male who signals his desire. Once she has become pregnant, she may not become sexually receptive again for years. Flo, the aging but sexy female studied by Jane van Lawick-Goodall, after delivering a baby, did not become sexually receptive again for five years. During this time, however, she attended carefully to her baby.

Figure 13.7 Chimpanzees move from place to place to find food and thus encounter many threats from other animals and the environment. (Courtesy: F. D. Schmidt from San Diego Zoo)

In the first months Flo carried her baby everywhere and diligently guarded it against danger. Later it was permitted brief forays on its own, but it was not allowed to stray far from her vigilant eye. During this time the baby would scurry back to her at any real or imagined sign of danger. As the young chimpanzee gradually became able to care for itself during those first few years, the association between mother and offspring lessened, until finally Flo was free to mate again and rear another baby. Among chimpanzees, the female does not maximize her reproductive success by giving birth to large numbers of offspring. Instead, few offspring are produced and these receive very careful attention until they become independent. Why has such a strategy evolved?

Chimpanzees have historically existed in a precarious and complex world. For centuries before humans decimated the African big cats in the name of fashion the chimpanzees were continually hunted by these vigorous and intelligent predators. Their lives were complicated by other factors as well. Chimpanzees feed on a number of kinds of food, including seasonal fruits, baby baboons, and small monkeys, as well as ants, which they pick up by using tools made of broken twigs. So their diets are varied and complex. Also, they must move from place to place to exploit new food sources. Each night they must build a new nest usually of bent branches or twigs and leaves. Thus they are constantly dealing with new and sometimes threatening situations.

To cope with such an environment the chimpanzee needs not only high intelligence, but a system of extended parental care to give the young chimpanzee time to learn about its complex world. Extended parental care, however, means that the parents can successfully rear only a limited number of young. So, in this species, selection has moved the animal not toward maximum production of infants, but toward efficient care of those which are produced. In other words, chimpanzees rear as many young as possible according to the demands of their particular kind of environment. So, we see that the number of offspring attempted by members of any species is likely to be set by natural selection and that the "strategy" of natural selection will depend upon a host of historical and environmental variables.

BIRDS

A quick look at some data from birds provides a clearer picture of how the dictates of a particular environment have influenced the number of offspring attempted by a given species. Again, for maximum reproductive success birds lay not as many eggs as possible, but the number of eggs that will enable them to rear as many young as possible. In most cases this is the number of young the parents can feed successfully.

Ornithologist David Lack found that females of the common swift, *Apus apus*, normally lay one to three eggs. Apparently, the number has not yet stabilized or is changing because of new ecological pressures. He found that in years when food was abundant the number of young which reached feathering age was 1.9 for nests with two eggs, and 2.3 for nests with three eggs. In the nests in which Lack placed extra eggs to bring the total to four, only 1.4 young survived until feathering age. If the food supply were to remain relatively constant, one might expect the number of eggs laid by females of this population to stabilize. Where?

The number of offspring attempted by a species may also reflect the vulnerability of the young to predators. For example, certain species of gulls nest on accessible beaches, where there is a high likelihood of the young being found by prowling foxes and other nest marauders. These gulls often lay three eggs, although they can usually feed only two young successfully. However, at least one of the young will probably be eaten, so apparently the three-egg nest is an adaptation to high predation. The kittiwake, a species of gull which nests on cliffsides, normally lays only two eggs, since the young are safe from predators and both offspring are likely to reach adulthood.

Figure 13.8 The Kittiwake gull nests on sheer cliffs which are inaccessible to predators. Other gulls nest on the ground. (Courtesy: Allan D. Cruikshank and Leonard Lee Rue III from National Audubon Society)

There are other factors that might have made small broods reproductively advantageous even when the parents could successfully feed larger ones. Conspicuous eggs, the activity of young, and the activity of parents tending the nest can all serve to attract predators, so fewer eggs means less activity around the nest.

Time may also be a factor in determining brood size. The same quantity of food can theoretically raise a small brood quickly or a large brood slowly. The rapid development of young is important in areas where the nesting season is short—such as the polar regions. In such areas there would be greater advantage in raising a small brood quickly than in portioning the same food out among a larger brood that might not mature fast enough to escape the winter.

One other fact which bird studies have helped to substantiate is the relationship between food availability and the production of offspring. It has been shown, for example, that when food is abundant, birds lay more eggs. In some cases the increase in egg production is accomplished by multiple nestings. A pair of Jamaican woodpeckers reared three successive broods when they were fed artificially. Conversely, it appears that when food is scarce birds lay fewer eggs. The mechanism here is obvious. Food is needed by the female in order to manufacture eggs, and with less food she simply forms fewer eggs. A reduction in the number of eggs may actually mean that more young will be reared, since each one will receive a larger proportion of whatever food there is.

There is less information on mammals, but wolves, for example, are reported to have no litters in years when food is scarce. In fact, during these periods they evidently do not mate. This may be because the hungry adults are more inclined to spend their time in foraging than in mating activities. It is possible that under extreme conditions some mammals might even by physically incapable of successful mating or of rearing young.

HUMANS

The reproductive strategies of any species evolve on the basis of its own particular environmental circumstances, so let's consider the factors that might have determined our own reproductive capacity. The human female has the ability to produce one child each year for over thirty years. However, does this mean that humans are physiologically and psychologically prepared to rear thirty children? Or is our reproductive ability an evolutionary adjustment to a historically rigorous life in which most, or even all, of the offspring were not likely to survive? In other words, the ability of the human female to reproduce prolifically may have evolved in response to a traditionally high death rate for children.

All this is highly hypothetical but the system might work like this. During its early years, the human is unequipped physically or mentally to deal effectively with the problems of the world. It is among the most helpless of mammalian offspring and one of the slowest to reach self-sufficiency. During this time it must be cared for by adults who strive to provide it with food, keep it warm, protect it from predators and see to it that it doesn't harm itself. But such care can only have minimal effects on reducing the incidence of disease and parasitism in these vulnerable years. The precariousness of its first years might be reflected in a high mortality rate. Even today, in highly stressful situations, the first to die are the very young, the very old, and the sick.

Should an infant die during its first year, the mother could be expected to replace the child rather quickly by becoming pregnant again. If the child continues to survive, however, she might not reproduce again quite so soon. One reason is that in many women, suckling somehow inhibits the secretion of FSH and LH from the pituitary, so pregnancy cannot occur as long as she is nursing. In some cultures the period of nursing is extended for as long as three years, since milk production will continue that long if the child continues to suckle.

The high reproductive capability of humans might also be a response to bleak prospects of an adult surviving through the entire reproductive period, a span of approximately 30 years (from 15 to 45). There is evidence that in some earlier civilizations, the life expectancy was only about 30 years (Table 13.2). In risky environmental situations in which there is a good possibility of an individual meeting death at any age, natural selection might place a higher premium on reproductive output even at the expense of the overall well-being of the individual.

Table 13.2—Estimated Average Life Span in Human Populations

Population	Years
Neanderthal	29.4
Upper Paleolithic	32.4
Mesolithic	31.5
Neolithic Anatolia	38.2
Austrian Bronze Age	38
Classic Greece	35
Classic Rome	32
United States, 1900–1902	48
United States, 1950	70

Table 13.3—Child Mortality Rates 1–4 Years, 1960–1962 (Average Annual)

Continent and country	Rate per 1,000 children (ages 1–4) per year	Continent and country	Rate per 1,000 children (ages 1–4) per year
Africa		*Europe*	
Mauritius	8.7	Austria	1.3
Réunion	9.6	Belgium	1.0
United Arab Republic	37.9	Bulgaria	2.4
		Czechoslovakia	1.2
North and Central America		Denmark	0.9
Barbados	3.7	Finland	1.1
Canada	1.1	France	1.0
Costa Rica	7.2	Germany, East	1.6
Dominican Republic	10.8	Germany, West	1.3
Guatemala	32.7	Greece	1.9
Mexico	13.8	Hungary	1.6
Puerto Rico	2.9	Ireland	1.2
Trinidad and Tobago	2.5	Italy	1.9
United States	1.0	Netherlands	1.1
		Norway	1.0
South America		Poland	1.6
Argentina	4.2	Portugal	8.0
Chile	8.0	Spain	2.0
Colombia	17.4	Sweden	0.8
Equador	22.1	Switzerland	1.2
Peru	17.4	United Kingdom—England and Wales	0.9
Venezuela	5.9	Yugoslavia	5.2
Asia		*Oceania*	
Ceylon	8.8	Australia	1.1
China (Taiwan)	7.2	Fiji Islands	3.7
Hong Kong	4.4	New Zealand	1.2
Israel	1.8		
Japan	2.2		
Kuwait	3.6		
Philippines	8.4		
Syria	8.3		
Thailand	9.1		

SOURCE: United Nations, *Statistical Series*, K/3, 1967.

But what about now? As Table 13.3 shows, a child born in developed countries these days is very likely to live through its first critical years. In humans today, mortality rate decreases after the first year and is very low through the teens, then it increases gradually until about age 60, when it rises rather rapidly. This is especially true in developed countries, such as the United States, but less true in certain poorer countries.

The means by which the human species has increased the likelihood of survival are almost universally viewed as good and desirable. For example, we have specific medicines to combat various maladies which, in earlier days, would have proven fatal. We also have therapeutic and corrective devices to aid the sick. If someone in our midst is unable to provide for himself, that person will usually be cared for, however minimally. We have laws which state that not only must the mentally deficient be allowed to remain in the population, but that their reproductive abilities must not be tampered with. Also, our society often provides for those who, in harsher days, would have been selected against. A person doesn't have to be keen of wit and physically agile in order to cross a busy street. He simply waits for a light—a light which means it's safe to cross leisurely. The result of our social care has been a negation of many of the influences of natural selection. However, at the same time that we have ceased selection for swiftness, strength and intelligence, we have increased the variation in our species. Thus we find among us myriad interests, talents, tendencies, and appearances. The point is that, with its highly developed social programs, our society is attempting to ensure every individual which is born that he will live and reproduce. With our reproductive potential so high, we may be placing our species in an untenable position by our uncontrolled breeding.

To sum up then, reproductive rates are subject to natural selection, and the reproductive potential of humans has been established through the eons of our development. We have recently altered our environment so that the direction and strength of natural selection has been changed, but we are left with a reproductive capacity which better fits our earlier situation.

If this is true, or even if it isn't, we might be led to ask ourselves a number of questions. Are large families with many children "natural"? Has the relaxation of natural selection resulted in our species increasing to the extent that we are physically or psychologically unable to deal effectively with the resulting swell of our own numbers? Do crowds make you uneasy? What might be the pressures on a social system in which parents are unable to carefully rear each of their children as individuals? Perhaps the most important questions involve the future. Where do we go from here? What can we expect? How much social change will be necessary to solve the problems resulting from an overbreeding population? How much social change can we tolerate? What are the odds we can pull it off? Those morose souls who spend time with such matters have imbued many of the rest of us with a deadening pessimism—partly for reasons we will consider in the next chapter. But, looking on the brighter side, let's see how populations may be kept in check by the deaths of their members.

CONTROLLING POPULATIONS THROUGH MORTALITY
Just as the numbers which are born to any species are partly a function of environmental factors, so are the numbers which are allowed to remain in the population. What are some of these factors

which tend to control populations by acting on the members themselves? What brings about death?

ABIOTIC CONTROL

Just as seasonal weather changes can alter population numbers, so can irregular or unusual weather. Drought may kill many kinds of plants and animals. Many birds perish in some years because they begin their northward migration in the spring only to be caught by a late cold spell. Such physical factors which depress population numbers are called *abiotic controls* (*a*, meaning without; and *biotic*, pertaining to life). A drought is not alive.

It should also be kept in mind that population-depressing influences such as severe weather are *density-independent*. In a severe drought, the parching sun doesn't care how many corn plants are struggling in the field below. It kills them all. In an area saturated with DDT, most of the insects die—whether there are few or many. Their numbers mean nothing (unless they somehow tend to protect or shelter each other). Thus mortality brought about by such means is said to be independent of the density of the population.

The awesome impact of the human species on the life-supporting world environment may soon bring the reality of density-independent population control into sharp focus. For example, we are building, along with other countries such as Japan, huge tankers which are designed to carry incredible amounts of petroleum products. It is disquieting to learn that, in one recent ten-year period, there were about 5,000 collisions of smaller, more maneuverable, tankers. What do you suppose might be the effect on the oceans' biota if a single supertanker were to sink or otherwise disgorge its cargo? We might expect to see wholesale destruction of oxygen and food-producing plankton over great areas of the ocean. At the time of this writing a supertanker has run aground and has already pumped a million gallons of its oil cargo into the Straits of Magellan in order to lighten its load to free itself. It's still stuck and still pumping. There have also been recent reports indicating that in certain sections of the country, industrial and automotive by-products emptied into the air are altering weather patterns. We know that the quality of air throughout the entire country has deteriorated sharply in recent years (although some people take solace in the fact that there has been a reduction of certain kinds of pollutants in some areas). We tend to sit in stoney complacency as televised reports warn our old people not to go outside, and our children to refrain from running at recess, because our air is dangerous to breathe. Should we continue to befoul and

Figure 13.9 Drought plagued Eastern Ethiopia for five years causing many deaths. These groups of rocks are actually graves which now litter the plains. The drought and subsequent famine threaten to wipe out whole tribes of people. (Courtesy: Wide World Photos)

Figure 13.10 The beautiful beaches of Santa Barbara, California, blackened with oil from a malfunctioning offshore oil rig. The cleanup effort was monumental and not totally successful. In spite of a massive campaign by citizens to save oil-covered diving birds, almost all died. (Courtesy: Dave Bellak, Photofind)

alter our environment with such recklessness, we may, indeed, come to understand more clearly just what density-independent population control really means.

BIOTIC CONTROL
Biotic population control simply refers to influences on populations

Figure 13.11 The classical Lotka-Volterra oscillation showing the relative fluctuation of predator and prey. (From G. F. Gause, *The Struggle for Existence*, Williams and Wilkens. Baltimore, 1934.)

Figure 13.12 Many whales are in danger of becoming extinct and yet some nations continue to hunt them. (From *N.Y. Zoological Society Newsletter*, November 1968. Reprinted by permission.)

which are alive. For example, the organism which causes bubonic plague can reduce populations. So can a tiger. But biotic influences can operate in subtler ways as well. If a territorial bird drives a competitor into an area where there is less food, when winter comes the underfed competitor may be more likely to succumb to the rigors of the season. Thus the territory holder has indirectly brought about the reduction of the population.

Biotic controls on populations are likely to be *density-dependent*. Density-dependence means that as population density rises there will be increasing pressures on it acting to reduce that density. As the density falls, then, the pressures on it will lessen, thus permitting the numbers to increase again. We can see how this works by considering the effects of predators.

Predation

One of the interesting facts about predation as a population regulator is that a predator normally doesn't eliminate its prey. For example, when the vigorous and intelligent dingo, a predatory relative of the dog, was introduced to Australia, it didn't kill off the primitive prey it found there. However, its hunting prowess proved so superior to that of its *competitors*, the Tasmanian devil and Tasmanian wolf, that those animals disappeared from the Australian continent. But why didn't it wipe out its prey?

Let's examine the classical theory which accounts for the density-dependent effect of predators. According to the theory (Figure 13.11), under undisturbed conditions prey numbers rise steadily, thus providing more food for predators. Then the predator numbers begin to rise. Their numbers do not rise immediately, however, since it takes time for the energy from food to be converted into reproductive efforts. When the predator numbers finally do rise, the increased predator population, of course, places increasing pressure on its prey. Then, as the prey are killed off, the predators find themselves with less food, and so their own members fall off. Because the fluctuation of predator numbers lags behind that of the prey, the prey may be well on the road to recovery before the predator population begins to rise again. In reality, the story is complicated a bit by such factors as prey-switching as the predators seek other, more available food to maintain their numbers.

Our own species is providing us with clear examples of how density-dependent regulation can fail. The great whales have been hunted to the brink of oblivion over the past few decades as modern whaling methods have reduced personal risk while increasing profits. Although there is nothing that whales provide which cannot be gotten elsewhere, the demand for whale products has not

World-wide total of whales killed (thousands)

Blue whales killed (thousands)

World-wide whale oil production (millions) of barrels

Fin whales killed (thousands)

Average gross tonnage of catcher boats (hundreds of tons)

Average horsepower of catcher boats (thousands)

Sei whales killed (thousands)

Average production of whale oil (barrels) per catchers boat per day

Sperm whales killed (thousands)

WWII

WWII

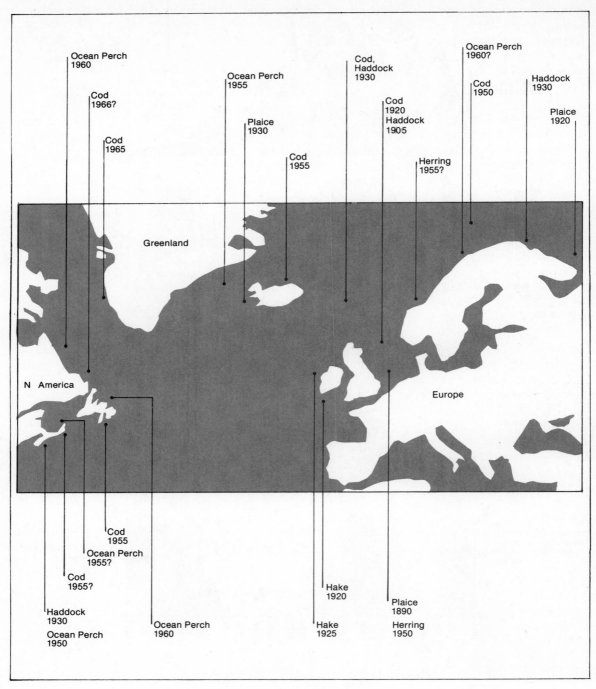

Ocean Perch
1960

Cod
1966?

Cod
1965

Ocean Perch
1955

Plaice
1930

Cod
1955

Cod,
Haddock
1930

Cod
1920
Haddock
1905

Herring
1955?

Ocean Perch
1960?

Cod
1950

Haddock
1930

Plaice
1920

Greenland

N America

Europe

Cod
1955

Ocean Perch
1955?

Cod
1955?

Haddock
1930

Ocean Perch
1950

Ocean Perch
1960

Hake
1920

Hake
1925

Plaice
1890

Herring
1950

diminished, especially in Japan. Thus, instead of man the predator relaxing his pressure and allowing the whale population to recover, whaling fleets continue to exert their depressing effect on populations of the great mammals. Then, as whales decrease in number, the price of whale products goes up, and the hunt becomes still more avid (Figure 13.12).

Parasitism

The relationship between parasite and host is, in a sense, much more delicate than that between predator and prey. After all, it is to the parasite's advantage to see that it does not kill its host or contribute to its death by weakening it. In many cases, as with tapeworms, the host is actually the habitat of the parasite, and who wants to destroy his habitat? Also, the offspring of the parasite will have greater likelihood of success if the host lives to reproduce.

In some cases the parasite and host are so well adapted to each other that they actually exist quite nicely together. In those cases where the reproduction or vigor of the host is seriously impaired by the parasite, the relationship can be considered to be a relatively new one in which the two species have not yet adapted to each other very well.

In other cases adaptation never takes place because of the disastrous effects of the parasite. For example, in 1904 the sac fungus *Endothia parasitica* was accidentally introduced into North America from China. In China the parasite was held in check by a variety of control mechanisms which do not operate in North America. Sadly, the majestic American chestnut tree proved defenseless to the fungus, and by the late 1940s the great tree which had dominated our Appalachian forests was virtually extinct. If you missed that drama, you can witness a similar tragedy as, even now, our elms are falling to the newly introduced Dutch elm disease.

Competition

As we have seen, populations may be held in check by what might be called "ecological" factors, such as competition for food. In some species, however, behavioral patterns such as territoriality or the formation of dominance hierarchies provide some members of the population with more access to commodities than others.

In *territorial* species some animals are excluded from certain areas held by other animals. If the habitat is variable, the result is that some animals will end up with "better" areas than others. If those areas hold more food, then in time of food shortage the most successful territory holders are more likely to survive. In some

Figure 13.13 Overfishing began in the North Sea about 80 years ago. Overfishing becomes apparent when increase in fishing for a species is not accompanied by increases in the catch. The dates show when this occurred for the species named. (From Holt, "The Food Resources of the Ocean." Copyrighted Scientific American 1969)

Figure 13.14 Many animals, including some birds, will fight to keep other members of their species out of their territory. (Courtesy: Phillip Strobridge from National Audubon Society)

species of birds the male must have a good territory in order to attract females. A splendid male holding an inferior territory will not find a mate. If there aren't enough good territories to go around, then not all individuals will be able to reproduce, and thus the population number is lowered.

In species which form *hierarchies*, or "peck orders," certain animals have freer access to commodities than others. In such a system, each animal must be able to recognize others individually and to respond to them according to their rank. Hierarchies are established in various ways for different species, such as by fighting or in the play which takes place as individuals grow up together. Hierarchies reduce conflict within groups by each animal acquiescing when confronted by a dominant individual. In times of shortage of critical commodities, then, the higher-ranking animals are more likely to survive as subordinates give way before them.

In both territorial and hierarchical systems (which are not, by the way, mutually exclusive), when certain animals have freer access to commodities than other animals, harsh conditions may quickly bring about the deaths of the "have-nots," thus increasing the likelihood of survival for the "haves." This doesn't mean that the have-nots accept their role for the "good of the species." Quite the contrary. It is to their advantage to remain in the breeding population at all costs. However, a low-ranking individual may have better luck if he searches for unclaimed commodities than if he tries to challenge a dominant individual.

Disease

Finally, disease can reduce populations. We will see in this chapter how the bubonic plague drastically reduced the human population at one time. Less dramatic diseases claim thousands of lives daily. The density-dependency of disease as a population regulator is obvious. The more closely individuals are packed together, the more opportunity there is for the transmission of certain diseases. Also, the more individuals there are in a population, the greater the number of potential reservoirs in which mutant strains of disease microorganisms can develop.

In some cases disease and predation may interact to reduce populations. For example, a two-week-old caribou fawn can outsprint a full-grown timber wolf. However, caribou are subject to a hoof disease which lames them before it infests other parts of the body, and it is these lamed animals which the wolf is likely to cull out of the herd. In areas where wolves were poisoned to "protect" the migrating caribou, this hoof disease spread unchecked and in a few seasons has decimated entire caribou herds.

DEATH

Have you ever wondered why death occurs at all? Why hasn't natural selection resulted in organisms which simply live forever? If "Life" is better than "Death," then why is there no marked selection for longevity? Why are there delicate insects which don't even have mouths with which to feed themselves, but which must instead spend whatever energy their frail bodies possess in finding mates before their few precious hours of life are gone? Why does man have so much trouble surviving past his "threescore years and ten"? Why death?

People with all sorts of perspectives on life have ruminated on the "meaning" of death. Ideas have been fiddled with by philosophers and theologists, as well as scores of novelists, dramatists, poets, and others. Even aboriginal societies dealt with the meaning of death in one way or another—some very casually, not fearing it at all, treating it as if it were simply the natural extension of life. So let's join the fray and consider death from a biological point of view.

We should begin by noting two things: first, the world changes and, second, it is limited. Not only have continents moved, islands formed, ice ages come and gone, and weather patterns shifted, but man has, in a very final sense, drastically altered the earth's surface in the past few thousand years. In a changing world the organisms which live there must also change. This is what adaptation is all about, adjusting to new situations.

As a population is confronted by changing conditions certain individuals can be expected to cope with them better than others. But the changes may well be too drastic for any individual, with his inherent limits, to be able to adjust. However, as new generations appear, their myriad genetic variations may provide material on which the new environment can have its molding effect through natural selection. Many in the new generations will fall by the wayside, their kinds of genes being separated out of the population. Individuals with other kinds and combinations of genes, though, may fit the new environment well. And these organisms will live and reproduce. The point is, the arrival of new and variable individuals is important in a changing world.

Now, in any system in which the commodities are limited, such as the planet earth, as the numbers of individuals increase, there will be competition for the available commodities, such as food, shelter, land and natural resources. So consider an organism which finishes its period of reproduction and then hangs around to compete with its offspring. Such an individual could be expected to leave fewer offspring than a parent which, after having re-

Figure 13.15 If this moth is captured by a bird, the bird is rewarded for its effort, has a better idea what this food looks like, and then will likely search for other similar moths in the area. (Courtesy: Louis Quitt from National Audubon Society)

produced, died, so as not to interfere with the success of its offspring.

David Blest has described certain moths with camouflaged markings in which the female, after laying her eggs, commits suicide. If she were not to commit suicide and stay in the area where her offspring are hatching, she might eventually be caught by a bird. That bird will then have a better idea of what the moths around there look like and will likely search for more of them— including the female's offspring. So when the female has laid her eggs, she flies around frantically, using up her energy reserves and fraying her wings beyond recognition. Then she either dies of her own efforts, or, because of her conspicuous behavior, she is captured by a bird, which is then unable to find another tattered morsel like that one.

This explanation of the moth's behavior is entirely speculative, of course, but the consideration of death as the ultimate form of altruism provides food for thought. Also, keep in mind that the moth may have no idea about the effects of her behavior. She doesn't "try" to save her young through self-sacrifice. Those moths which behave in such a way, however, leave more offspring, each of which, perhaps, carries a "suicide gene."

If new generations help to ensure success in a changing world, and a parent that remains after reproducing may be a detriment to its offspring, then the phenomenon of death can theoretically be associated with the end of the reproductive period. Why, then, isn't there selection for long lives with extended reproductive periods in all species? Since we're dealing in pure conjecture here, your own answer might be as good as anybody's. You might want to consider such matters as the limits of variability in the offspring of a single parent. Which would you expect to be more different from a parent, its children or its grandchildren? In natural selection is such difference good or bad?

When we consider death as a mechanism which helps to ensure perpetuation of a particular set of genes, we might be left with the rather irreverent idea that any organism is simply a vehicle utilized by its chromosomes to enable them to reproduce themselves. Most of us, however, prefer to feel that we have other raisons d'être than simply as carrying cases for our gonads.

We've tossed around some heavy concepts fairly loosely here, as you are aware. However, these are legitimate questions for biologists. Perhaps we are not in a position to uncover whole truths but we can begin to accumulate certain kinds of evidence. Anyhow, this sort of approach makes as much sense as dealing with the concept in subjective, artistic, abstract, or mystical terms.

HUMAN POPULATIONS

Now let's consider some points that relate specifically to the human population of the earth, something of the history of our species, and what our future looks like from here.

EARLY MAN

Somewhere between 1 and 2 million years ago one might have noticed among the different kinds of living things on the earth a small population of primates which differed somewhat, but not much, from other animals. Strangely enough, the appearance of these animals, ambling in little bands over the vast African plains, was one of the most momentous events in the history of the planet. Who could have guessed that virtually every other kind of life, as well as great mountains, endless oceans, vast plains, immense forests, and even the air itself were to give way, change their form, before something as seemingly insignificant as that hunched figure sitting in the shade, picking at the soft parts of an insect.

We have no idea how many individuals comprised the human species in those early times. In fact all the evidence before about

Figure 13.17 Human population growth during the last half million years. If the Old Stone Age were drawn to scale, the line would extend about 18 feet. (Annabelle Desmond, "How Many People Have Ever Lived on Earth?" *Population Bulletin,* vol. 18, no. 1 Washington, D.C.: Population Reference Bureau, February, 1962)

1650 B.C. is entirely conjectural. We know, for example, that agriculture was almost unknown before about 8000 B.C., or possibly a bit earlier in Malaysia and the Middle East. Before that time humans must have lived by hunting and gathering their food. Considering the inefficiency of such methods, and given a land area of about 58 million square miles, it is estimated that the earth could have supported no more than about 5 million people in 8000 B.C.

It is believed that these early human populations were separate, but interbreeding, groups, possibly divided into yet smaller bands. From studies of present-day hunters and gatherers, it has been calculated that the average size of such groups would have been about 500 men, women and children, of whom about 200 were of breeding age. Population size was probably relatively stable and density-dependent. (Keeping in mind the food sources, can you tell why?) It is also generally believed that the number and size of the groups approached the carrying capacity of the environment.

At about 8,000 B.C., however, that carrying capacity became elevated by the advent of agriculture. Men began to cease roaming and began to grow and store their food and thus they added a relatively stable element to their food supply. The lean seasons no longer marked the deaths of so many children and old people, since food could be stored. The result of the shift toward agriculture was a slow steady increase in human populations.

How could the shift to agriculture have raised the environment's carrying capacity? We have already mentioned that the ability to store such products as grain meant a more dependable food supply and a greater likelihood of surviving the harsh seasons. Something else happened as well, something which may come to be increasingly important in coming years. Agriculture permitted man to shift more toward the level of a *primary consumer* or herbivore. Let's see what this means.

THE FOOD PYRAMID

The concept of consumer levels may be understood when we consider the relative energy efficiencies of taking different kinds of food. Let's compare the relative efficiencies of eating plant and animal food. Some animals eat plants, but it should be apparent that there is no way that all the energy stored in plant tissues can be recovered by those animals. Furthermore, animals which eat the plant eaters can't recover all the energy stored in the bodies of those herbivores. Thus much of the food stored in living things is wasted. But the wastage varies depending on how far one's food supply is removed from the level of the primary producer, the plant.

The principles have been demonstrated in what is called an *energy pyramid* (Figure 13.18). Note that of the great amount of energy stored in the tissues of the plants, less than half (8833/20810) is available to animals. The reason is twofold, as we have mentioned. First, plants have their own things to do; they aren't in the business of supporting animals. Thus, much of the food they make in photosynthesis is used to produce energy to enable them to do the things that plants do. Only a part of it is turned into "plant." Also, a herbivore uses much of the energy it retrieves from plants for its own activities. This means that only a portion of the energy it gets from plants is stored in its own tissues and thus made available to the carnivores which eat it. So, much of the energy stored by organisms at any food level is used in metabolic activities by those organisms, or is lost as heat or in other ways. It is simply not made available to the next consumer level.

Second, animals are not 100 percent efficient at recovering energy stored in the things they eat, so much of the energy which *is* available to them is not utilized. The result of all of this is apparent when you notice in Figure 13.18 how many producers are necessary to maintain even a few herbivores and even fewer carnivores. A massive amount of food energy must be produced by plants to maintain only a few secondary carnivores.

With this in mind, then, would you say it is more efficient for humans to eat the fish which ate the fish which ate the fish which ate the algae, *or* is it more efficient to eat the algae? Should we feed grain to beef cattle or should *we* eat the grain? The improved diet of North Americans places *four times* the demand on agricultural resources than does the poorer diets of the less fortunate. The reason is because of our interest in eating from higher levels of the food pyramid. With such wastage involved, can you see why steaks are so expensive (politics and market manipulations aside)?

Back to our earlier point, that beginning about 10,000 years ago man began to raise the carrying capacity of his environment

Figure 13.18 An energy pyramid calculated for Silver Springs, Florida. The upper numbers for each tier designate the stored energy (expressed as kilo-calories) for each square meter investigated over a year. Not all the stored energy at any level is available to the level above it so the kilo-calories which are available as food are shown in parentheses, and as the shaded areas of each tier. Also, each tier is much smaller than the one below it because it is not completely efficient at recovering all the energy which *is* available. P = producers (plants); H = herbivores; C = primary carnivores (animals which eat herbivores); SC = secondary carnivores (animals which eat carnivores). (From H. T. Odum. "Trophic structure and productivity of Silver Springs, Florida." *Ecological Monographs*, 27:55–112. 1957.)

through agriculture by insuring a more stable food supply, and by increasing his reliance on lower food levels. Man was now able to make more efficient use of plants and thus to support a larger population. All this is not to say that the agricultural revolution has been the only factor which has changed the carrying capacity of our environment. The advent of tool making also had its marked effect, and we will learn shortly of the phenomenal impact of the recent scientific-industrial revolution.

As you can see from Figure 13.17, the human population continued its slow rise through the beginning of the Christian era. It is believed that at the time of Christ, the earth's population was between 200 and 300 million and that it increased to about 500 million by about 1650, with only one period when it dropped significantly, because of the impact of the bubonic plagues.

GROWTH RATE

It is interesting, and tremendously significant, that not only has our population steadily increased, with few irregularities, in the last 10,000 years, but so has our *rate* of increase. Growth rate increase can perhaps best be shown by considering "doubling times"—the time necessary for a population to double in size (Table 13.4). To go from 5 million people on earth in 8000 B.C. (the present number in only three of New York City's five boroughs) to 500 million in 1650 A.D. took between six and seven doublings over a period of 9,000 to 10,000 years. So, during that time, the human population doubled on an average of about every 1,500 years.

Table 13.4—Doubling Times of the Human Population

Date	Estimated World Population	Time Required for Population to Double
8000 B.C.	5 million	1,500 years
1650 A.D.	500 million	200 years
1850 A.D.	1,000 million (1 billion)	80 years
1930 A.D.	2,000 million (2 billion)	45 years
1975 A.D.	4,000 million (4 billion)	35 years*

*Computed doubling time around 1972.
SOURCE: From *Population, Resources, Environment: Issues in Human Ecology*, Second Edition by Paul R. Ehrlich and Anne H. Ehrlich. W. H. Freeman and Company. Copyright © 1972.

A glance at the right-hand column of Table 13.4 will show that, all other things being equal, in only thirty-five years we will need two cars, two schools, two roads, two houses, and two cities throughout the world for every one that presently exists. And that will only maintain our status quo as far as material goods are concerned. The problem is that not all things can remain equal over that time. Can India or Guatemala double its food? Can we double the population of the Eastern Seaboard without the inner cities falling apart? Can California stand another Los Angeles? Where will the water come from? If we double the number of our oil wells, can we double the amount of the earth's oil?

If we *should* manage to withstand environmental resistance and double the earth's population in thirty-five years, it will be precious few years after that before the population would double again! And then again and again!

Those with a great faith in technology have suggested that the answer lies in increasing the carrying capacity of the earth through agricultural, technological and medical "revolutions." They maintain that it is possible to accommodate billions more people. We are an aggressive and innovative species and there is no doubt that we *can* successfully withstand population increases, for a time. The question is, do we want to? How old will you be when the earth has 7 billion people? How old will your children be when there are 14 billion people on earth at the current rate of population increase? How hard is it to get away by yourself right now with the earth's population at only $3\frac{3}{4}$ billion? (The world's population in mid-1972 was estimated at 3.78 billion by the Population Reference Bureau.) In the past 300 years the human population has increased an average of 10 million per year. But the present rate of increase is 80 million per year! How many is that each day? Each hour?

POPULATION CHANGES SINCE THE 1700s
Somewhere around the seventeenth century the world's human population suddenly began to shoot upward. Such an abrupt rise has heralded population crashes in other species, and until we have some sort of evidence that humans are immune from natural laws, we might want to carefully examine the factors behind our sky-rocketing numbers.

The sharp rise in the population of Europe between about 1650 and 1750 is attributed to innovations in agriculture and the beginning of the exploitation of the Western Hemisphere. The accompanying rise in Asian populations is harder to explain. India, for example, was in turmoil following the demise of the Mogul Empire. In China, after the fall of the Ming Dynasty in 1644, the Manchu emperors brought political stability and new agricultural policies, which may account for the increase in the Chinese population.

Box 13.1 Birthrates and Deathrates   Birthrate may be expressed as the number of babies born per thousand people in the population per year. At the year's end the total number of births is divided by the estimated population halfway through that year. The deathrate is calculated in a similar fashion. For example, if 200 babies were born to a population of 10,000 people, the birthrate would be 20 per thousand. If 180 people died in the same period, the deathrate would be 18 per thousand. The rate of increase for that period, then, would be 20 per thousand per year. This is an annual growth rate of 2 percent.

The world's growth rate in 1972 was 2 percent. Thus the population will double in 35 years if the rate remains constant. The chart below describes doubling times as related to annual percent of population increases.

Annual percent increase	Doubling time (years)
0.5	139
0.8	87
1.0	70
2.0	35
3.0	23
4.0	17

The growth rate in the United States presently approaches 1 percent, but in several Latin American countries, the 4 percent figure is reached.

Not much is known about Africa in this period, but its estimated population was about 100 million in 1850 when European medicine and technology were introduced. In the next hundred years, then, its population doubled. Between 1750 and 1850 the populations of Asia increased only about 50 percent, but in Europe the numbers of people swelled much more rapidly. The factors permitting such growth included newer agricultural techniques, better sanitation, and advances in medicine, such as the smallpox vaccine.

In the next hundred years the world population grew from over 1 billion in 1850 to about $2\frac{1}{2}$ billion by 1950. Since the Korean War we have added another $1\frac{1}{4}$ billion to our numbers. Between 1850 and 1950 the populations of Asia, Africa, and Europe approximately doubled, the population of Latin America increased about five times, and that of North America increased sixfold.

Table 13.5—Populations in Millions

Date	World	Africa	North America	Latin America	Asia (except USSR)	Europe and Asiatic USSR	Oceania
1850	1,131	97	26	33	700	274	2
1950	2,495	200	167	163	1,376	576	13

SOURCE: Walter F. Willcox: *Studies in American Demography.* Copyright 1940 by Cornell University.

AFTER 1850

Between 1850 and 1900 the death rate continued to decline in the face of advances in industry, agriculture and medicine. The lot of the European previous to this time had been incredibly bad. The cities were garbage-filled and overrun with rats. The stench was devastating. Life in the country was not much better, with the scattered hamlets being little more than isolated slums inhabited by people of numbing ignorance. But, by the beginning of the twentieth century, the situation of these people improved as the likelihood of crop failure decreased. In addition, transportation advances meant that food could be imported to locally famished areas. Also, about this time the role of bacteria in infection became known, and so more lives were saved. Death rates in Europe dropped from about 24 to 20 per 1,000 in those fifty years.

At about the turn of the century a new and highly significant trend appeared in the world's populations. It seemed that, for some reason, industrialization was followed by a decreased birth rate. The phenomenon, which has been highly speculated upon, has never been entirely explained, but there are some good guesses at why it happened. First of all, in agricultural societies, children are a source of labor and a social security measure. They provide assistance before they leave home and they care for their aged parents later. In an industrial society, however, they are less likely to aid in production, but they maintain their status as "more mouths to feed." Also, mobility is important in industrial societies and mobility is reduced by the presence of children. Given these reasons, it is interesting that, in the early twentieth century, as birth rates dropped off in the industrial cities they did the same in rural areas. One reason was the mechanization of farming which reduced the need for labor. So, it seems that industrialization tends to depress both death rate *and* birth rate.

About the time of World War II many underdeveloped countries experienced a dramatic reduction in death rates. The cause was apparently the rapid import of public health programs, including new drugs, from the more developed countries. (Among the most important changes was the control of malaria through the widespread use of DDT.) Such importation of "death control" produced the most sweeping changes in human population dynamics in the history of the species. It is important to realize that the changes in death rate in these underdeveloped countries were not accompanied by the social and institutional changes which marked the slower reduction in death rate in the more developed countries. The changes were much more rapid and were brought about from the outside.

AGE PROFILES

In the consideration of how any population changes, it is important to consider the ages of the people in the group. Figure 13.19 shows the "age profiles" of two kinds of populations. Mauritius, a small island in the Indian Ocean, is typical of many rapidly growing countries with high birth rates and declining death rates. The United Kingdom, on the other hand, is representative of older developed countries which have had low birth and death rates for many generations.

The profile of Mauritius is produced by a high birth rate, combined with a sudden recent decline in the deaths of infants and children. There hasn't been enough time for those benefited by imported "death control" to have moved into the older age groups. Furthermore, the children of these young people will further inflate the youngest end of the age spectrum. However, Mauritius is an island, just as the earth is, and this kind of growth can't go on forever. A program of population control will have to be initiated, or environmental resistance in the form of famine or

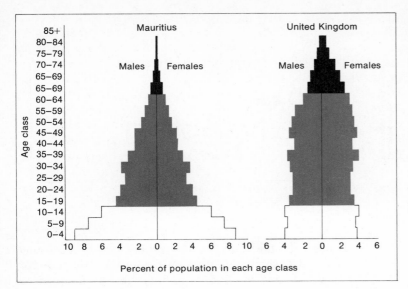

Figure 13.19 Age profiles of Mauritius and the United Kingdom in 1959. Mauritius is a rapidly growing under-developed country with a very young population. The United Kingdom is an old developed country in which the people are more evenly distributed over the age spectrum. Males are shown to the left of the center line, females to the right. The working ages are shown in green, the young in white, the old in black. (From Annabelle Desmond and Judy K. Morris, "The Story of Mauritius— From DoDo to Stork" *Population Bulletin*, Vol. 18, no. 5, Washington, D.C.: Population Reference Bureau, August, 1962)

some other check will eventually alter the shape of the age pyramid. What would the age pyramid look like for a rapidly developing country ten years after it had suddenly instigated strict birth-control programs? Judging from the age profile of the United Kingdom, at about what age would you say that death-control programs begin to lose their effectiveness?

FERTILITY RATE

It has been pointed out by demographers, the people who study human populations, that *fertility rate* is a better predictor of the behavior of populations than birth and death rates. Fertility rate is calculated as the number of births per thousand among women of child-bearing age, established for statistical purposes as fifteen to forty-four.

The importance of considering fertility rate was apparent in the erroneous interpretation of the widely heralded drop in birth rate in the United States during the 1960s. The news media pounced on this phenomenon as indicating the end, at last, of the population explosion in this country. But, a closer look at the data showed that although couples were having fewer children, a more important reason for the drop was that at that time there were simply fewer women of child-bearing age in the population. A glance at the age profile of the United States would have shown that there was a large number of girls under fifteen in the population, just waiting to move into the reproductive brackets.

Figure 13.20 Fertility rate in the United States from 1910 to 1971. (From *Human Ecology: Problems and Solutions*, by Paul R. Ehrlich, Anne H. Ehrlich and John P. Holdren. W. H. Freeman and Company. Copyright © 1973)

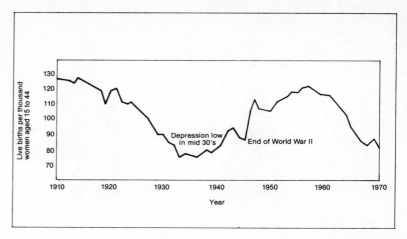

What has been the course of the fertility rate in the United States? Figure 13.20 shows that it dropped between 1959 and 1968. Both the birth rate and the fertility rate declined to about 85 per thousand, but still failed to reach the low of the depression years. Then in 1970 it rose to 87.6, fell in 1971 to 82.3 and in 1972 it dropped to below 80.

Does this mean the end of the population explosion in the United States? The experts are unanimous! They don't know. Our population has historically followed erratic routes. The optimists like to point to such factors as an enlightened public, the extension of family planning services to more people, the trend toward male and female voluntary sterilization, liberalization of abortion laws and the impact of the feminist movement. On the other hand, pessimists are quick to point out that high unemployment and a tight job market for prospective fathers has traditionally depressed population growth, and that if the economy should regain its strength, we will be on our merry way once again.

WAR

INTERGROUP AGGRESSIVENESS AND COMMODITIES

At this point we're going to consider a form of human interaction which has had important effects in reducing small populations but which has had less effect on the overall population of the world. The topic here is *war*. Because of its complexities we can only discuss it in a limited and rather superficial way, and perhaps there are reasons why it could better be fitted in somewhere else. However, it seems that war must be considered, at least in part, as a *population* phenomenon. It has traditionally been a population *interaction*. In essence, what we will say is that war may be one expression of a long history of competition between different groups as each seeks to advance its particular types of genes over those of other groups. In other words, we will consider war as a biological phenomenon. Of course, and by *all* means, war is not simply that. We are much too complex and involuted creatures for any of our behavioral traits to be considered a result of "simply" anything. Perhaps, however, competition for commodities is an *element* of what causes wars. In any case, let's see if we can develop a few ideas along this line. Hopefully, the material will at least serve as a springboard for more extensive discussions in this and related areas.

WAR'S EFFECT ON NUMBERS

A quick look at a world population curve will show that there has been no great depression in our numbers during the period of any war. In fact, it has been calculated that the number of *all* people killed as a result of all recorded wars is presently made up in only a few days population growth. After almost 20 years of war in Vietnam, with millions of casualties, the populations of North and South Vietnam have increased by two million each. Smaller populations, on the other hand, have been virtually decimated by warfare. We have been provided recent examples of such tragedies in Nigeria and Bangladesh.

THE EVOLUTIONARY IMPLICATIONS OF WAR

Let's consider our topic on an evolutionary basis. We are sure to fall into argument with some sociologists, political scientists, anthropologists and probably some biologists. But that's all right.

In order to develop our case, we must be allowed three premises, ones which were established in previous discussions. First, we must recall that in a system of limited commodities which occasionally fall into short supply, there may be strong competition between the ones who rely on the commodities.

Second, within groups in which the members are likely to be related, there may be reproductive advantages in cooperative behavior. A helpful and cooperative individual is likely to be advancing his own kinds of genes by aiding relatives. Thus "in-group" members are likely to receive favorable treatment from each other.

But, third, those members of the species which are designated "out-group" individuals (not members of the reproductive group, perhaps of another group) are likely to be strong competitors and unrelated. So aggressiveness might be expected to be primarily directed toward such individuals.

Now, let's consider what might have been early man's situation. You will recall that the carrying capacity of the environment for hunters and gatherers is rather low at all times and that environmental resistance can be expected to increase seasonally. In the hunting and gathering groups, then, there would have been times, such as in winter, when there were severe shortages of commodities. At such times, then, an individual might be likely to share food with his immediate family (think of the reproductive advantages), but less likely to share with more distant relatives. Not only would he not be likely to share hard-won food with out-group members, but he could be expected to fall into strong competition with them for whatever food was available. Such strong competition, of course, could be expected to result in outright aggression. Even when the carrying capacity of the environment was raised by agriculture, after the population had increased to its new limits, competition for land or water could be expected to increase accordingly.

So, in such a system of limited commodities, what sort of individual would be most likely to leave progeny carrying his genes —a friendly, cooperative individual who was willing to share with everyone, or an aggressive, competitive person who took for himself and his immediate family all he could? What do you suppose, then, would be the disposition of most of the individuals in the next generation?

The aggressive spirit, of course, would be tempered by the

Box 13.2 Out-Group Aggression   The principle of identifying individuals on the basis of their group or subgroup status and then behaving toward them according to their affiliation is well represented throughout social animals. Hyenas live in *clans*, or groups of families, and rarely will they permit a member of another clan to enter their group or share their kills. Wolves may attack any strange wolf which enters their area. Jackals drive away other jackals which they do not recognize. Lions, amiable enough within their own pride, may kill a stranger which they encounter within their hunting area. Rats and mice are notorious killers of strangers.

   Out-group hostility is not universally demonstrated among animals, as would be expected, since we're dealing with a diverse and complex kingdom. For example, chimpanzees are highly social, but they do not usually greet strangers with hostility. (They also eat a variety of abundant food types and have no permanent or semi-permanent nest sites. Why might this be a factor?) The general point can be made, however, that there must be a great advantage to out-group hostility in certain cases as revealed by its prevalence among social animals.

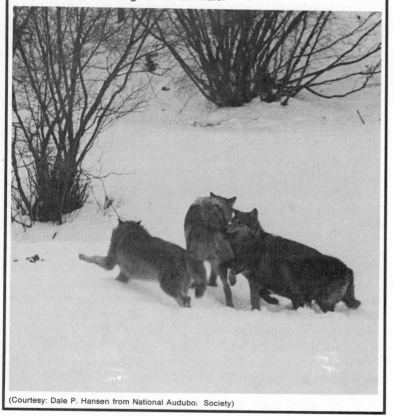

(Courtesy: Dale P. Hansen from National Audubon Society)

need to get along with in-group members. Aggressiveness could also be reduced by behavioral mechanisms, such as hierarchical systems which would effectively designate rank and thus reduce in-group strife. Among humans, aggression might also be reduced through such cultural phenomena as laws or religion.

If any of this is correct, then, it seems that there would be a tendency to identify individuals according to their group affiliation (or potential relatedness). Within groups, such as the breeding populations of 500 or so individuals which roamed the earth about 10,000 years ago, there would be subgroups, such as families, in which relatedness was higher than in the group as a whole. Aggressiveness within subgroups could be expected to be lower than between subgroups, with the highest aggression existing, potentially, between different groups. The principle of identifying out-groups as targets of aggression has not escaped political leaders, who, seeing their position jeopardized by strife among their followers, initiate war or other actions toward another, or "out" group. The result is that the followers are pulled together, at least temporarily, as their aggression is directed toward the designated group. See Box 13.2.

COMMODITIES AND WARFARE

How well does the human condition fit the general evolutionary scheme described above? It matters little for purposes of our discussion whether our behavior in these areas is indelibly stamped into our nervous systems as a result of natural selection or whether our actions are the result of learned patterns, a "fault" of our culture. (If we want to discuss *changing* our condition, however, these questions become critical.) The questions we must ask here are, do we generally show out-group aggression? Are we in competition for commodities? Does our competition often result in aggression? Is war an extension of this sort of aggression?

It seems, on an intuitive basis, that the competitive nature of humans can hardly be denied. Apparently, all that is necessary to stir vague feelings of animosity toward any group is to identify it as "them." We compete at all levels. Not only are we quietly aware of whether we do any particular thing better than our friends, but our competitive nature has been directed and channeled culturally by such institutions as little league baseball, school football teams, and the system of academic grading. Needless to say, the merchants among us have very successfully capitalized on the tendency as millions of TV viewers watch athletes, with whom we somehow manage to identify, compete on the playing field each week. Then, of course, periodically there is the great international competition of the Olympiad. And millions of people who have utterly no idea of what the game is about will devotedly follow the world championship chess tournament if one of "ours" is playing one of "theirs."

But if this competitiveness exists, does it function in gaining commodities through the more serious practice of waging warfare? At this point we might take a look at some recent events.

A Recent Lesson

In 1969 two tremendously overpopulated countries in Central America, Honduras and El Salvador, declared war. Elements in the American press played up the notion that it started over a soccer game, but the issues were a bit more deeply rooted. El Salvador, with a population of 3.3 million, had 413 people per square mile (782 per square mile of arable, or cultivatable, land). Honduras had a population of 2.5 million, with only 57 people per square mile (155 per square arable mile). The doubling time of both populations was only 21 years. Obviously El Salvador was under greater pressure. The pressure was partly relieved as about 300,000 Salvadorans moved to Honduras in search of land and jobs. They were greeted as "outsiders" by the Hondurans and friction developed between the Salvadorans and their hosts. The problem escalated into a brief but nasty war which was ended by the intervention of the Organization of American States. This international body, in an unprecedented move, acknowledged that population pressure had been at the cause of a war—too many people seeking too few commodities.

A More Recent Lesson

As another, perhaps more painful, example of competition for commodities precipitating war we might consider our own policy in Southeast Asia. Untold thousands of people have died there as a result of programs funded by American citizens. The death toll of our own young men is set at about 50,000, but this figure ignores the paralytics and amputees who were returned to us. It also says nothing of the emotionally scarred, both those who went to war and those who watched the misery each evening before beginning a large meal. One wonders, on what rational basis would the world's most powerful nation exert such a monumental effort to rain destruction on what one recent president termed a "raggedy-ass little third-rate country"? Perhaps clues can be found in the statements of some of our political leaders. Richard Nixon, in 1953, when he was vice-president, said, "If Indo-China falls, Thailand is put in an almost impossible position. The same is true of Malaya

Figure 13.22 The populations of some Latin American countries are doubling in less than twenty years. Their resources, however, cannot double. (Courtesy: David Glaubinger from Jeroboam)

Figure 13.23 Throughout human history wars have been fought for a variety of reasons. Not only are the young and healthy combatants the victims, but the very young and elderly die and are scarred psychologically as well as physically. (Courtesy: Authenticated News International)

with its rubber and tin." In 1963 President Eisenhower wrote, "The loss of all Vietnam together with Laos in the west and Cambodia on the southwest . . . would have spelled the loss of valuable deposits of tin and prodigious supplies of rubber and rice." Earlier, in 1954, Mr. Eisenhower had thought to include tungsten as an important commodity found in Indo-China. He made the statement in an interview published in the April 16, 1954 edition of *U.S. News and World Report*. The article was entitled: "Why U.S. risks war in Indo-China" and the subheading states:

One of the world's richest areas is open to the winner in Indo-China. That's behind the growing U.S. concern. Communists are fighting for the wealth of the Indies. Tin, rubber, rice, key strategic raw materials are what the war is really about. The U.S. sees it as a place to hold—at any cost.

Is it possible that, in the name of "liberty," our 19-year-olds actually died for tin, rubber, rice, petroleum, tungsten, iron, manganese, zinc, coal and antimony—the commodities listed in another article entitled "The Big Prize—Southeast Asia"? It is interesting, and perhaps a little sad, that while people on both sides were dying those horrible deaths we've witnessed so many times, and while homes, farmland and hospitals alike were being reduced to rubbish, American oil interests were busy mapping out and arranging oil leases in the waters off South Vietnam. Possibly economist Percy W. Bidwell, writing for the Council on Foreign Relations, summed up the American position when he stated:

Our purpose should be to encourage the expansion of low-cost production and to make sure that neither nationalistic policies nor Communist influences deny American industries access on reasonable terms to the basic materials necessary to the continued growth of the American economy.

Recently, Washington columnist Jack Anderson, a tough critic of American politics, was cautioned by one of the generals serving under Secretary of State Henry Kissinger not to publish a "sensitive" report on our relationship with the African country of Burundi. Anderson, not surprisingly, chose to publish the report. In it he described the normalization of American relations with Burundi, and why it came about.

In early May, 1972, the ruling Tutsi tribe of Burundi began the systematic slaughter of their ethnic rivals, the Hutus, who had attempted a coup. The scale of the slaughter was incredible. By late 1973 over 200,000 Hutus were dead, and another 100,000 had fled the country. The U.S. government showed its disapproval of the events by adopting a "minimal relations" policy with Burundi. Then suddenly, the U.S. stand warmed, almost overnight.

The thaw corresponded with the discovery of vast deposits of nickel in Burundi. Since that time, agents for U.S. corporations have swarmed into Burundi, and the country has become a much-courted friend of the U.S. government.

One top White House official wrote in a memorandum that the normalization would provide "opportunities for American corporations that are interested in exploiting the major new mineral discovery." He also wrote that resumption of U.S. aid (from our tax money) might also earn "concessions for American companies who believe there may be hydrocarbon deposits (oil) on the eastern shore of Lake Tanganyika." According to Anderson's report, then, our foreign policy can be altered in a number of ways, depending on where the *commodities* lie.

Of course, all of this sounds rather cynical, as perhaps it should. But on the other hand, it is argued that if the United States did *not* vigorously seek to insure its access to the world's goods, other countries would soon outstrip us economically and militarily. We might then be placed in the position of eventually being denied essential commodities.

Perhaps the real question should center over what should be regarded as essential. Do we really *need* the oil which lies off the coast of Vietnam? The answer is assuredly yes, if we continue to cultivate our appetite for petroleum products. Perhaps the next question is, then, will we be able to reduce the likelihood of war by controlling our appetites? This is a complex question, but one which we will approach in the next chapter.

THE PROXIMAL CAUSES OF WAR

In an effort to determine the proximal causes of war, Robert C. North and others at Stanford University have correlated involvement in wars in modern Europe with population growth, rising Gross National Product (in 1972, President Nixon posed proudly for photographers below a sign showing a record GNP for the United States), and expanding military budgets. These were only correlations and, as such, were subject to debate over their interpretation. However, a later, more complex analysis, showed a causal chain which involved population growth in areas with unchanging or slowly growing resources, high technological development, increases in the demands of consumers, and a tendency to invest energy beyond normal limits. On the other hand, North found that countries in nonaggressive phases, such as Sweden, are likely to have relatively small and stable populations, high and developing technology and good access to resources, either domestic or as a result of favorable trade.

THE PROSPECTS

It's hard to guess how long we will be able to keep ourselves out of war. The prospects do not appear very good. The next conflict, for example, could well be related to oil. We are desperately in need of prodigious amounts of the slippery stuff. The TV commericals are not wrong; we have an oil-based economy. Powerful and self-serving oil lobbies have, over the years, managed to make us utterly dependent on their product.

Now that we are beginning to understand the meaning of "finite resources," there is some anxiety over our continued access to foreign oil. In an effort to reduce our present need for imported oil, we are currently preparing to ravage some of our most beautiful and irreplaceable Western mountains in an effort to squeeze the last drop out of their oil-bearing shale. We are also in the process of building an environmentally hazardous pipeline to funnel oil from the North Slope of Alaska to our waiting machines. To reduce our dependence on oil we are building potentially hazardous nuclear power plants (see Chapter 14). *Yet, currently, our demands for energy are doubling every eight years.* We continue to respond faithfully to the imploring merchant. Each family is seeking to own more goods, newer cars and niftier gadgets as they continue to measure their value as citizens according to the list of their possessions. Unrestrained, these good citizens will continue to propel their government to perform abominable acts to provide "things" for them.

However, restraint, by definition, implies loss of freedom. Freedom has long been important to our society. The basis of the American economy is even called "free enterprise." So here is a dilemma. If individuals continue freely in their search for more and newer goods, they will be brought into more intense competition as a result of dwindling resources and burgeoning demand. We cannot expect wars to be averted as shortages in one commodity are followed by the realization that yet another is becoming scarce. The answer must lie in some system of effective international law. All such attempts thus far have not been encouraging but perhaps even self-serving men can soon be made to see that worldwide control of commodities is in their own best interest. We might also consider the creation of a world body which could effectively enforce a lowering of the destructive capabilities of nations; but here, the problems seem great also.

NEW APPROACHES TO THE OLDEST PROBLEMS

At least part of the problem in achieving worldwide solutions is that national leaders are almost always old men—even in youth-

Box 13.3 Disarmament "The basic need is evident; once again it is a change in human attitudes so that the 'in-group' against which aggression is forbidden expands to include *all* of humanity.

If this could be accomplished, security might be provided by an armed international organization, a global analogue of a police force. Many people have recognized this as a goal, but the way to reach it remains obscure. The first step might involve partial surrender of sovereignty to an international organization. It seems probable that as long as most people fail to comprehend the magnitude of the threat, that step will be impossible. At the very least we must learn to weigh the risks inherent in attempting controlled disarmament against those run by continuing the arms race. An attempt at disarmament could lead to a war, or the destruction or domination of the United States through Chinese or Soviet 'cheating.' But, if we were successful at disarming and if we achieved an international police force, the reward would be a very much safer world in which resources would be freed for raising the standard of living for all humankind. Few problems deserve more intensive study. The dynamics of disarmament appear to be even more complex then those of arms races. In spite of this, in 1970 the Arms Control and Disarmament Agency (ACDA), the only U.S. agency charged with planning in this area, had a budget of only a few million dollars (contrasted with $80 billion for 'defense'). Moreover the ACDA is heavily influenced by the State Department, a stronghold of cold war bureaucracy."

(From *Population, Resources, Environment: Issues in Human Ecology*, Second Edition by Paul R. Ehrlich and Anne H. Ehrlich. W. H. Freeman and Company. Copyright © 1972.)

Box 13.4 Values "We have but one explicit model of the world and that is built upon economics. The present face of the land of the free is its clearest testimony, even as the Gross National Product is the proof of its success. Money is our measure, convenience is its cohort, the short term is its span, and the devil may take the hindmost is the morality.

Perhaps there is a time and place for everything; and, with wars and revolutions, with the opening and development of continents, the major purposes of exploration and settlement override all lesser concerns and one concludes in favor of the enterprises while regretting the wastages and losses which are incurred in these extreme events. But if this was once acceptable as the inevitable way, that time has passed.

The pioneers, the builders of railroads and canals, the great industrialists who built the foundations for future growth were hard-driven, single-minded men. Like soldiers and revolutionaries, they destroyed much in disdain and in ignorance, but there are fruits from their energies and we share them today. Their successors, the merchants, are a different breed, more obsequious and insidious. The shock of the assassination of a President stilled for only one day their wheedling and coercive blandishments for our money. It is their ethos, with our consent, that sustains the slumlord and the land rapist, the polluters of rivers and atmosphere. In the name of profit they preempt the seashore and sterilize the landscape, fell the great forests, fill the protective marshes, build cynically in the flood plain. It is the claim of convenience for commerce—or its illusion—that drives the expressway through neighborhoods, homes and priceless parks, a taximeter of indifferent greed. Only the merchant's creed can justify the slum as a sound investment or offer tomato stakes as the highest utility for the priceless and irreplaceable redwoods.

The economists, with a few exceptions, are the merchants' minions and together they ask with the most barefaced effrontery that we accommodate our value system to theirs. Neither love nor compassion, health nor beauty, dignity nor freedom, grace nor delight are important unless they can be priced. If they are non-price benefits or costs they are relegated to inconsequence. The economic model proceeds inexorably towards its self-fulfillment of more and more despoliation, uglification and inhibition to life, all in the name of progress—yet, paradoxically, the components which the model excludes are the most important human ambitions and accomplishments and the requirements for survival."

(From *Design With Nature*, Copyright © 1969 by Ian McHarg. Reprinted by permission of Doubleday & Co., Inc.)

worshipping America. This is probably good since, presumably, age brings wisdom. But these leaders can be expected to be steeped in the principles which seemed to operate so well in the world of only a few decades ago, and they can be expected to be reluctant to accept the sorts of drastic measures which are now called for. This is not to say that being young automatically imbues one with new wisdom, especially if the young manage to remain ignorant. However, answers must lie in new ways of thinking.

In most political systems, including our own, the problem is complicated by the nature of the systems themselves. From among a generally competitive populace, it seems that those who are able to rise to high office must be among the most competitive and aggressive of all. For every election winner, there are scores of losers—people who competed and lost. In other words, these kinds of political systems automatically select from among the most hard-driving competitors. It is these people, then, who make the decisions regarding the course their nations will take. What sorts of decisions can we expect from them?

Also, particularly in the United States, political campaigns are costly and those driven souls who reach high office are likely to be beholden to special-interest campaign contributors. In many cases these contributors are in the position of needing resources which cannot be furnished from within our own boundaries. Thus the likelihood of foreign policy dedicated to providing these resources increases—and so does the risk of war. It seems apparent that we must either change our system of values so that "gain" loses its status as our loftiest ideal—or we must realize that it is perhaps more important to gain freedom and a sense of peace, than a faster car and a bigger house.

14
Resources, Pollution, and Values

(Courtesy: Navajo Reservation, 1966 and today, Gail Weingart, Hap Stewart)

The unprecedented swell of the human population is placing new kinds of demands on the world's resources. It seems that each new member of the species is born with a burning desire to drive a Porsche. However, in most cases throughout the world, the more immediate demand is not for the exotic symbols of status, but for the "simpler" things, such as food. It has been estimated that as the world's population grows, two people join the ranks of the hungry for each one which will be adequately fed. So we might consider the human species as being composed of two more or less distinct groups, those seeking the things necessary to maintain their very existence and those who have the necessities of life and are seeking other things.

We might also divide the world's resources into two groups, those which are renewable, such as crops and timber, and those which are nonrenewable, such as oil and tin. We will explore the critical nature of this difference later, but it should be made clear that no matter what the nature of the resource, it is subject to steadily growing demands. These demands are beginning to reach such proportions that we can no longer be content with "asking ourselves questions" regarding how our resources will be used. We had *better* begin providing clear, carefully chosen answers. It may be that these answers will be unwelcome or even dreaded. Nevertheless the questions cannot be avoided. A mountain climber seeing a rockslide above him is in much the same predicament; he must make a decision or be lost. His decision to move in a certain way may save him, but he has no guarantee of that. He may be lost anyway; still, he *must* make his move.

But exactly what is the nature of our own predicament and what kinds of decisions have been made so far? Let's begin by considering how we are doing with one of our most basic necessities: food.

FOOD

You have eaten recently or you wouldn't be sitting here reading;

Figure 14.1 Much of the problem with the world's commodities lies in their distribution. It is difficult for most Americans to grasp the level of poverty endured by most of the rest of the world. A starving child in Biafra has little to do with us at home. (Courtesy: Action and United Press International)

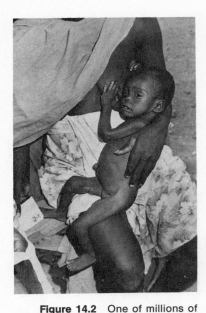

Figure 14.2 One of millions of babies who have died of starvation in the past few years. The swollen belly and scrawny limbs are common in such cases. One reason for this tragedy is that the mother cannot produce milk unless her own diet is adequate. One of the most recent examples of mass starvation occurred in the drought-ridden Saheal area of Africa.

During this drought, it was reported by news sources that American milk producers had dumped about a quarter million gallons of skim milk on the ground in order to raise prices for their product. (Courtesy: Action)

you would be out hustling food. It may be hard for most of us with time to read books to envision the hunger and starvation which exists on our planet. But perhaps 10 or 15 people who were alive when you started this paragraph are now dead because they didn't have enough to eat.

The sight of starving people is a grim one. Once, when I was working in the West Indies, I saw Haitian children, black skinned, with swollen bellies, tottering about in the dust as they slowly died of *kwashiorkor*, a disease caused by protein starvation. This strange sickness, which causes the hair to become reddish and brittle, usually appears when the child is weaned. I was told by a group of missionaries at dinner one night that even if these children should begin a good diet, they would probably be mentally retarded. I found out later that this is because the brain grows rapidly in the very early years and requires high levels of protein for its normal development. According to N. S. Scrimshaw and J. E. Gordon, in their book *Malnutrition, Learning and Behavior,* if the growing brain is deprived of protein at this time *irreversible* abnormality will result. These are great human tragedies but they are not confined to small, isolated areas of our world.

Of the 60 million people who die each year, it has been estimated by French food specialists René Dumont and Bernard Rosier that 10 or 20 million of these deaths are due to starvation or malnutrition. This figure is regarded as too high by some authorities but it should be remembered that starvation (due to the lack of food) or malnutrition (due to an unbalanced diet) may *indirectly* bring about death. The coup de grace may be administered by some disease or parasite after the person has been weakened by an inadequate diet. Environmental biologist Paul Ehrlich considers deaths by starvation to include all those which would not have occurred if the person had been well fed.

THE DISTRIBUTION OF FOOD

Why are people starving? Isn't there enough food to go around? In 1965, the President's Science Advisory Committee placed the world average caloric needs at 2354 per person each day. The United Nations Food and Agriculture Organization has estimated that there were, in 1965, 2420 calories per person available in the world's marketplaces. So, in terms of calories (ignoring protein, vitamin and mineral needs), there probably is enough to go around —barely. The fact that people are starving means the problem is partly one of distribution. The rich have more than their share, the poor have less.

Another problem involves storage. At least 10 percent of the

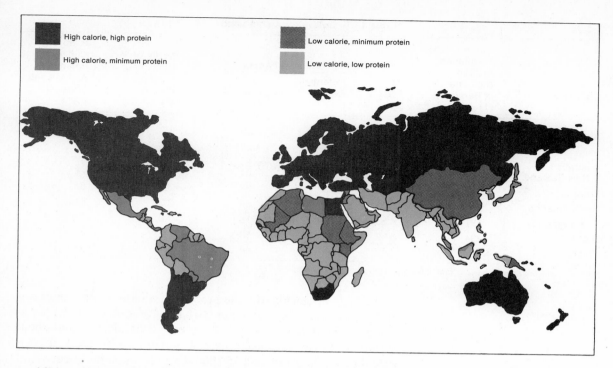

| High calorie, high protein | Low calorie, minimum protein |
| High calorie, minimum protein | Low calorie, low protein |

world's available food is destroyed by pests, waste, and spoilage somewhere between the marketplace and the stomach of the consumer. Look at Figure 14.3 to see how the world's food is distributed.

In the best of worlds, it would seem that there would be a net flow of food from richer areas to poorer ones in an effort to alleviate the plight of the hungry. But is this the case on *our* small planet? It is true that the developed nations send about 2.5 million tons of gross protein to the undeveloped countries each year. The gifts are well-publicized. It is less well known, however, that the developed countries *take* about 3.5 million tons of *higher quality* protein from the poorer countries in the same period. Much of our shrimp, for example, comes from India. Protein-deficient Latin America provides us with much of the fish which we convert to dog and cat food. (I must confess to a particular ill-feeling toward the producers of those commercials of imbecilic families following finicky cats around with cans of various high-protein foods until they gleefully find one it will accept.) Sadly, it seems inevitable that sooner or later we are going to have to place strong restrictions on the ownership of pets. I say this reluctantly since I live with fish, a turtle, three mice, two chinchillas, a hamster, a cat, a dog and a horse. Actually,

Figure 14.3 Protein-calorie comparison throughout the world. (From *Human Ecology: Problems and Solution* by Paul R. Ehrlich, Anne H. Ehrlich and John P. Holdren. W. H. Freeman and Company. Copyright © 1973)

Table 14.1—Cultivated Arable Land

| Continent | Population in 1965 (Millions of Persons) (1) | Area in Billions of Acres | | | Acres of Cultivated* Land per Person (5) | Ratio of Cultivated* to Potentially Arable Land (Percent) (6) |
		Total (2)	Potentially Arable (3)	Cultivated* (4)		
Africa	310	7.46	1.81	0.39	1.3	22
Asia	1,855	6.76	1.55	1.28	.7	83
Australia and New Zealand	14	2.03	.38	.04	2.9	2
Europe	445	1.18	.43	.38	.9	88
North America	255	5.21	1.15	.59	2.3	51
South America	197	4.33	1.63	.19	1.0	11
U.S.S.R.	234	5.52	.88	.56	2.4	64
Total	3,310	32.49	7.88	3.43	1.0	44

SOURCE: President's Science Advisory Committee, *The World Food Problem* (1967).

the horse lives outside. But the problem is a serious one. For example, it is estimated that in Santa Barbara, California, with its population of about 75,000 people, there are 60,000 dogs in and about the city. Each day the 500,000 owned dogs in New York City deposit 90,000 gallons of urine and 150,000 lbs. of feces on the streets. And it seems that about half of this ends up on someone's shoes! Much of the rest of the protein which is imported to developed countries is fed to poultry and livestock, since the wealthy can afford to eat from the top of the food pyramid.

INCREASING THE PRODUCTION OF FOOD

Because of our growing numbers, we must greatly increase the world's food production just to maintain our present level of food abundance. So how are we doing? One effort has involved opening more land to cultivation. Table 14.1 shows the amount of land presently under cultivation as compared to potentially arable land. The term *potentially arable* is somewhat misleading, since much of this land cannot be used without irrigation. The immense problems of finding water and providing irrigation conduits mean that, in Asia for example, almost all the land that *can* be cultivated *is* being cultivated.

Then there is the problem of cost. It has been calculated that on the very conservative assumptions that one acre of land will support one person, and that it takes only $400 to develop each acre, the world would need $280 billion in the next ten years just to feed the people who will be *added* to the population!

Box 14.1 How Hunger Kills   What happens when someone starves to death? What does starvation do? The first thing is that the person loses weight. The living organism must metabolize and if there are no nutrients entering the gut, the victim will burn his own tissues for fuel. The body literally digests itself.

A person is considered to be starving when he has lost about 33 percent of his body weight. When he has lost 40 percent, death is almost inevitable. As starvation progresses, various organs begin to lose their efficiency. The liver, kidneys and endocrine system may cease to function properly. Lack of carbohydrates may affect the mind so that the person becomes listless and confused, often seemingly unable to understand his plight. Kwashiorkor and vitamin-deficiency diseases can set in. Soft bones may appear in children who lack vitamin D. Lack of thiamine may cause memory lapses. Lack of niacin causes skin inflammation, diarrhea, insanity and death.

Children are often permanently affected since improved diet cannot straighten bones or build normal brain cells, but adults can often approach starvation and then recover to a degree, as we know from the rescue of concentration camp inmates. But these people usually die sooner than their better fed contemporaries.

It was once suggested that the food problem could be greatly alleviated simply by opening up the great jungles of the world to cultivation. Brazil made such an effort in setting up an agricultural colony in the Amazon Basin. The program failed dismally, however, because tropical soils are desperately poor in nutrients. You may find this surprising, since jungles seem to be so lush and rich. There are great stores in tropical foliage, to be sure, but when that foliage falls to earth and decays the nutrients are quickly washed from the soil by the torrential rains. This difficulty is compounded by the problem of clearing an area in the face of almost overnight encroachment by rapid jungle growth. And then there is the problem of trying to till this soil. In many areas when tropical soil is exposed to sunlight and oxygen it forms *laterite*, a hard rock-like substance. The beautiful and enduring temples of Angkor Wat in Cambodia are made of laterite and sandstone.

This is not to say that tropical jungles *cannot* be made to produce food for human populations, but the problem will have to be approached much more carefully than it has been thus far. We're going to have to make concentrated efforts to develop new farming techniques and to learn more about how indigenous peoples have been able to successfully cultivate such areas in the past. Whatever efforts are made to make land more productive will have to be considered very carefully. Important mistakes have been made when developers failed to consider the full impact of their schemes.

Figure 14.4 The rice at left is a traditional variety. When carrying a high yield, the tall, lean plants tend to "fall over." Therefore, the rice at right was developed. It is shorter and can hold up to three times that of the traditional plants. "Miracle grains" such as this rice will help relieve the world's hunger. (Courtesy: UPI)

MIRACLE CROPS

In the late 1960s there was a great fanfare heralding the impending "green revolution." New kinds of grains had been developed which yielded plants heavy with seed. For example, one strain of dwarf rice, IR-8, produced over twice as many rice grains per plant as the conventional strains. The new miracle crops do, indeed, hold great promise for the world's populations. However, they have their problems.

First of all, the new strains require special handling. For the nations most in need of food to be able to use these crops, farmers must be carefully trained in new agricultural methods. However, there has traditionally been a reluctance of people steeped in their traditions to adopt new ways. There is also a problem in under-developed countries of providing trained technicians to teach the farmers. In some cases the technology itself is useless. How can controlled-temperature farming be implemented in areas which have no electricity?

Miracle crops also must be heavily fertilized, and fertilizer of course costs money. Then there is the problem of the fertilizer run-off. As rains wash fertilizer from the fields into bodies of water, the nutrients may cause the water to *eutrophy*, or "age," prematurely. Nutrient-rich water may also be subject to great blooms of microscopic algae which can interfere with the balance of life and cause massive fish kills. The "red tides" of the oceans, when the waters

Box 14.2 The Aswan High Dam   The Aswan High Dam, built in Egypt with Russian support, was supposed to provide hydroelectric power and to increase Egypt's food supply by controlling the unpredictable Nile River. The project meant that great art treasures were flooded as submerged land was drained for cultivation. However, only one tenth of an acre of land was made available for each person added to Egypt's population during the period of construction. One result of the dam was that the Nile no longer flooded the delta farmlands annually. These annual floods served to restore the farmland fertility with deposited silt. This no longer the case, the quality of the farmland decreased. The dam also cut off the nutrients which had been washed to the Mediterranean sea as a result of the annual floodings. Because of this, or the change in the salinity of the sea which the dam produced, the sardine catch dropped from 18,000 tons per year to 500 tons per year. The stable lake created by the dam allowed aquatic snails to flourish. The snails serve as an intermediate host to a blood fluke which bores into humans causing the dreaded disease, schistosomiasis. The construction of the dam had important political implications at the time. These tombs had to be moved to be saved from the dam's waters. The political scene has changed now. So has the environmental one.

(Courtesy: Monkmeyer Press Photo Service)

strangely turn red and fish die from lack of oxygen, are the result of such blooms. An increase in agricultural production, then, may reduce the catch of fish in offshore waters.

World grain production currently amounts to about 1.6 pounds per person per day, with wheat, rice, and corn accounting for about three-fourths of the total yield. Since grains are concentrated sources of energy and are easily stored, they probably represent a good focus for continued research effort.

Ecologists such as Guy Bush at the University of Texas have warned of the consequences of relying too heavily on artificial strains of grains with low genetic variation. One of the dangers cited is that wild populations of insects or disease organisms, replete with their wide genetic variation, might through the processes of natural selection quickly accommodate themselves to any single strain and wipe it out in short order. Thus in the long run we might be better off paying the price of lower productivity for increased genetic variation in our crop species.

This point was dramatically illustrated by an unusual event in 1969. Specially developed high-yield corn plants were attacked by a new type of fungus called the southern corn blight. By 1970 the epidemic was so severe that the United States corn crop was reduced by 15 to 20 percent (a loss calculated at more than a billion dollars). The tragedy jolted many agriculturalists into a new appreciation of how food supply is a function of biological realities. The lesson was a simple one: extreme uniformity of our food crops makes them particularly vulnerable to the appearance of new kinds of pathogens or predators.

DOMESTIC ANIMALS AS FOOD

One of our biggest problems is not in just providing more food, but in providing the essential kinds of food. As you recall, each time the food produced by plants is transferred to a higher level in the pyramid, most of the energy of the preceding level is lost. However, the use of animal protein from high on the pyramid is advantageous for a number of reasons. Since we are animals, the flesh of other animals provides us with the eight or ten essential proteins in proportions which are relatively close to our needs. To get all the various proteins we need from plant tissue, we would have to consume large volumes of plant bulk and a wide variety of plants. Also, for many of us, meat tastes better than plant food.

It is important, however, which animals we choose to raise as sources of food. For example, goats are a major source of milk, but the goat is a destructive grazer and is best suited for marginal areas that have no other use. In their overgrazing, goats have contributed to ecological disasters by stripping the soil of its protective plant layer. Sheep also tend to overgraze; in addition, sheep are not universally popular as food items and because of their heavy wool coats they are not well suited to tropical areas. Hogs, on the other hand, can be reared in a wide range of habitats and are extremely efficient in converting plant protein to animal protein. The trouble is that their dietary habits make them direct competitors with man for the available plant food. In fact, like man, hogs are *omnivores*—that is, they eat about anything—animals as well as plants. Hence whereas they compete very directly with man, they can also subsist on man's refuse.

Cattle, because of their complex digestive arrangement, are able to survive on plants which are not food sources for humans. Since they ordinarily don't wander from their grazing areas, they don't require constant attention. Moreover, they travel well, are large enough to be used both as beasts of burden and for other work, and they provide leather as well as beef. In addition, the milk of cattle is a food staple in certain areas whereas not many people think much of pig milk.

FISHING

Perhaps fishing should be included in the section on nonrenewable resources, because, in a very real sense, we do not replace the fish we take. Our fishing practices are more akin to mining.

There are many paradoxes associated with our fishing practices. The world fish catch more than tripled between 1938 and 1968. At that time the Food and Agriculture Organization estimated the catch at 64 million metric tons (Figure 14.5). Since then, there have been some alarming new trends. In spite of increasing worldwide efforts by the fishing industries, the fish catches are actually diminishing in many areas. However, since the oceans are a prime source of food for the world's burgeoning human population, the pressure on fish populations is *not* being relaxed. Instead, the fishing industries are diligently studying fish behavior and developing their technical arsenals as they continue their drive to meet the demands of our own hungry numbers, although fish play a relatively minor role in the world's total consumption of protein (Figure 14.6).

The management of the oceans is complicated by what biologist Garrett Hardin refers to as the "law of the commons." For the most part the decisions made by various nations are not compatible with optimal productivity. For example, the decision to add another vessel to the fleet may be marginally productive to the country, but in the long run may contribute to a diminished catch because of overfishing. Yet if that nation chooses not to add another vessel,

Figure 14.5 From 1938 to 1968 the world's fish catch more than tripled. By far the greater part consists of marine fishes. Actually, humans directly consume only about half the catch. Much of the rest is fed to live-stock in an incredible display of nutritive inefficiency. (From *"The Food Resources of the Ocean"*, S. J. Holt. Copyright © 1969 by Scientific American. All rights reserved.)

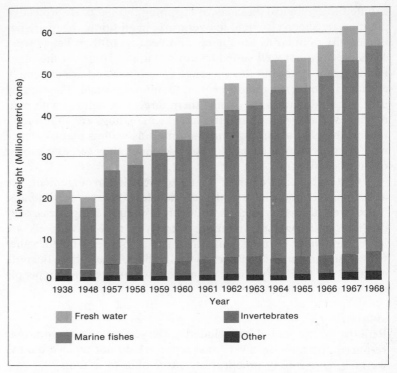

it may find itself on the short end as other nations continue to add vessels. The first nation comes up short both ways. So what do you think its decision might be? What will be the long-range effects of such decisions? Can you think of some better way of apportioning the world's resources?

OTHER IDEAS

There are also other ideas regarding how to feed the world. Some are conceptually fascinating, but as yet unproved. For example, it is known that single-celled organisms, which are high in protein, can be grown on petroleum. If these organisms could be produced in large enough amounts and purified economically, they could conceivably provide us with high-grade protein. Other scientists are busy trying to produce yet newer grains with higher quality proteins (which means containing a better balance of amino acids). A nutritious mixture of corn and cotton seed meal enriched with vitamins A and B has been available in Central America for over ten years but unfortunately has not been accepted by the people. This is one of the great problems of food scientists—how to get people to eat what is being newly developed. Historically, the hungriest people

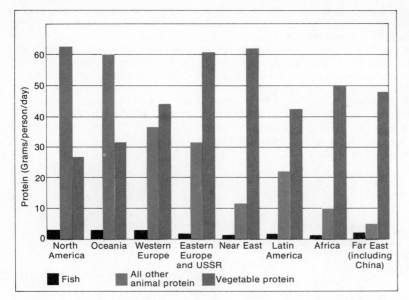

in the world are those which recognize the fewest items as food.

And then there are a variety of other ideas being considered, such as culturing algae, ranching various rodents or antelope, converting weeds to cattle feed, extracting protein from plants or small fish, converting wood to cattle feed, and using sewage to grow edible slime. It's likely, however, that quite a few people will have trouble accepting most of these as food unless they can be made palatable through processing techniques, or unless the hunger level rises high enough.

It is interesting that of the thousands of species with which we share this planet, only a handful have been designated food items. In fact, it turns out that modern man, with his vast abilities and imagination, has virtually discovered *no* new crop plant or domestic-able animal. Of course, food may be imported into one area from another. Captain Bligh was directed to take the *Bounty* to bring back a "new" plant which had long been cultivated in the South Pacific. New kinds of fish meals are still made of fish. And the miracle grains are simply variations of plants which we have used for thousands of years. Recently, a promising crop "species" was developed which is a hybrid of wheat and rye, old standbys.

One is tempted to ask whether our energies *should* be directed toward frantically trying to provide a minimal diet to teeming millions of people as their numbers grow increasingly faster. Should this be the role of our scientists and the lot of modern man? Or should we begin to put strong pressures on the swell of our num-

Box 14.3 Triage During World War I a "triage" system was put into effect to determine which wounded would receive priority in field hospitals. Medical attention was based on the condition of the wounded. Those who were so injured that it seemed they would probably not recover even if treated, were sedated and left to die. Time and energy were spent only on those which, with help, could probably survive.

Some scientists are now advocating such an approach on a worldwide basis. The idea is that food or other supplies should only be sent to those countries which could benefit from such assistance. The poorest countries with the highest birthrates and least likelihood of being able to survive would be left to collapse of their own weight. The suggestion is based on the belief that if an attempt is made to distribute resources to all nations the demand will be too great and no nations will be able to survive in anything like their present form.

(Courtesy: The American Red Cross)

450 *Biology: The World of Life*

bers? Should we stop asking how we can manage to keep ever greater numbers of people barely alive, and begin to ask how we can improve the quality of life for those already living? Some of the answers being considered are grim indeed.

But food is only one problem and one type of problem. Let's take a look at some of our nonrenewable resources.

NONRENEWABLE RESOURCES

The line between renewable and nonrenewable resources is a fine one. For example, water is considered cyclic in that it evaporates and falls again as rain and thus is "renewable." However, much water is tied up in various manufactured and natural products and thus is not easily retrievable. Some water also becomes so polluted that it is no longer usable. For example, the atomic wastes dumped by the AEC onto bare earth will undoubtedly seep into the water tables below, rendering that water unusable for many thousands of years. It is thus nonrenewable.

For purposes of this discussion, we will consider nonrenewable resources to be those naturally occurring materials, such as minerals, whose amounts are fixed and can therefore be depleted.

In our consideration of these commodities we must remember that they do not occur uniformly throughout the earth. Hence, their use may be manipulated by nationalistic policies, thereby potentially heightening the chance of conflict over their distribution. Consider the imbalances in metals. Asia is rich in tin, tungsten, and manganese. In fact, over half the world's tin is located in Indonesia, Malaya, and Thailand. North America, on the other hand, is poor in these elements, but well-endowed with molybdenum. Gold, of course, occurs mostly in South Africa; so does platinum, but not much silver is found there. Over half the world's nickel is in New Caledonia and Cuba, and most mercury is found in Spain, Italy, and the Sino-Soviet areas. Because of the implications of such resource distribution, let's look at resource reserves both for the world and for the United States.

It should be apparent from Figure 14.7 that there are certain problems we can expect if we continue to "develop" (and "exploit") these resources on a "devil take the hindmost" basis. In another twenty-five years there will be only eleven of them left in the world, and only four in the United States. (How old will you be by then?) By the year 2042 the world will have only eight of these left, and the United States three. Also, barring deliberate population-control measures or some devastating catastrophe, there may well be 15 billion people on earth by that time.

Of course, these figures are only estimates. Some will prove too

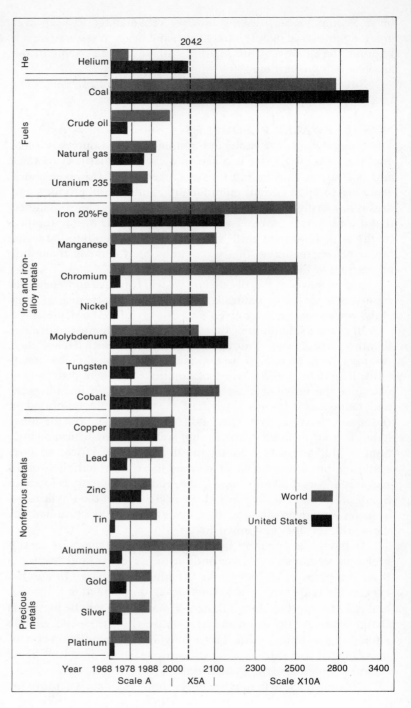

Figure 14.7 Estimated natural resource reserves for the world and the United States.

small and others too large. We will undoubtedly begin to recycle more materials as we find new deposits and develop new technology. At the same time, the population will probably continue to boom for some time, each new human replete with his needs and propensity to acquire.

Geologist Preston Cloud points out that we can make the best predictions for those substances which are found associated with specific rock layers. These include the fossil fuels—coal, natural gas, and oil—and to a degree, iron and alumina. The reserves, grades, locations, and recoverability of many other critical metals are much harder to estimate. Thus the problem of developing long-term policies regarding their use is magnified. Should we continue to drain each known deposit, confident that others will be found because they *must* be found?

The approach to such problems taken by the United States is particularly important for a number of reasons. Since the United States has been technologically developed for a relatively long time, we have exploited our own resources to a greater extent than most other countries. Whereas a great wealth of undiscovered resources may exist in Russia, for example, we are probably much more knowledgeable regarding our own reserves. It seems that we would therefore be more acutely conscious of the limits of natural resources and would take a rather cautious approach to their removal or exploitation. For some reason, though, it doesn't work that way. Recall the power plays of the oil industries as they gained governmental approval to extract the oil from the North Slope of Alaska—to transport it by ecologically hazardous pipelines to loading facilities at the port of Valdez. Also, at the time of this writing, hearings are being held in Santa Barbara regarding whether new offshore oil wells will be permitted in the seismically unstable Santa Barbara Channel, where the water continues to be polluted from the seepage left from the great 1968 spills. (I will make the prediction that by the time this book goes to press, the oil companies will have the approval they seek and will be merrily and profitably drilling away in the geologically unstable channel. This is a comment on the power of the oil companies, not the speed of my publishers.) Now the oil executives are stroking their chins and wondering how much oil they can get from under the waters of the Eastern Seaboard.

It seems that, in spite of increased public awareness of the importance of developing ecologically sound policies, there has been almost no reordering of priorities by our government, or the government of any other developed country. Big business goes on as usual. But it is simplistic to lay the entire blame for policy errors at any single door. Certainly, the oil companies would not so vigorously search out the earth's oil if there were no public demand for it. We need that oil. But then, the demand exists, at least in part, because the powerful "highway lobby" has such inordinate influence in Washington. Thus Americans have only woefully inadequate means of public transport and must depend heavily on automobiles, automobiles which use petroleum products. Such circularity can only be broken, it seems, by a powerful national effort to free ourselves from the sinking vortex, and such an effort is not likely to be made in the absence of a large-scale shift in values. Unfortunately, there is no sign of such a shift on the horizon.

Another reason that the decisions made by the United States are especially critical has to do with our national appetite. The United States, with something over 5 percent of the world's people, currently uses about 30 percent of the world's resources! And that figure is expected to go much higher within the next ten years. (Other developed countries also consume much more than their "share." The U.S., Canada, Europe, the U.S.S.R., Australia and Japan consumed about 90 percent of the energy and the steel produced in the world.) Considering our own limited reserves, then, it is apparent that much of what the United States uses must come from outside its borders (Table 14.2).

Table 14.2—Net U.S. Imports as a Percentage of the Country's Consumption

Resource	1960 Imports	1970 Imports
Chromium	94	100
Nickel	72	65
Manganese	92	94
Tin	73	69
Petroleum	16	22

SOURCE: Interim Report of the National Commission on Materials Policy, April 1972.

To see how our consumption compares to some other countries, consider the following: in 1968 our steel consumption per person (production plus imports minus exports) was about 342 times that of Indonesia, 86 times that of Pakistan, 68 times that of Ceylon, 23 times that of Colombia, 9 times that of Mexico, twice that of Switzerland and France, 1.5 times that of the United Kingdom and the Soviet Union, 1.4 times that of Japan, and 1.1 times that of Sweden. In 1968 we used over one-third of the world's energy output, over one-third of its tin, one-fourth of its phosphate, potash, and nitrogenous fertilizers, and about one-eighth of its cotton. All

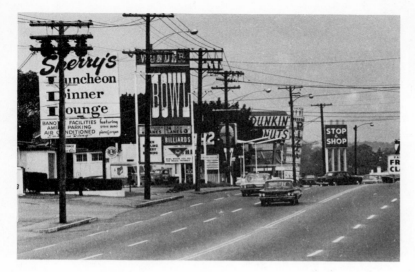

Figure 14.8 Signs like these are seen in every city in America and are becoming more numerous around the world. Besides being unsightly, they waste valuable energy. (Courtesy: Wide World Photos)

this by one-twentieth of the world's population. Thus the United States has an environmental impact far in excess of its numbers.

Not only is what we do important on the basis of our staggering appetite, but also because we are often imitated by other countries. I was once told by an emissary from a West African nation that the burning desire of his countrymen was to build another Los Angeles. Imagine that!

ENERGY

Energy is simply the capacity to do work. The problem of energy supply and utilization is, of course, a common denominator for all living organisms. Just as plants must acquire energy from the sun and animals must derive energy from plants, humans must also find energy to be able to rearrange trees and oil and iron to build houses so that they may be shielded from disruptive elements in their environment. Now, if the humans demand more or larger houses, then they need not only more resources, but more energy to rearrange the resources.

Some of the demands placed on energy resources are "legitimate" in that they relate to very basic needs such as food production or environmental shields. We may be reaching the point, however, where it will be necessary to distinguish such essential energy uses from nonessential demands. Do we really need elaborate neon-lit billboards, automatic can openers, oversized electronic amplifiers, electric toothbrushes, and overpowered six-passenger cars which are usually used to carry one person to work? But you could compose your own list of such nonessentials just as well. So, let's take a look at our energy resources and some of the demands being placed on them.

We should first make it clear that, at present, we are not running out of even the current forms of energy. The United States, for example, has vast deposits of coal. However, the utilization of these resources involves considerations other than availability. There are heavy costs to our environment both in producing energy and in consuming it, and these costs must be carefully weighed in calculating the net benefit of energy use. We will see again the enormous impact of the industrialized countries, particularly the United States, on energy resources (Table 14.3).

Table 14.3—Sources of Energy in 1968
(excluding food and the burning of wood and dung)

Source	Percentage of World Consumption	Percentage of U.S. Consumption
Coal	36.6	22.5
Petroleum	42.7	43.0
Natural gas	18.3	33.0
Hydroelectric	2.1	1.3
Nuclear	0.3	0.2

We will also consider, in terms of dollars, our efforts in the research and development of various energy sources. In a special report to the White House in December 1973, a government task force proposed a monumental increase in such efforts. In fact, it requested $10 billion be spent over the next five years. By way of gaining perspective, this is about one-eighth of the Pentagon's budget for 1974 alone.

WATER, WIND, EARTH, AND SUN

Moving water has been harnessed as a source of energy for many years. Waterfalls turned paddlewheels to grind corn in early America. Later, moving water was backed up behind great dams and then released to turn huge generators for *hydroelectric* energy. In a few places *tides* have even been used to produce hydroelectric energy. Since rivers and tides are likely to persist we can probably always count on hydroelectric energy to some degree. At the present, the world's *potential* production of hydroelectric power is roughly half that *now* produced by fossil fuel.

There are serious problems, however, in the use of hydroelectric power. To begin with, for a country to make full use of such a power source it must *already* be industrialized, and much of the potential hydroelectric power lies in nations that are now underdeveloped. Another problem involves dams, which are after all only temporary structures. They must be carefully monitored, and in time they must be removed or replaced. Also, they tend to "silt in," so that they can lose their control of water flow. These problems, of course, ignore a central question: do we really want to dam up all our free-flowing rivers?

How about *wind*? Wind has provided cheap and pollution-free energy throughout much of Western Europe and the American Southwest, but windmills are presently often treated as quaint relics. Part of the problem is that wind is a rather local phenomenon and is, as any sailor knows, extremely unpredictable in most places. Theoretically, however, wind might be able to provide significant amounts of energy if the proper technology could be developed for utilizing and storing its power. Although no such large-scale effort is underway, a federal research grant has recently been awarded to at least one educational institution to design and build an effective wind-powered generator.

One of the most promising sources of power is *geothermal energy*. This means using the steam that is produced when the molten masses deep below the earth's surface heat the water trapped above. A few countries, such as Italy, Iceland, and New Zealand, are already using geothermal energy effectively. However, critics say that such sources can be developed only in a few locations around the earth, and that there is no way to estimate how long the superheated water will last in these areas.

And then there is *solar energy*. It has the obvious advantage of being as common as sunlight. The amount of sunlight striking different parts of the earth varies, of course, depending on the latitude and the local weather patterns, but with our present technology it is feasible to utilize solar energy in most parts of the United States.

Because sunlight is a rather low-level energy source, a large surface is needed to collect it. In fact, a collecting device designed to meet the energy needs of a city of 750,000 people would require an area of about 16 square miles (assuming a solar-to-electric energy conversion efficiency of about 10 percent).

Sunlight is already dispersed, so that there is no need to distribute it from a central source. Therefore, the roofs of individual houses may provide surfaces large enough to run individual heating and cooling units and to heat and cool water.

Solar heating may be one of the most realistic solutions to the current energy problem. However, for some reason, it is not receiving the research attention it warrants. Although $13.2 million in federal funds was allocated to solar-energy research and a total

Figure 14.9 The John Day Dam. Such dams devastate large natural areas. However, they house large generators which provide a low cost source of electrical energy. (Courtesy: U.S. Army Photo and U.S. Bureau of Reclamation, photo by J. D. Roderick)

Figure 14.10 This windmill once used the wind to grind wheat. Other windmills use the wind to generate electricity. Once considered obsolete, the concept is being reconsidered as a source of modern power. (Courtesy: Wide World Photo)

Figure 14.11 Geothermal power plants at Geyserville, California. The development of such plants, at present, has received little attention in the United States. The "smoke" is actually harmless steam. (Courtesy: National Science Foundation)

Figure 14.12 A solar heating system (far right) is used to heat this junior high school in Minneapolis, Minnesota. (Courtesy: National Science Foundation)

of $200 million is earmarked for such use through 1979, for purposes of comparison, this is less than $\frac{1}{40}$ the military budget for 1974 alone.

FOSSIL FUELS

The supplies of petroleum, coal, and natural gas will not be quickly depleted, but they are certainly not going to hold out indefinitely. Coal, the most abundant of these, will probably be largely consumed within the next few hundred years, but petroleum and natural gas will go much sooner. Geologist M. King Hubbert, who has formulated the most reliable estimate to date, states that these latter two resources will be substantially depleted in about 100 years. This does not mean that all the earth's fossil fuels will be gone, but the residual amounts will, for the most part, be so small that their extraction would be economically unfeasible. It is possible that mining for such products might continue if a tiny vial of petroleum should come to have the approximate value of a rare and exotic perfume. Plastics, clothing, vitamins, phonograph records, and carbonated water would, of course, be priced accordingly.

Before we reach that stage, however, we will undoubtedly have ravaged the great oil-laden shale deposits in the vulnerable mountains of Colorado, Utah, and Wyoming. Proposals are already under consideration to strip mine these mountains in what sometimes seems to be a diligent, and certainly profitable, effort to remove the last of the American oil as soon as possible.

Future shortages may temporarily be diverted by converting coal to oil and gas. It is expected that economically feasible means will be developed in the 1970s. But, then, such a program would decrease the life expectancy of coal.

So how goes the funding? Coal research, in 1973, received $167.2 million, about $400 million in 1974 with the governmental task force proposing that $2.18 billion be spent by fiscal 1979.

THE ENERGY OF THE ATOM

One of the most talked-about solutions to our energy problems is *nuclear power*. Presently, nuclear reactors produce less than 1 percent of the energy used in the United States. However, this percentage is doubling about every two years.

At present nuclear power is based on *fission* processes. This means that when the nuclei of uranium atoms are split, part of the mass is released as energy. Today, we are able to utilize only 1 or 2 percent of the energy produced, but the fuel is rather cheap and enormous amounts of energy can theoretically be realized.

Box 14.4 Shale Oil   Shale oil is potentially an ideal fuel. We know where it is. There is no need for costly searches with hit-or-miss exploration techniques. It is present in immense quantities (estimates range up to 3 trillion barrels). It contains little sulfur. It's right here in the United States, 70 to 80 percent of it on government-owned land. And the price of such oil would be no higher than that of oil from other sources. The problem is, the shale land is among the most beautiful that the country has to offer. It lies in the mountains of Colorado, Utah and Wyoming. The decision has to be made, then, whether to devour the mountains to satisfy our demand for oil. The Interior Department has announced an experimental program to tap these reserves, "in the national interest."

In the extraction methods which are proposed, the shale is literally chewed up and cooked. The rock then yields a heavy oil similar to crude petroleum. Of the 8 million acres of oil-bearing shale, only six tracts of about 5,000 acres each will be involved in the initial tests. In two of the tracts, oil will probably be directly boiled out of its underground reservoir. A chamber will be blasted, then natural gas will be injected into it and set afire. The oil would then be extracted by special wells.

Another method may be the conventional "pillar-and-room" tunneling which is presently used in coal mines. But in the case of the shale the project would be gargantuan. Perhaps the most destructive method involves open-pit mining, the third alternative. This is the strip-mining that has so devastated coal-laden areas of Kentucky and West Virginia, and is presently being used to get at the copper near Butte, Montana. Here, enormous earth-moving machines scrape the earth bare of billions of tons of its natural cover to expose the shale beds 100 to 850 feet below the surface. These beds are then gouged out, forever destroying the contour of the land.

Environmentalists are concerned for a number of reasons. For one thing the waste shale occupies more space *after* it is boiled. Where will it all be put? The oil companies suggest they might fill in canyons with it, fertilize it, water it and plant it.

(Courtesy: United Press International)

In the future, much of our energy may come from "breeder reactors" which will permit the extraction of 40 to 70 percent of the energy of new uranium. These reactors, then, could theoretically supply our anticipated demands for electrical energy for thousands of years. But here lies an important point. Nuclear power can produce only *electrical* energy, and electricity accounts for only one-fourth of the current energy budget in the United States. This figure is expected to go up to one-half by the year 2000, but even so, nuclear power is not being considered as a replacement for fossil fuels in such areas as private transportation. Nor can it be, unless there are great changes in our concept of how we should be getting from one place to another.

Nuclear fusion may, in the long run, prove to be our best energy bet. Fusion involves the joining of heavy hydrogen nuclei under superhigh temperatures, forming helium nuclei and releasing huge quantities of energy. It is this process which produces the energy of the sun and the hydrogen bomb. We haven't yet learned to control fusion reactions for the sustained production of energy, but such an ability is projected by the end of the 1970s. The great advantage of nuclear fusion over fission is that very low radioactivity would be produced and hence the system is inherently much safer. In the discussion of radiation pollution we will consider some rather alarming reasons why we should proceed with the development of nuclear fission plants with the spirit of mating porcupines—very, very carefully.

Figure 14.13 The 330 megawatt San Onofre nuclear reactor in California. (Courtesy: Southern California Edison)

POLLUTION

A *pollutant* is generally defined as a harmful product which enters the environment. Such products may be in the form of noise, heat, chemicals or garbage. In this discussion we will consider the pollution produced by human activity, although nature may produce some of the same substances. For example, there is some natural seepage of oil from undersea deposits which has been entering the oceans for thousands of years. Oil companies accused of polluting oceans and beaches are fond of bringing up the possibility of natural seepage to account for the mess. And then, much of Los Angeles is lost to aerial view on days when weather dictates that the city should breathe its own effluent; but the eruption of volcanoes can also emit prodigious amount of ash and fumes into the air.

The present environment has been molded by such natural occurrences throughout the earth's history, just as the life on earth has adapted to such events. The kind of pollution which concerns us here, however, is that which, in kind or degree, is new to the earth and its creatures—that which has only recently begun to be pro-

Figure 14.14 Air pollution may actually dissolve buildings and statues and rain may become so acid as to be dangerous to many forms of life.

duced by the brainiest of the earth's forms. The reason for the concern is that the new pollutants may alter critical aspects of our environment so rapidly that much of the earth's life will not have time to adapt to the changes before it is destroyed.

Of course we're not simply concerned about how to continue living or *existing* on our planet, but we might also legitimately have an interest in protecting the *quality* of our lives. The once-beautiful Lake Erie is now so polluted that many cities along its shores require boaters to have typhoid shots in case they should accidentally fall in! But, because of the factories there we now have more paper cups and more brightly painted cardboard boxes to package toothpaste in. And such goods may define life's quality to some people. I once saw an old man fishing in New York City's East River as traffic honked, screeched and roared only a few yards behind him. I wondered if he remembered quieter days there when one could catch something besides sludge-eating bottom dwellers. I also couldn't help wondering what the term "quality of life" might mean to him.

But let's get on with the consideration of a few specific types of pollution.

AIR POLLUTION

If you live in a large city, you might be particularly concerned about air pollution because you can *feel* it. It burns your eyes and nose and may give you terrific headaches on the worst days. It may even make your days darker in a literal sense. Air pollution, for example, cuts by 40 percent the amount of sunlight reaching Chicago. You may also find it mildly "irritating" that in many cities certain air pollutants dissolve nylon stockings, windshield wipers and statues. You may be even more moved by the fact that air pollution can also kill crops and people.

Air pollution comes from a variety of sources, but principally from the automobile (Table 14.4). Fuel burned in stationary sources, such as in heating units for buildings, also largely contributes to foul air.

We should realize, also, that air pollution is not just the problem of big cities, nor is it restricted to industrialized countries. Meteorologists are now describing a thin veil of pollutants which hangs over the entire earth. Polluted air has even been found over the North Pole! Generally, however, the foulest air is found in the industrialized countries where it is particularly concentrated over large cities.

How serious is the threat of air pollution in such areas? Pretty serious. During one great smog in London in 1952, 4000 deaths

Table 14.4—Air Pollution Sources in the U.S. in 1968 (millions of tons)

Source	Carbon Monoxide	Particles	Nitrogen Oxides	Sulfur Oxides	Hydrocarbons
Fuel burned for transportation	63.8	1.2	8.1	0.8	16.6
Fuel burned in stationary sources*	1.9	8.9	10.0	24.4	0.7
Other industrial processes†	9.7	7.5	0.2	7.3	4.6
Solid waste burning	7.8	1.1	0.6	0.1	1.6
Agricultural burning, forest fires, and other	16.9	9.6	1.7	0.6	8.5
Totals	100.1	28.3	20.6	33.2	32.0

SOURCE: Reprinted from Man's Impact on the Global Environment,'' SCEP, W. Matthews, editor, p. 296, by permission of the M.I.T. Press, Cambridge, Massachusetts. Copyright © 1970 by M.I.T. Press, Cambridge, Mass.
*Major components are electricity generation, industrial process heat, and space heating for homes, commerce, and industry.
†Major contributors are pulp mills, smelters, refineries, and cement plants.

were *directly* attributable to the quality of the air. Indirectly, probably many more people were affected, because air pollution works it's way in subtle and insidious ways. Since air pollution kills slowly, it is difficult to attribute human deaths to pollution levels or specific contaminants. It has been shown, however, that death rates are higher in smoggy areas. The most likely to be affected are the very old, the very young and those with respiratory ailments.

Carbon Monoxide

Carbon monoxide is the most prevalent air pollutant, as you can see in Table 14.4. It is biologically important because of its tendency to combine with the blood's hemoglobin in the place of oxygen. The effect is to cut down the blood's supply of oxygen to the tissues, thus causing the heart to have to work harder to oxygenate the body. The increased demands on the heart place a severe strain on people with heart or respiratory ailments. Looked at another way, carbon monoxide can have the same effect as loss of blood. In fact, spending eight hours in an atmosphere with 80 parts per million of carbon monoxide has the same effect as losing over a pint of blood. It may interest you to know that in a traffic jam the air may contain nearly 400 parts per million of carbon monoxide. The people breathing this air may experience headache, loss of coordination, nausea, abdominal cramping, and even partial blindness.

Nitrogen Oxides

Nitrogen oxides have much the same effect as carbon monoxide in reducing the oxygen-carrying capacity of the blood. But nitro-

Figure 14.15 A view of a street in New York City. Traffic such as this has become a severe problem in our cities. Aside from the nuisance of traffic jams, the amount of carbon monoxide in the air over such areas is dangerous to the health of those who spend time in such areas. (Courtesy: Wide World Photos)

Box 14.5 The Effects of Air Pollution   Because the effects of air pollution are slow to show and because such effects often contribute to an already weakened condition, other detrimental influences which set in may receive the blame. Therefore, it is hard to pin anything on air pollution. But some hard evidence does show air pollution as being harmful. For example, it is known that the amount of head colds and pneumonia is higher in areas of high air pollution. Cigarette smokers from polluted St. Louis have four times the incidence of emphysema as do smokers from Winnipeg, Canada, where the air is cleaner. Chronic bronchitis is more serious among British postmen who serve high-smog areas than those who work in the more pollution-free districts. There have generally been sharp increases in emphysema deaths in areas where air pollution has worsened.

Lung cancer deaths in America and England have been correlated with the density of smoke in the air. In Staten Island, New York, 55 men in every 100,000 have lung cancer in the smoggiest areas, but a few miles away where it is less smoggy, the incidence is only 40 in 100,000. It has been argued that the point that air pollution contributes to death can only be made statistically, but those familiar with statistics are often alarmed at the findings. In 1969, sixty members of the faculty of the Medical School of the University of California at Los Angeles released a statement urging anyone who didn't have compelling reasons to remain, to move out of the smoggier areas of Los Angeles to avoid respiratory illnesses.

(Courtesy: Wide World Photos)

gen oxides have other surprises as well. They may act as poisons in the lungs, producing a variety of respiratory illnesses.

Sulfur Oxides

Sulfur oxides, produced mainly by burning coal, are thought to exert particularly severe reactions during heavy smog. These chemicals are very harsh elements which produce coughing, wheezing, and choking. They are believed to figure importantly in the increased incidents of bronchitis, asthma, and emphysema which are so frequently suffered by people in high smog areas. Emphysema is a progressive lung disease and is considered fatal.

Hydrocarbons

Hydrocarbons, as you know, are a diverse and variable group of elements, some of which are believed to be a causative factor in respiratory cancers. The hydrocarbons emitted in automobile exhaust are rather easily trapped by the present crankcase antismog devices. However, nitrous oxides, with which the hydrocarbons would normally bond to form relatively harmless compounds, are not filtered out by these devices. Thus, partly because of improper antismog devices and partly because automobile manufacturers have increased the compression ratios of their engines, nitrous oxide pollution has soared in areas with heavy automobile traffic.

Particulate Matter

Certain industrial processes release dusts, minerals, vegetable matter, or other waste particles into the air which are large enough to be filtered. Some of these particles are organic compounds which contain chemically active groups of sulfates, nitrates, fluorides, or ammonia. Others are essentially metallic, containing cobalt, iron, lead, or copper.

Because of the difficulties in pinpointing causal relationships in subtle processes, it has been difficult to determine the extent to which air pollution can cause deaths. This task may become easier, however, if current trends are allowed to continue. According to the U.S. Public Health Service, annual sulfur dioxide emissions are expected to increase from 20 million tons in 1960 to 35 million tons by the year 2000. The quantity of nitrogen oxides will triple to almost 30 million tons in that period, and particulate matter will increase by 50 percent to about 45 million tons. And by then, unless we make some changes, we will have four times the number of automobiles we have now. The concomitant rise in gasoline usage will not be the only problem. How long did it take you to find a parking place this morning?

Temperature Inversions

It is likely that some spectacular tragedies due to polluted air will occur in large cities before long. Such crises are most likely when *temperature inversions* occur. Normally the air temperature decreases with altitude, but sometimes a layer of warm air moves above the cooler air, thus limiting the normal upward flow of the air from below. The result, as you see in Figure 14.16, is that the polluted lower air is not allowed to escape into the upper atmosphere. This means the foul air is trapped and the people below must breathe their pollutants. When temperature inversions occur over cities, the number of deaths from respiratory ailments usually rises sharply.

Part of the problem of air pollution in cities lies with the design of the cities themselves. It is unfortunate that we have managed the growth of our cities so miserably, especially since most people in this country live in urban areas. At present 70 percent of our population lives on only 1 percent of the land, and the outlook is for even greater shifts to metropolitan areas in the next fifty years. Such trends give us a great opportunity to devise economical ways of living, since compactness can result in manageability. However, our approach to urban development has not changed noticeably in the past few years, as evidenced by the fact that we continue to develop sprawling megalopolises (such as the one which is presently emerging as a single metropolitan area extending from about Richmond to Boston). Our cities are, at present, grossly inefficient and often dangerous places to live, and it is not likely that simply increasing their sizes will help matters much.

Those who appreciate irony may glean moments of grim humor by traveling on the New York City subway system. The cultural center of the world asks those seeking its pleasures to roar and screech about in antiquated old cars which reek with a scent suspiciously urinesque. And to wait for the privilege on dirty, sodden platforms where the sounds of passing trains can impart a permanent high-pitch deafness to the frequent traveler. It is true that such transportation methods cause far less pollution per passenger mile than the myriad honking taxis lurching above. But is this the best we can do?

One might argue that, at least, an effort has been made, at least a lip-service to rationality. After all, Los Angeles has almost no system of mass transit other than an inadequate bus system. The citizens are almost entirely dependent on automobiles. In 1974 they voted down a one percent sales tax which would have provided funds for the beginning of a mass transit system. And when the wind is right, lettuce crops 90 miles away wilt and die.

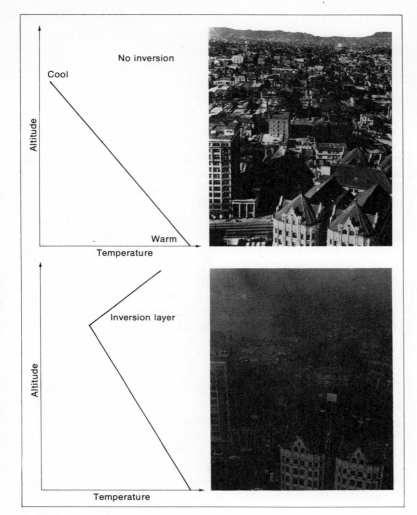

Figure 14.16 In a temperature inversion, a warm layer of air lies over a cooler layer, trapping air pollution below it. These photographs were taken in Los Angeles in the early 1950's and demonstrate that smog is not a recent phenomenon. (Courtesy: Wide World Photos)

The Prospects

The prospects for cleaning up the air on a worldwide basis are, at present, nor particularly promising. Many countries seem to associate effluence with affluence (with good reason if the United States serves as a model) and are willing to put up with a fouled environment in exchange for physical conveniences and manufactured goods. In certain less-developed nations, the environmental movement in the United States is viewed as a preoccupation of a pampered society which can afford such luxuries.

Undoubtedly, we can make great gains toward cleaner air in the United States by the initiation of a forceful program, and the beginning of such an effort is already under way. In 1970 Congress passed a new Clean Air Act which set strict standards for both automobiles and factories. Hydrocarbon and carbon monoxide emissions from automobiles must be reduced by 90 percent by 1976. Of course there have been powerful industry efforts to relax the standards.

There is some concern that our present fuel problems, whether real or designed, may result in lowered air quality standards so that "unclean," or high-polluting, fuel can be burned in industrial operations and automobiles. Another result of a "fuel shortage" might be to relieve automobile manufacturers from the responsibility of developing low-polluting engines which, with present engine size and design, will burn more gasoline.

Other methods which might improve the quality of our air include the prompt development of good mass-transit systems and the encouragement of smaller automobile engines. This encouragement might be provided in the form of horsepower taxes—which would mean that only the wealthy could afford the luxury of pollution—or the strict regulation of engine sizes. Eventually, however, we must come to rely less on the internal-combustion engine. And there we have a marvelous example of understatement.

WATER POLLUTION
Sewage
It has been said that the tap water of New York City contains many interesting ingredients—not the least of which is chow mein. A more precise, but no less fascinating statement is that in some localities in the United States the water one drinks has already passed through the bodies of seven or eight other good people. Under the best of conditions that water will have been filtered and treated with small amounts of deadly chlorine in an effort to poison microorganisms, so that the water will be relatively safe to drink. (Chlorine is sometimes questioned as a water additive because certain chlorine compounds can cause mutations, but the consensus is that, considering the risk of impure water, chlorination is the lesser of two evils.)

In many locales the water is not safe to drink, as evidenced by the recent outbreaks of infectious hepatitis in the United States. Infectious hepatitis is believed to be caused by a virus carried in human waste, usually through a water supply which is contaminated by sewage. There is some disturbing evidence that this virus may be resistant to chlorine, especially in the presence of high levels of organic material. Despite our national pride in indoor plumbing and walk-in bathrooms, sewage treatment for many communities

Box 14.6 Sewage Treatment When our population was smaller, sewage could be safely emptied into moving rivers where bacteria and fungi (decomposers) could digest it. Those large organic molecules from excretion, fallen leaves and other by-products of life were broken down into their elemental constituents. These constituents could then be used as building blocks by primary producers in the recycling of life's materials. When the numbers of waste-producers along such rivers became too high, the natural processes of decomposition proved too slow and sewage-treatment plants were devised. These plants are, in effect, places where the natural processes are speeded up within the confines of a restricted area.

Ideally, a modern two-stage sewage treatment plant collects the wastes discharged into our waters, gets rid of the waste, treats the water and pumps it back into our supply. First, the largest pieces of waste are screened out. Then the water is pumped into settling tanks where finer material sinks to the bottom as sludge. The supernatant (fluid over the sludge) is pumped into tanks where air is bubbled through it to provide oxygen for bacteria and fungi which break down the organic molecules. Sludge is again allowed to settle out and the supernatant is chlorinated to kill the bacteria and it is ready to drink again.

Unfortunately, many of our cities lack facilities for complete processing of sewage. This means that partly treated or untreated wastes are pumped into our waterways. Even those plants with adequately treated sewage present certain environmental problems. For example, the effluent from the best sewage plants is high in phosphorus, nitrogen and certain other elements. These can act as fertilizers by enriching the water into which they are dumped. (Their effect is added to that of phosphate from detergents which accounts for about half the phosphate overload.) The result can be clouded, smelly water covered with algae scum. Many species of algae which respond so well to inorganic fertilizers are not used as food by zooplankton and thus do little to initiate a food chain. Also, it has been determined that where pollutants raise the overall "productivity" of water, in terms of mass, it decreases the diversity of species present. Certain blue-green algae produce poisons which have even killed cattle and which can produce rashes and vomiting in humans.

in the United States is grossly inadequate, and waste which has been only partially treated is discharged into waterways. Recently the news services carried a story announcing that the New Orleans water supply may be dangerous to drink. However we have been assured that there is no cause for alarm—a committee has been appointed to study the problem!

Chemical Effluents
As we provide for our expanding population increasing amounts

Box 14.7 A "Dead Sea" Moves on New York   The ability of man to defile his environment is being starkly demonstrated by the sea of fetid filth which is moving inexorably in Atlantic waters toward the beaches of Long Island.

For four decades, New York and its surrounding communities have been dumping their sewage sludge some 12 miles offshore where, out of sight and out of mind, it was supposed to sink and dissipate miraculously with the wash of the sea. But the heavy and continuous influx of diseased muck—five million cubic yards a year—has proved too much for the waters around it. A "dead sea" was created which hangs like a pestilential nightmare just beyond the city's horizon. And in 1970 that sea began to move. Its drift course was back toward the communities which in their innocence and ignorance had created it.

The presence of this hovering sea of ooze has been no heavily guarded secret among Atlantic seaboard officials. It has been the observable and devastating result of a rapidly conceived policy of somehow meeting the problem of an excess of human and industrial effluence. So recently as Earth Week of 1971, in fact, it was listed as one of the top 10 ecological disasters in the world.

What has been unexpected, however, is the slow surge of this mass toward the coastal civilization that spawned it. To date the drift defies scientific explanation. Its contamination could reach the Long Island beaches and those of the Fire Island National Seashore within three years. The consequences of that surge, with its fecal coliform bacteria, hepatitis, encephalitis and other diseases, have so far eluded description.

Man's inhumanity to man manifests itself in many ways. But in the drift of the "dead sea" toward the New York shoreline, he must come to understand that the recuperative powers of the seas are not inexhaustable—that the sea, too, and the men who live along its inspiring borders must pay enormous environmental prices for the abuse of its tidal blessings.

(Excerpted from "A 'Dead Sea' Moves on New York," *Santa Barbara News Press*, January 15, 1974, Section C, page 12. Reprinted by permission.)

of fertilizers, pesticides, radioactive particles, and strange new industrial by-products are finding their way into our water supplies. In many cases, we not only do not know the effects of the chemicals, we don't even know what the chemicals are. In the instances in which the chemicals are known, we rarely have any idea of the effects of long-term, low-dosage ingestion. The continued need for energy, food, and manufactured goods, however, ensures the entrance of a number of potentially dangerous substances into our water supply. Under these circumstances we desperately need to initiate studies on the effects of chemical contaminants as soon as possible.

The danger of our present ignorance is clear from the belated

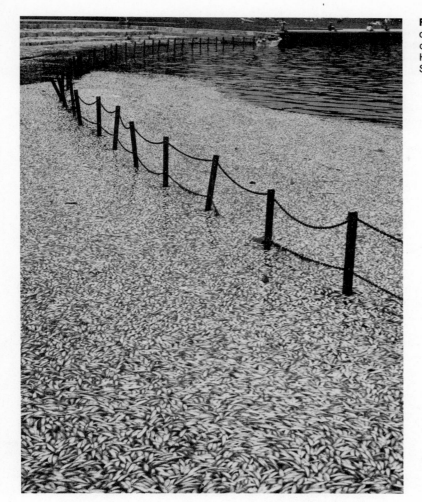

Figure 14.17 Alewives which have died as a result of our pollution of their habitat. (Courtesy: John Hendry from National Audubon Society)

information on certain long-familiar contaminants. For example, we learned only around 1950 that nitrates, a common constituent of agricultural fertilizers and water supplies, could alter hemoglobin so as to impair its oxygen-carrying capacity. More recently it was found that certain bacteria in the digestive tract can convert nitrates to highly dangerous nitrites. Nitrate water pollution is especially dangerous in certain areas of California, Illinois, Wisconsin, and Missouri. As another example, we have known for some time that low concentrations of cadmium can be dangerous to human health, but its allowable limit in water was not set until 1962.

In many cases water pollution is so obvious that it has been the subject of wry humor, such as in cartoons of small fish begging

Box 14.8 **Spilled Oil**   There has been a lot of discussion recently over the effects of oil spilled into the ocean from the operations of petroleum companies. Actually, it is estimated that the 10 million tons of oil which reaches the oceans by such means each year, is matched by natural seepage. However, an important difference exists in pollutants from the two sources. Much of the man-made contribution is partly refined and therefore its effect on living systems is much more drastic. Just how drastic isn't known. The ocean presents such severe problems to researchers that only the most obvious and flagrant effects of oil pollution can be determined. Much of the research which has been done so far also presents some rather puzzling data. It is known that short-chain hydrocarbons damage marine larvae, and that simple ring systems are generally poisonous to living things, while complex ring systems are known to cause cancer. In spite of great efforts by local citizens, sea-birds which become covered with oil almost invariably die. Even flamingos and penguins have been killed by oil. In Alaskan waters great rafts of oil-covered corpses of birds, sea lions, hair seals, killer whales and sea otters have been found (but in the interest of fairness to oil companies, no one wanted to say the oil had caused the deaths).

Oil contamination does not always kill. In certain cases, barnacles have grown back on oil-slicked rocks with apparently little difficulty. Some tiny isopods and stunted goose barnacles may live within the tar balls which form from oil slicks. Certain bacteria are known to digest such tar balls, but they do it so slowly, even under ideal conditions, that the oil residue is likely to enter the food chain unchanged through another route before the bacteria have their way.

If the oil reaches the ocean floor where little oxygen exists, it is likely to persist for long periods. Inshore refined-oil spills are highly toxic and may kill worms, crustacea, mollusks and fish for months afterward and may impart objectionable tastes to denizens of scallop and oyster beds.

(Courtesy: Dave Bellak)

the fisherman not to throw it back. In some cases the level of pollution is even startling. For example, areas of the Cuyahoga River in Ohio have been declared a fire hazard because of the high levels of combustible materials it carries. A recent report showed only *traces* of water in parts of the Houston ship channel. Such cases are extreme, but our country also abounds with less dramatic examples of pollution. Certain winding rivers passing through the beautiful Maine countryside are piled high with yellow foam from upriver paper mills. The smell emanating from such water is rather remarkable in its own right. Joggers along the beaches of southern California, running within sight of great offshore oil rigs, find their feet covered with heavy layers of tarlike substances. Backpackers in Colorado drink freely from remote mountain streams, but they must purify water as they get closer to "civilization."

Figure 14.18 The effect of temperature variations on populations of water fleas. The largest dots represent populations at 33.6°C. The medium dots; 24.8°C. The smaller dots; 19.8°C. Note that after ten days, only the populations grown at 19.8°C survived, despite the more rapid increases in the others.

Heat Pollution

A form of water pollution which is even subtler than chemicals or bacteria is *heat*. Water absorbs heat when it is used as a coolant or lubricant in manufacturing or in energy production. When it is returned to its source at this higher temperature, it may alter the biota of the water to which it is added. The temperature increase need not be great to exert profound effects. A difference of only a few degrees can markedly alter the growth rates of small aquatic organisms. Such a temperature change may also cause changes in the species composition of communities of larger organisms such as fish. Thus delicate balances which have been established as the result of long periods of evolution may be disrupted.

Among the greatest potential sources of heat pollution are nuclear power plants. The nuclear process generates enormous heat, and the water used as a coolant, if it is dumped offshore, could drastically alter the marine life in that area. Of course, it is in coastal waters that many of the tiny organisms at the base of the ocean's food pyramid begin their existence.

But why is the addition of only a little heat so critical? How does it exert its effect? Basically, heat pollution works in two ways, (1) by increasing the rate of oxygen consumption by aquatic organisms while, at the same time, lowering the oxygen-carrying capacity of the water (resulting in a lowering of the ability of the biological system to decompose waste), and (2) by increasing the rate of evaporation of the water, thus raising the concentration of pollutants left behind.

Enforcement

We have long had legislation which limits industrial water pollution

Figure 14.19 Unfit for human consumption? (Courtesy: Photo by Erika)

—the Refuse Act of 1899. But it has only recently begun to be enforced, under the joint authority of the Environmental Protection Agency and the Army Corps of Engineers. The act was bolstered in late 1972 by the Water Pollution Control Act, which has as its goal the end of water pollution from industrial and municipal sources by 1985. So there are laws, of sorts, against polluting our waterways. But water pollution continues to become worse, partly because of lax enforcement and partly because of inadequate surveillance.

The drinking water for municipalities is monitored by the U.S. Public Health Service, but this agency is empowered only to set standards relating to the initiation or spread of disease. It has had, until very recently, little to say about the introduction of chemicals, such as pesticides, into our drinking water. The PHS is presumed to be currently revising its monitoring procedures to broaden the scope of its surveillance.

It is apparent that we need some system of firm national standards to replace the varying and flimsy set of regulations which now govern what we allow in our waters. Any such system should probably include criminal penalties for violators. Perhaps *individuals* should be held accountable for infractions instead of faceless corporations, since individuals make the decisions. One recent example of corporate punishment was the $5,000 fine imposed on an oil company responsible for a major oil spill attributed to negligence. The payment was about equivalent to a man who earns $20,000 a year being fined a nickel. It has also been suggested, perhaps too facetiously, that each member of the board of directors of any corporation which discharges wastes into waterways be required to drink a glassful of their effluent at a public ceremony every six months. A less facetious suggestion is that board members be required to pay the corporation's fines from their own pockets.

PESTICIDE POLLUTION

Pesticides are biologically rather interesting substances. They have no known counterpart in the natural world, and most of them didn't even exist just thirty years ago. Today, however, a metabolic product of DDT, called DDE, may be the most common and widely distributed man-made chemical on earth. It has been found in the tissues of living things from the polar regions to the remotest parts of the oceans, forests, and mountains. Although the permissible level of DDT in cow's milk, set by the U.S. Food and Drug Administration, is 0.05 parts per million, it often occurs in human milk in concentrations as high as 5 parts per million and in human fat at levels of more than 12 parts per million.

Box 14.9 Pest Control   Ranchers have persuaded their representatives in Washington that some kinds of animals were either preying on their stock or eating the stock's feed. They argued that their profits were being reduced. As a result the government sent men, employed at taxpayers' expense, to trap, shoot or poison the designated offenders. These efforts have been aided by a bounty system which enables any citizen to get a piece of the action. As an example of one such effort, men are hired to stroll through prairie dog towns spreading poisoned grain. As the tiny prairie dogs lie in their death throes, they become easy prey for various predatory birds as well as the rare black-footed ferret. The poisons in the bodies of the prairie dogs are then transferred to other plains animals such as coyotes.

The government has also regularly employed men who place meat over a short-barreled device so that when a coyote or other animal tugs at the meat, cyanide is discharged by a shotgun blast into the animal's head and mouth. (The same economic reasoning undoubtedly contributed to recent slaughters of eagles by midwest ranchers shooting from private planes. Their targets included the bald eagle, our endangered national symbol.) There are little hard data to show the economic effects of such predators. However, many conservation biologists feel the ranchers' attacks are unwarranted, but are, instead, attempts to gain favorable consideration when losses are reported for tax purposes.

NOW, THIS'N HERE IS THE BALD EAGLE, OUR NATIONAL BIRD AND SYMBOL — I WANT YOU TO KILL EVERY ONE OF THE VARMINTS YOU SEE!

Figure 14.20 Leaf damage by Japanese beetles. (Courtesy: A. W. Ambler, National Audubon Society)

Pesticides, of course, are products which kill pests. But what is a pest? Biologically the term has no meaning. The Colorado potato beetle, for example, was never regarded as a pest until it made its way (as a result of human activity) to Europe, where it began to seriously interfere with potato production. Perhaps this points up the best definition of a pest: it is usually something that competes with man.

It seems that the greatest pesticidal efforts have been directed at insects and much of it has been beneficial. The heavy application of DDT since World War II resulted in sharp decreases in malaria and yellow fever in certain areas of the world. But DDT and other chlorinated hydrocarbons have continued to be spread indiscriminately any place in which insect pests are found.

There are several risks involved in such use of insecticides. For example, many such products are unselective in their targets; they simply kill *all* the insects they contact. Many insects, of course, are beneficial, and as a group they form an important part of ecosystems. Also, chemical insecticides move easily through the environment and can permeate far larger areas than intended. Another particularly serious problem with pesticides is indicated in Table 14.5: many of them persist in the environment for long periods. In other words, the chemicals are very stable and it is difficult for natural processes to break them down to their harmless components.

The most persistent of insecticides is DDT. Since it just doesn't break down very easily, it remains intact in the environment, poisoning every susceptible organism it encounters. And it does this for years. This persistence assumes major significance when you consider the history of heavy DDT use.

Table 14.5—Upper Limits of Longevity of Chlorinated Hydrocarbons in Soils

Pesticide	Percent Remaining
	14 Years Later
Aldrin	40
Chlordane	40
Endrin	41
Heptachlor	16
BHC	10
Toxaphene	45
	15 Years Later
Aldrin	28
Dieldrin	31
	17 Years Later
DDT	39

SOURCE: "Persistence of Chlorinated Hydrocarbon Insecticides in Soil," Nash, R. G. and Woolson, C. A., *Science*, Vol. 157, pp. 924–927, Table, 25 August 1967.

Box 14.10 Insecticides   The majority of synthetic insecticides falls into two groups: *chlorinated hydrocarbons* and *organophosphates*. The chlorinated hydrocarbons include DDT (and its still-potent breakdown product DDE), benzene hexachloride (BHC), dieldrin, endrin, aldrin, chlordane, lindane, toxophene, and isodrin. DDT is by far the most studied of these. DDT is also rather typical for the group, although other compounds may be more or less toxic, water soluble, or persistent (hard for nature to break down). These chemicals somehow work on the central nervous system of insects causing excitability, convulsions and death. They are soluble in fats and tend to collect in the fatty tissues of animals rather than being excreted. Chlorinated hydrocarbons are rather broad-spectrum poisons and affect a wide variety of animals in addition to insects, including vertebrates. They are also very stable (especially DDE), and mobile. The threat from chlorinated hydrocarbons is thus based on four properties: their affinity for living systems, their range of biological activity, their stability and their mobility.

The organophosphates include parathion, malathion, diazinon, azondrin, TEPP, phosdrin, and others. These poisons are descended from a nerve gas called Tabun developed by the Nazis in World War II. Basically, they function as inhibitors of cholinesterase activity. Cholinesterase breaks down the neural transmitter substance, acetylcholine. The poison thus causes increased activity of the nervous system and the victim dies, twisting and jerking wildly. Organophosphates may be more immediately poisonous to vertebrates than chlorinated hydrocarbons, but they are unstable and nonpersistent and thus do not accumulate in food chains.

Persistent chlorinated hydrocarbons such as DDT, aldrin, dieldrin, and benzene hexachloride have another interesting feature. They not only remain in the environment for long periods, but they also persist in the bodies of animals which ingest them. They are thus passed intact through the food chain. This is why human milk often has higher DDT levels than cow's milk. The cow eats the grass which has been sprayed, and the fat-soluble DDT moves into the cow's tissues. Then, when a human eats the cow, the DDT which has been accumulating for years in the cow's tissues is passed along to the human. If that person is fond of beef, the DDT level in his tissues will be quickly magnified. Other animals which occupy the top of a food pyramid may also concentrate DDT.

The tendency of DDT to be magnified in food chains has been particularly disastrous for those birds which feed high on the food pyramid. Reproductive failures in peregrine falcons, the brown pelican, and the Bermuda petrel have been attributed to this phenomenon. The problem is that DDT interferes with the birds' ability to metabolize calcium. As a result, the birds lay eggs with

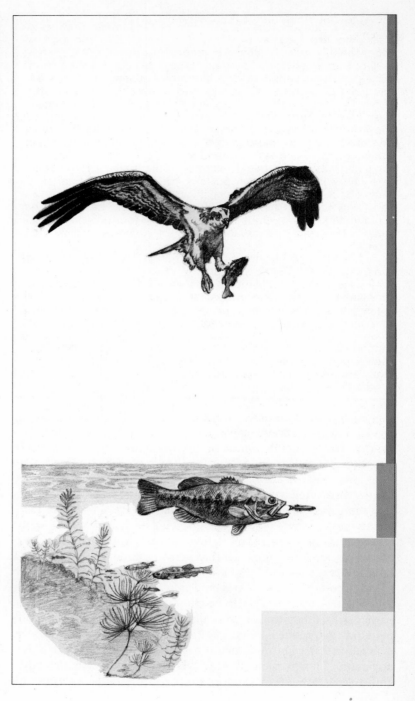

Figure 14.21 Once DDT enters the food chain it travels from one food level to the next where its concentration in animal tissue increases (shown by increased color density). The top carnivores contain the highest DDT levels. The width of the boxes in the pyramid represents the amount of biomass at each level.

shells so thin that they cannot support the weight of a nesting parent.

Manufacturers of DDT have tried to demonstrate its harmlessness to humans by publicly swallowing spoonfuls of the pesticide. The idea is an ad-man's dream, but it actually proves nothing. What we're concerned about are the effects of long-term, low-level exposure to the chemical, at such dosages that occur throughout the environment. There are a host of unanswered questions regarding DDT. What overall effect is the insecticide having on the complex life systems on the earth? How serious is the evidence that it inhibits photosynthesis? In a more immediate sense, what effect does it have on mammals, particularly humans? Are there alternatives to its use?

Much of the recent evidence of the effects of DDT on mammals is not encouraging. For example, it has been shown to increase the incidence of cancers, especially liver cancer, in mice. In rats DDT causes an increase in certain liver enzymes which render many drugs useless. It also induces an increase in the deposit of dextrose in the uterus of female rats. DDT can make rats sterile, and it has similar effects on the sex hormones of birds. In the seals of San Miguel Island, off California, high levels of DDT in the tissues of females are correlated with premature births and abnormal behavior of mothers toward their young.

We don't yet know whether these effects operate in humans, since there would probably be a problem in rounding up experimental subjects. However, certain alarming findings have come to light regarding the effects of DDT on humans. For example, concentrations of DDT, DDE, DDD (another breakdown product of DDT), and dieldrin were higher in the fat of patients who died of softening of the brain, brain hemorrhage, hypertension, cirrhosis of the liver, and cancers than in a group of patients who died of infectious diseases. Those with the highest DDT levels in their body were those who had used the chemical most freely at home.

In summary, we know that the earth's living things are being exposed to long-term low dosages of chlorinated hydrocarbons. We don't know what the effects of such exposure are on living systems, or on individual species, including our own. Even in the face of such questions, DDT continues to be used in large quantities on a worldwide basis, each application being added to the still active residue of earlier years. Thus, whether we like it or not, we are part of a grand experiment. The results should prove interesting.

Figure 14.22 This pelican egg was crushed by the weight of its parent. In the brown pelican which nests on the Channel Islands off California there have been such great losses of eggs that the species was considered endangered. The losses were due to high levels of DDT in the tissues of the mother which caused her to produce very thin shells. (Courtesy: Joseph R. Jehl, Jr.)

RADIATION

Life on earth has always been exposed to the effects of ionizing

Box 14.11 Integrated Control of Insect Pests   It has been suggested by promoters of chlorinated hydrocarbons that all that stands between us and starvation because of insect-borne diseases are insecticides like DDT. Not so. We are aware of short-lived insecticides which bear very careful handling by humans because they also may kill mammals. And there are ways to reduce threats from insects which involve means other than chemical. It has been found that sterile males released in large numbers will virtually wipe out populations of some insects. The females can breed only once. If they mate with a sterile male, of course, no progeny are produced. If the population density should become so low that even potent individuals are unlikely to find each other, the population is doomed.

In the 1880s the cottony cushion scale threatened the developing California citrus industry. The growers introduced vedalia beetles which attacked the scale and the problem was solved. In the 1940s, however, DDT was used on the citrus. It killed off the vedalia beetles and the scale insects blossomed again. When DDT use was discontinued and the beetle reintroduced, the scale insect was controlled once again. So the introduction of *very carefully* chosen predators may reduce pests.

Probably the system called "integrated control" is our, at present, best bet. Integrated control involves a multifaceted attack which simply keeps numbers of pests below levels where they pose a threat. For example, mosquitos might be controlled by draining swamps, introducing fish which eat mosquito larvae, spreading biodegradable chemicals on the surface of water to suffocate larvae, the judicious use of short-lived insecticides, and luring adults to traps. This way, no dangerous pesticides are used.

radiation in one form or another. In fact, we live in a veritable sea of radiation. One source of radiation is the earth's crust, where radioactive substances undergo spontaneous nuclear disintegration, emitting both high-speed atomic particles and penetrating electromagnetic rays (similar to X rays). In addition, we are continually bombarded with radioactive products from extraterrestrial sources in the form of cosmic rays. Other radioactive substances, such as potassium 40, circulate through living systems. Such naturally occurring radiation is called *background radiation*. We do not know at present the degree to which background radiation is responsible for cancer or birth defects in the human population. We do know, however, that ionizing radiation causes mutations and cancer under experimental conditions.

A much more serious problem is the radiation level to which we are exposed through our own activities. The use of radioactive substances has greatly increased in the last thirty or forty years, and now poses severe problems for us. More than 90 percent of man-made radiation exposure occurs in medical and dental uses, and there is some evidence that such use should be sharply curtailed. For example, health physicist Karl Z. Morgan has argued that unnecessary medical radiation may presently be causing 3,000 to 30,000 deaths annually in the United States alone.

We are also exposed to the results of fallout from nuclear weapons. In the United States, however, fallout presently accounts for only 1 or 2 percent of our environmental radiation. The rate of increase of radioactivity was rising sharply until the United States, Britain, and Russia stopped atmospheric testing in 1963. Of course, the Chinese and the French have continued surface testing of atomic weapons and we are all exposed to the radioactivity. Mark Twain would have new things to say about the French if he were here to witness their nuclear policies.

Nuclear Power Plants

In the United States an argument has been raging for the past few years over the risk involved in the proliferation of nuclear power plants. There can be no question that with our electrical power "needs" doubling every eight years, we cannot rely indefinitely on the earth's remaining fossil fuel supply. The question is, can we safely shift our reliance to nuclear fission power plants—considering the present state of our knowledge and technology?

It is not likely that an explosion of the type produced by atomic bombs can occur in the kinds of nuclear reactors being used today. But other types of tragedies are more likely. Radioactivity could be released into the environment from activities related to mining and processing the fuel, from the transportation and recycling of the fuel and from storage of the radioactive wastes. In addition, accidents can occur while the plants are in operation. Should the reactor core melt, or be disrupted by some event such as an earthquake or sabotage, radioactivity equivalent to that which would be produced by hundreds of bombs the size of those that devastated Hiroshima and Nagasaki could be released into the environment. Such accidents are said to be unlikely, but the mere possibility of such an event renders "odds" meaningless.

Even on the basis of the calculated risk, insurance companies, which are presumably well-advised in such matters, will not underwrite more than about 1 percent of the potential liability of such an accident. Instead, by act of Congress, the public would receive up to $478 million in public funds and assume the rest of the damages themselves. Figure *that* one out! A 1957 report by the Atomic Energy Commission concluded that a single major accident might cause 3,400 deaths at distances up to 15 miles and 43,000

Box 14.12 Radiation   The nuclei of some atoms are unstable. Of the more than 320 isotopes that exist in nature, about 60 are unstable, or radioactive. In addition, man has created about 200 more. Such creations are called *radioisotopes.* The substances in which they exist are *radioactive.* Some of these unstable isotopes occur in natural systems; others are manufactured by man. When a radioisotope decays (or decomposes) certain particles (positively or negatively charged) and rays are emitted from its nucleus. The radioisotope is then changed from one element to another. The new atom may also be radioactive, or it may be stable.

Radioisotopes decay at predictable rates. These rates are usually expressed as the element's *half-life.* Half-life refers to the time which must expire in order for half the atoms in any amount of the isotope to decay to another kind of atom. Half-lives may be very long. The half-life of natural uranium (U_{238}) is 4,500,000,000 years! However, as a rule of thumb, the time it takes for a substance to become nonradioactive is considered at 20 half-lives.

When radioisotopes decay ("explode" may be a better word), the particles and rays may strike other atoms in the environment in such a way as to tear electrons away from them. Thus the process is called *ionizing radiation.* When living tissue is thus disrupted, the effects can range from reduced life expectancy through cancer, leukemia or other means, to instant death. There is *no* threshold below which radiation is genetically harmless. Also, the genetic effects of radiation are accumulative in the exposed individual and his offspring. A "snowball effect" operates so that an increased radiation level which has little effect in the first few generations may, if maintained, produce disasters, even extinction, in following generations.

Radioactivity of any substance can only be reduced by allowing the radioisotopes to decay naturally. There is no means for hastening the process.

The irradiated onions are the ones that are not sprouting (Courtesy: University of Michigan)

Figure 14.23 Diagram of a nuclear power plant. The fuel is uranium bars encased in metal sheaths. As the uranium decomposes, heat is produced and is picked up by water circulating past the fuel elements. Steam is produced which then turns turbines and operates electric generators.

injuries at distances up to 45 miles, with property damage up to $7 billion. That report is almost twenty years old. Today property is higher priced, populations are denser, and nuclear power plants are larger. Of course, these figures refer only to the immediate destruction. The residents of Nagasaki are still dying from the progressive effects of radiation damage incurred thirty years ago. Also, today, there is another factor which was not operative in 1957. Today *I* own a house and *I* live near a nuclear power site! The issue is suddenly more than academic for millions of people.

Completely apart from the possibility of accidents, there is the unsolved problem of what to do with the radioactive wastes generated in the course of normal nuclear plant reactions. The problem of how to store nuclear wastes is a tough one since such wastes can only be rendered safe by the passage of time. The waste radioactivity is generated in the fuel system of the reactors because only a part of the fuel is fissionable, and for technical reasons not all of the fissionable elements are spent. Much of the spent fuel materials removed from the reactor core can be reused. However, some of the radioactive fuel in the spent elements cannot be recovered, and this material comprises the radioactive waste.

And then there is a problem with coolants. The coolant, whether it is water or gas, becomes radioactive as fission products manage to escape through the metal casing of the uranium. Water coolants may also become radioactive by a process called *neutron activation*, which occurs when one of the hydrogen ions in the water molecule is replaced by a tritium ion. The present practice is to dilute this radioactive water with plain water and then to dump the mixture onto the ground, where, of course, it can seep into the water table below. The enthusiastic developers of this process have

Figure 14.24 This Greek drinking vessel is typical of those used by the Greeks around 430 B.C., which were often lined with sweet-tasting, but deadly, lead. (Courtesy: Los Angeles County Museum of Art, Kylix, The Marlay Painter, William Randolph Hearst Collection)

even proposed that water with too high a radioactive level for human consumption be used for such purposes as irrigating crops and watering livestock. A single course in general biology should dispel such foolish ideas soon enough. The frightening thing is that competent, highly-trained scientists have proposed other ideas which are potentially just as dangerous.

Highly radioactive liquid wastes are presently being stored in steel tanks at several sites around the country. It was recently found that 200,000 gallons of this waste had leaked from a tank farm holding 80 million gallons at Hanford, Washington! Because such wastes remain dangerous for thousands of years, there is a problem in building containers which will safely hold the material for that length of time. Certainly, steel holding-tanks are not the answer. The Atomic Energy Commission has suggested reducing high-level wastes to solid form and then storing these, perhaps permanently, in salt beds, deep underground. Technical problems and an increasingly skeptical public have stalled these plans for the present.

HEAVY METAL POLLUTION

Believe it or not, lead-lined food containers, according to some historians, may have contributed to the fall of both the Greek and Roman civilizations. Because lead is noncorrodible and has a sweet taste, it was used to line their bronze vessels. However, lead can cause brain damage. Thus, it is believed, widespread lead poisoning may have caused mental debilitation in the citizenry of these great societies.

Today, we prefer our lead other ways, such as in our gasoline (so our high-compression cars don't knock), pesticides, paint, pipes, ceramics, glassware and the solder in food cans. We're very familiar with the hazards of lead, but unleaded gasoline in mid 1971 accounted for only 5 percent of all gasoline sales. Part of the reason is that unleaded gasoline is from two to six cents higher in price per gallon. Since 1970, however, cars have become more tolerant of unleaded gas, and all 1975 cars must be able to use it; that is, unless we relax our new regulations. The strictures on depositing lead into the environment should be strengthened because of the severe effects of lead poisoning. The symptoms include general apathy, weakness, loss of appetite, loss of coordination, and miscarriage.

Mercury provided the news media with a new toy recently as they projected the dangers of this metal into the public consciousness. The story involved finding high levels of mercury in the flesh of swordfish. Swordfish sales immediately plummetted, despite statements from the fishing industry that prodigious amounts of swordfish would have to be eaten for the mercury to pose any danger to humans. Because of the apparent threat, studies were immediately launched to see whether human activities had raised the mercury levels in the oceans. The findings indicated that man had not added significantly to natural levels of the oceans' mercury.

Mercury pollution in fresh-water bodies is a different story. It seems that man has sharply raised mercury levels in rivers, lakes, and streams. The Saskatchewan River, which runs through Edmonton, Alberta, has only half the amount of mercury before it enters the city that it has when it leaves. The usual sources of fresh-water mercury are industrial processes (especially those involved in the production of plastics and caustic soda), pulp and paper mills, seed fungicides, and fossil fuels. The relatively small amount of mercury in fossil fuels is a threat because of the prodigious volume of such fuels which is burned.

Interestingly, mercury is relatively harmless in its metallic form. However, certain microorganisms present in water are able to convert it to methylmercury (CH_3Hg) and dimethylmercury (($CH_3)_2Hg$), highly poisonous substances which are water soluble, and hence easily dispersed. The startling part of the story is that even if we immediately stopped dumping mercury into water on a worldwide basis, there may already be enough metallic mercury in the environment to pose a serious threat to later generations, as it continues to be slowly converted to the methylmercuries.

Mercury, like DDT, is amplified as it moves through the food chain, which accounts for the high mercury levels in the large fish at the top of the food pyramid. It is also likely to appear in higher concentrations in human populations which rely heavily on fish in their diet. The symptoms of severe mercury poisoning include blindness, deafness, loss of coordination, insanity, and death.

Other heavy metals are also increasing in our environment and are being viewed suspiciously, although we do not fully understand the implications of long-term, low-dosage exposure. Such metals include cadmium, chromium, arsenic and nickel.

NOISE POLLUTION
The real meaning of noise pollution is known to anyone who has ever retreated in solitude to a remote area, only to have the silence suddenly shattered by the blast of a trail bike. That person's eyes may turn to mere slits while he wistfully fondles an oak branch as the biker roars by, supremely confident in the sporting image cultivated by paunchy advertisers.

The same animosity is often described by people exposed to

the raucous spattering of snowmobiles, power lawnmowers or any number of other machines which seem to be designed with disturbance in mind. For sailors, nothing can ruin a day easier than sharing the water with a noisy powerboat. Then there are the "camper vehicles," that come replete with televisions and dishwashers, utterly dependent on outside power sources, and which increasingly disturb our national parks, requiring the development of more and more facilities to cater to their monumental needs. (You may detect faint personal prejudices on my part here and there in spite of my efforts to be objective.)

We need not enumerate other well-known major sources of noise, especially since you undoubtedly have your own list, but two deserve special mention. There is some evidence that the high amplification of rock music causes *permanent* hearing damage. Thus, the advent of hard rock in the '60s may have contributed to an entire generation of the hard of hearing. Also, there is great scientific concern over the potential effects of supersonic transport planes (SSTs), should they finally find legislative approval for commercial use. Several powerful lobbying organizations quietly continue to work rather effectively in Washington in an effort to secure approval for the development and use of the SST in this country.

Noise is usually measured in decibels. A tenfold increase in the volume of a sound adds 10 decibel units. A 100-fold increase adds 20, and so on. Table 14.6 gives the decibel units of some familiar sounds.

Table 14.6—Decibel Levels of Sounds

Noise Level	Decibels
Threshold of hearing	0
Normal breathing	10
Whispering	30
Home noises	45
Conversation	60
Food blender	80
Heavy automobile traffic	100
Jet aircraft taking off	120

In addition to hearing loss, noise may cause several stress-related diseases such as hypertension and peptic ulcers. There is also growing evidence that noise may be associated with irreversible changes in the nervous system. Much of the evidence is circumstantial, and researchers are presently engaged in analyzing more specific data on the effects of noise.

THE FUTURE

On the basis of the magnitude of some of our problems, you might be touched by traces of pessimism regarding our future. However, that spirit simply cannot be allowed to prevail. Some of our problems demand immediate, vigorous, and intelligent action, and they must be boldly addressed if we are going to buy the time to consider long-range solutions.

If we can deal symptomatically, through technology and stopgap efforts, with our most pressing problems, perhaps we can buy that time. However, we need to clearly differentiate those kinds of solutions from the long-range ones. For example, bolstering food production, at least with present methods, can only be considered stopgap. The population will eventually catch up. There is no question, in fact, that overreliance on technology at this point will ultimately lead to quite a different kind of world than the one which we could build if we shifted the *direction* of our development.

The ultimate answer does not lie in finding newer and better ways to do what we're doing. It can only be found, it seems, in a change in our behavior, a wholesale redirection of our course. Such shifts, of course, demand new thinking, new ways of looking at things, new values. This is the hard part. What if we thought through our options one more time and decided that welfare rolls must be drastically *increased* to pull people *out* of production, and that, in order to support people who choose not to work at all, we must all pay much *higher* taxes; that industrial management is somehow akin to prostitution; that aquirers without direct need are socially maladjusted; that a pleasant storyteller is a more socially valuable individual than a manufacturer; that patriots are those people who refuse to fight, or support fighting, on any soil other than their own. Spring those arguments on someone you know in order to see how entrenched some of our ideas are.

But aren't there things we can do which we know would be beneficial? Things which are more realistic? Something that we might be physically and psychologically prepared to do at present? Some changes, after all, would not greatly alter our life style. Some are even absurdly obvious. We could cut back on our use of disposable goods. We could reduce our reliance on the automobile by developing good mass-transit systems. We could decrease our use of oil and plastic. We could reduce our bizarre fascination with packaging. (Notice how many non-essential goods you end up with when you simply buy a shirt.) It is argued that such measures would reduce production and that would lead to unemployment. The answer is to put these people to work making things we really need.

So, what if people were hired to produce things we need rather than things which are used once and thrown away. Undoubtedly the initial reorganization and retooling would temporarily reduce profits. In the long run, however, we might benefit from changing from material-centered production to labor-centered services. People who work in automobile production, a trade which demands new steel from our earth, new plastics, new cloth and immense amounts of energy, could switch to automobile maintenance. Other people could be put to work finding better ways for us to travel.

The new trend, then, might be toward services. We need mechanics, repairmen, doctors, lawyers, regional planners, plumbers, environmental architects, city planners, and even teachers. There is plenty of work to be done. And, although you may disagree, I think we need artists, entertainers, performers and musicians. It may be worthwhile to shift a greater amount of our financial resources in their direction. One of the many lessons I learned from living in New York is that there are uncounted, virtually unknown artisans among us, people who truly excel at their trade. There are musicians who can leave us spellbound and writers who can tell stories that reflect our spirit as accurately as any mirror. Many of these people can barely keep body and soul together. Would we be willing to pay higher taxes to sponsor such people? The point is that perhaps we can come to place less importance on *things* and to place more importance on services. We might eventually come to elevate the value of sheer experience. If many of us felt this way, we might cultivate new Nureyevs, Picassos, Croces, Stravinskys and Hendrixes among us.

One would hope that with some new insight, life could come to be more than a series of accumulative events. This is not to say that any new perspective or information will produce a utopian society. Perhaps that was possible at one time, but not now. Now we must engage in trade-offs, partly because there is no chance of pulling our population down to a reasonable size for many years. Driving a well-maintained 1957 automobile is not the same as having a new car. Cardboard comes from trees; toothpaste tubes without boxes are harder to stack. If we should de-emphasize production, we will have less goods. If we opt for subsidizing nonproducers, we will have less money.

Economist Kenneth Boulding has suggested that our problems of resources and pollution could be greatly helped by a single philosophical shift—from a "cowboy economy" to a "spaceman economy." The cowboy attitude involves a reckless, exploitive philosophy. The American cowboy could always assume that more resources were waiting just over the horizon. His charge was to press on, find those resources and leave the garbage behind in the untracked wilderness. It may have made sense—then. Human welfare was most likely to be improved through economic growth—*any* kind of growth, with no thought for the wastes. The important thing was to grow, develop, and surge onward to seek new ways to grow.

But that's over now. At least it should be. Now we've covered the entire land and have exploited most of the resources which could be reached. Our numbers are so great that our efforts to grow threaten the system with collapse. Our waste is poisoning us, our animals and our crops. There is no place to put it—no way to leave it. We are spacemen in cowboy hats. The sooner we realize that our planet is merely a spaceship with only limited resources to last us through our entire journey, the sooner we may be able to shed our cowboy mentality and see ourselves for what we are: travelers on the Spaceship Earth.

If you are reading this book, chances are that you are young so you have the most to lose. The decisions will be particularly important for you, whether you make them yourself or allow them to be imposed on you. There are those who say that yours is the last generation which will live out its lifetime. Whether they are right or wrong depends on what you do, what you allow and what you won't allow.

Appendix

CLASSIFICATION

Living things are classified in a number of ways, and the diversity indicates the arbitrariness of the present system. People who classify living things are called *taxonomists,* and taxonomists disagree. What is one man's family may be another man's genus. Basically, however, the idea is simple enough. For example, most of us can look at the living things we usually encounter and tell whether they should be considered part of the plant kingdom or the animal kingdom. Dogs are classified in the animal kingdom.

But within the animal kingdom are kinds of animals that are similar in certain respects and that differ from other animals. These, then, can be placed into a subgroup on the basis of their similarity. For example, there is a wide range of animals that develop notochords in some period of their life, and so all these are called *chordates.* But the group is a diverse one that includes both acorn worms and dogs. So the phylum Chordata is divided into those animals that eventually develop backbones (like dogs) and those that do not (like acorn worms). The acorn worms are placed into the subphylum Hemichordata, whereas dogs fall into the subphylum Vertebrata. Dogs and a number of other vertebrates have hair and give milk, so all these are placed into a class called Mammalia. Organisms continue, then, to be separated into yet finer and finer groups until all the organisms within the group have so many qualities in common that they can, and normally do, successfully interbreed. These are called species. From kingdom to species, organisms are classified according to:

kingdom
  phylum
    subphylum
      class
        order
          family
            genus
              species

Depending on the taxonomist with whom one wishes to agree, there are various super- and subgroups within each division. We have already indicated one: the subphylum. Within species, also, there may be great diversity. For example, great danes and beagles are varieties of the same species. We can get an idea of the taxonomy problem by noting

that wolves and dogs are usually considered different species, yet they can, and do, successfully interbreed.

The following briefly reviews the classification of organisms.

Kingdom Viruses. Submicroscopic, noncellular parasites.

Kingdom Monera. Procaryotic cells (see Chapter 5). About 1600 species of these bacteria and bacterialike organisms are recognized. Many more species are involved if the class Rickettsia (very small, bacterialike organisms) is included. Also included is the phylum Cyanophyta, the blue-green algae. This is a highly complex kingdom.

Kingdom Protista. Eukaryotes (see Chapter 5). They are nonmotile, form pseudopods, or move by flagellae or cilia.

  Phylum Protozoa. Unicellular or simple colonial animals. About 30,000 species.

    Class Mastigophora. The flagellates.

    Class Sarcodina. The pseudopod formers.

    Class Ciliophora. The ciliates.

    Class Sporozoa. The spore-forming parasites.

  Phylum Chrysophyta. Golden algae and diatoms. 6,000 to 10,000 species.

    Class Bacillariophyceae. Diatoms with double siliceous shells. 5,000 to 9,000 species.

    Class Chrysophyceae. Golden algae. Over 1,000 species.

  Phylum Pyrrophyta. Golden-brown algae, mostly biflagellated. About 1,100 species.

    Class Dinophyceae. Dinoflagellates with a peculiar mitotic process. Over 1,000 species.

  Phylum Euglenophyta. The euglena. Contain chlorophyll but may be secondarily heterotrophic. A single flagellum. About 450 species.

  Phylum Gymnomycota. The slime molds. Heterotrophic, amoeboid. Form sporangia at some stage.

    Class Myxomycetes. Plasmodial slime molds that creep along as a mass. About 450 species.

    Class Acrasiomycetes. Cellular slime molds. Separate amoebas clump together to form a mass. About 26 species.

    Class Protostelidomycetes. Amoebas that may remain separate or massed. About 15 species.

Kingdom Fungi. Single or multicelled eucaryotes. Heterotrophic. Reproduction may be sexual or asexual. At present, one phylum: Mycota. About 100,000 species known. Many

more not yet classified.

    Class Oomycetes. Mostly aquatic species. Motile cells at some stage. Several hundred species.

    Class Zygomycetes. Terrestrial fungi such as black bread mold. Several hundred species.

    Class Ascomycetes. Terrestrial and aquatic fungi including mildews, truffles and yeasts. About 30,000 species.

    Class Basidiomycetes. Terrestrial fungi including mushrooms and toadstools. About 25,000 species.

    Class Fungi imperfecti. The last word refers to our knowledge, not to any fungal deficiency. The sexual cycle has not been observed in members of this group. As more is known about a species, it is usually placed into another group. In this group are our penicillium, the source of athlete's foot, and certain of the molds that give some cheeses their flavor.

    Class Lichens. Mostly comprised of about 17,000 species of Ascomycetes which occur symbiotically with certain algae.

Kingdom Plantae. Multicellular eucaryotes and certain related one-celled forms. Autotrophic (photosynthetic). Cell walls contain cellulose. Reproduction primarily sexual showing alternation of generations, with the gametophytic stage generally reduced.

  Phylum Rhodophyta. Red algae. Marine plants. No motile cells present at any stage. Body made of closely packed filaments. No differentiation of roots, stems, or leaves. No conducting elements. About 4,000 species.

  Phylum Phaeophyta. Brown algae. Marine. Motile stages are biflagellate. May show considerable differentiation, especially in the kelps. Some genera have conducting elements. No leaves, roots, or stems. About 1,100 species.

  Phylum Chlorophyta. Green algae. Unicellular or multicellular. Motile stages biflagellate. Multicellular stages do not show complex differentiation. Probably well over 7,000 species.

  Phylum Bryophyta. Mosses, hornworts and liverworts. Multicellular. Biflagellate sperm. All stages show complex multicellular patterns of development. Conducting tissues are absent or not well differentiated. Most photosynthesis carried out by the gametophyte stage. Over 23,500 species.

    Class Hepaticae. The liverworts. Gametophytes are undifferentiated or leafy; simple sporophytes. 9,000 species.

    Class Antherocerotae. The hornworts.

differentiated gametophytes. About 100
ecies.

Class Musci. The mosses. Leafy
metophytes. About 14,500 species.

Phylum Tracheophyta. The vascular
ants. Terrestrial, with leaves, roots, and
ms. Male cells of some species are motile,
opelled by cilia. Well-developed conduct-
g elements in the vascular plants. Reduced
metophyte stage (autotrophic in ferns,
terotrophic in the rest). Seed bearers.

Subphyla Lycophytina. The clubmosses.
ry diverse homosporous and heterosporous
ants. Motile sperm. About 1,000 species.

Subphyla Sperrophytina. The horsetails.
mosporous. Noded stems. Scalelike leaves.
otile sperm. About 24 species.

Subphyla Pterophytina. Ferns, gymno-
erms, and flowering plants. About 260,000
ecies.

Class Filicineae. The ferns. Game-
ohyte. Generally free-living and photo-
athetic. Multicellular gametangia and free-
imming sperm. About 11,000 species.

Class Coniferinae. The conifers, such
pines and spruces. Seed plants having
aple leaves. Ovules not enclosed. Sperm
flagellated. About 550 species.

Class Cycadinae. The cycads. Slow
owers. Compound pinnate, palmate, or fern-
e leaves. Ovules not enclosed. Motile sperm.
llen tube formed. About 100 species.

Class Ginkgoinae. Ginko. Fan-
aped leaves with dichotomous venation.
ules not enclosed. Pollen tube formed.
otile sperm. One species.

Class Angiospermae. The flowering
ants. Ovules enclosed in carpel in most
ses. Seeds borne within fruits. A diverse
oup. Gametophytes greatly reduced. About
0,000 species.

Subclass Dicotyledonae. The di-
ts. Flower parts usually in fours or fives. Leaf
nation usually netlike, pinnate, or palmate.
scular cambium usually present. Two
tyledons. Vascular bundles (xylem and
loem) usually organized in a ring. About
0,000 species.

Subclass Monocotyledonae. The
onocots. Flower parts usually in threes.
ually parallel venation. One cotyledon.
scular bundles scattered. Includes palms
d grasses. About 60,000 species.

ngdom Animalia. The animals. Multicel-
ar heterotrophic eucaryotes. Generally
ck cell walls. Complex embryology. Repro-
ction primarily sexual. Between 1 million
d 10 million species.

Subkingdom Parazoa. The only multicellular group with no digestive cavity.

Phylum Porifera. The sponges. Largely
marine. Stiff skeletons. Porous bodies. About
4,200 species.

Subkingdom Metazoa. Animals that consist
of many cells.

Phylum Coelenterata. The coelenterates.
Radially symmetrical. No mesoderm. Reproduction sexual or asexual. Stinging cells in the
outer of their two layers. Aquatic, mostly
marine. About 9,600 species.

Class Hydrozoa. Hydra. Obelia and
other hydralike animals. Sometimes colonial
with alternation of generations. Polyp stage
dominant.

Class Scyphozoa. Marine jellyfishes.
Medusa stage dominant. True muscle cells.

Class Anthozoa. Sea anemones and
colonial corals. No medusa stage.

Phylum Ctenophora. The comb jellies
and sea walnuts. Spherical and free-swimming.
Translucent, in some cases bioluminescent.
Possess eight bands of cilia. About 80 species.

Phylum Platyhelminthes. The flatworms.
Bilaterally symmetrical. Three germ layers.
Saclike gut when present. No coelom or
circulatory system. Hermaphroditic. Excretion
by flame cells. About 15,000 species.

Class Turbellaria. The planaria and
other nonparasitic flatworms.

Class Trematoda. The flukes. Parasites having a saclike digestive tract.

Class Cestoidea. The tapeworms.
Parasites that absorb their food through the
body surface.

Phylum Nemertea. The ribbon worms.
Nonparasitic. Usually marine. Tubelike gut.
Protrusible proboscis with a hook. About
550 species.

Phylum Aschelminthes. The roundworms
and nematodes. Includes free-living forms
and parasites. Elongated, cylindrical, non-
segmented bodies. About 80,000 species.

Phylum Acanthocephala. The spiny-
headed worms. Parasites. No digestive tract.
About 300 species.

Phylum Chaetognatha. Arrow worms.
Free-swimming, coelomate marine worms with
a tubelike digestive tract. Sickle-shaped hooks
around mouth. About 50 species.

Phylum Nematomorpha. The horsehair
worms. Slender, up to 3 feet long. Adults
free-living, but larvae parasitic in insects.
About 250 species.

Phylum Rotifera. "Wheel animals." Microscopic, wormlike or spherical. Tubelike
digestive tract. Circle of cilia around the

head which, when beating, suggests a wheel.
Males unknown, minute or degenerate in many
species. About 1,500 species.

Phylum Gastrotricha. Microscopic, wormlike. Move by cilia. About 140 species.

Phylum Bryozoa. Moss animals. Microscopic, aquatic. Usually in branching colonies.
Have tubelike digestive tract. Larvae kept in
brood pouch. About 4,000 species.

Phylum Brachiopoda. The lamp shells.
Marine animals having two shells.

Phylum Phoronidea. Elongate wormlike
animals that live within a leathery tube which
they secrete. Marine. About 15 species.

Phylum Annelida. Segmented worms such
as the earthworm. Well-developed coelom.
Tubelike digestive tract. Closed circulatory
system. Appendages nonjointed when present.
About 7,000 species.

Class Archiannelida. Small primitive
marine worms. About 35 species.

Class Polychaeta. Marine worms
primarily, such as *Nereis*. Distinct head, palps
(mouthparts), tentacles, bristled appendages.
About 4,000 species.

Class Oligochaeta. Terrestrial, fresh-
water, and marine annelids. Includes the
earthworm. About 2,500 species.

Class Hirudinea. Leeches. Posterior
sucker, anterior sucker in some species.
Marine and terrestrial. Free-living or parasitic.
About 300 species.

Phylum Mollusca. Unsegmented, with a
head, muscular foot, and a mantle that may be
involved in secreting a shell. Mostly aquatic,
soft-bodied. Three-chambered heart. About
100,000 species.

Class Amphineura. The chitons.
Elongate body covered by mantle. Eight dorsal
shell plates. About 700 species.

Class Pelecypoda. Two-shelled mollusks, such as clams and oysters. About 15,000
species.

Class Scaphopoda. The tooth shells.
A tubular, conical shell. About 350 species.

Class Gastropoda. Asymmetrical
mollusks, including snails and slugs. A distinct
head with tentacles. About 80,000 species.

Class Cephalopoda. "Headfoot"
animals, such as octopus and squid. Possess
tentacles and mouth with two horny jaws.
Well-developed eyes and nervous system.
About 400 species.

Phylum Arthropoda. Segmented, joint-
legged animals with paired appendages and a
hard exoskeleton. Grow by molting. Dorsal
brain. Complete digestive tract. Ventral nerve
cord. Paired ganglia in each segment. Over

750,000 species—the largest phylum.

Class Merostomata. The horseshoe crabs. Aquatic. "Book" gills. Called "living fossils."

Class Crustacea. Includes crayfish, lobsters, and crabs. Mostly aquatic. Two pairs of antennae, one pair of mandibles. Thoracic segment with appendages. Abdomen segmented and with or without appendages. About 25,000 species.

Class Arachnida. Spiders, scorpions, mites. Mostly terrestrial and air-breathing, with 4 or 5 pairs of legs, the first pair used for grasping. About 30,000 species.

Class Onychophora. Terrestrial arthropods. Simple body plans. One genus, *Peripatus*, has many short, unjointed pairs of legs. About 75 species.

Class Insecta. The insects. Most are terrestrial and have trachea, one pair of antennae, three pairs of legs. Body divided into head, thorax, and abdomen. Most have two pairs of wings. About 700,000 species.

Class Chilopoda. The centipedes. Trunk divided into 15 to 173 segments, each with a pair of jointed legs. About 2,000 species.

Class Diplopoda. The millipedes. Abdomen has 20 to 100 segments, each with two pairs of appendages. About 7,000 species.

Phylum Echinodermata. The spiny-skinned animals. Marine. Radially symmetrical as adults. Well-developed coelom. Endoskeleton of chalky particles and spines. Peculiar "water vascular" system. Tube feet. About 5,700 species.

Class Crinoidea. Sea lilies and feather stars. Sessile, often attached to ocean bottom by jointed stalks. Ten slender arms. Well represented among fossils.

Class Asteroidea. The starfishes. Five to 50 arms. Mouth directed downward.

Class Ophiuroidea. The brittle stars and serpent stars. Elongated, highly flexible stars.

Class Echinoidea. Sea urchins and sand dollars. Skeleton (or test) formed of rigid plates holding many moveable spines.

Class Holothuroidea. Sea cucumbers. Sausage-shaped, sluggish, sometimes over 18 inches long.

Phylum Chordata. Animals having a notochord, pharyngeal gill slits, and a hollow, dorsal nerve chord at some stage in their life cycles. About 45,000 species.

Subphylum Hemichordata. Marine animals with a notochordlike structure in the head, gill slits, solid nerve cord. Includes the acorn worm. About 91 species.

Subphylum Tunicata. The tunicates. Saclike animals. The adults usually sessile. Marine. May form branching colonies. Adults have gill slits, reduced nervous system. Larvae have notochord and well-developed nervous system. About 1,600 species.

Subphylum Cephalochordata. The lancelets. *Amphioxus* and related forms. Fishlike, marine, permanent notochord. No bone or cartilage. Only about 13 species.

Subphylum Vertebrata. The vertebrates. The supportive function of notochord taken over by cartilage or bone that forms a segmented backbone. Skull surrounds a complex brain. Tail usually present. About 43,000 species.

Class Agnatha. Lampreys and hagfishes. Aquatic. Eellike. No limbs or jaws. Sucking mouth. No bones, fins, or scales. Single nostril.

Class Chondrichthyes. Fish with cartilagenous skeletons, such as sharks, skates, and rays. No air bladders.

Class Osteichthyes. Fish with bony skeletons. Includes many marine and almost all freshwater fishes. Usually with an air bladder, or (in rare cases) a lung. There are about 23,000 species of fishes in this and the two preceding classes.

Class Amphibia. The amphibians, such as salamanders, frogs and toads. In most cases, breathing is by gills in the larval stage and lungs in the adult stage. Legs are usually present in adults. Eggs unprotected by a shell. About 2,000 species.

Class Reptilia. Includes the turtles, snakes, lizards, alligators, and the dinosaurs. Have lungs, scales, are cold-blooded. Except for snakes, legs are present. Primarily terrestrial, a few aquatic. Embryo develops in an egg with an amnion and allantois. About 5,000 species.

Class Aves. The birds. Feathered, warm-blooded, winged. Embryos develop in eggs outside the mother's body (as do many reptile and all amphibian eggs). About 9,000 species.

Class Mammalia. The mammals. Warm-blooded, usually hairy. Mothers secrete milk. Four limbs, which may be legs, wings (bats), or fins (whales). Respiratory diaphragm. Three bones in middle ear. Mandible formed of single pair of bones. Seven vertebrae in neck. About 4,500 species.

Subclass Prototheria. The monotremes. Egg-laying mammals with poor temperature regulation. Two living species are the platypus and the spiny anteater.

Subclass Metatheria. The marsupials. Live-bearers, with the placenta absent or poorly developed. Young develop primarily in the mother's pouch. Include kangaroos and opossums. Most are found in Australia.

Subclass Eutheria. The placental. Comprises most mammals.

ORDERS OF MAMMALS

Insectivora. Inappropriately designated the "insect-eaters," such as shrews, moles, etc.

Edentata. The toothless mammals such as anteaters and armadillos.

Rodentia. The rodents such as rats and squirrels.

Artiodactyla. The even-toed ungulates such as cattle, llamas, deer.

Perissodactyla. The odd-toed mammals such as horses and rhinoceroses.

Proboscidea. The big-nosed animals, such as the elephants.

Lagomorpha. Rabbits and hares.

Sirenia. The large aquatic mammals. Forelimbs, fins, and hindlimbs are absent. The manatee, dugong, and sea cow.

Carnivora. The flesh-eaters. Dogs, cats, seals, ferrets.

Cetacea. Aquatic mammals with finned forelimbs; hindlimbs absent. The whales, porpoises, dolphins.

Chiroptera. The flying mammals such as the bats.

Primata. The lemurs, monkeys, apes, and you.

uggested Readings

apter 1: Yesterday and Today
rwin, C. R. 1962. *The voyage of the* Beagle. Garden City, N.Y.: Doubleday.
bzhansky, T. 1962. *Mankind evolving.* New Haven, Conn.: Yale University Press.
seley, L. 1956. Charles Darwin. *Scientific American* 194(2):62–72.
———. 1958. *Darwin's century.* Garden City, N.Y.: Doubleday.
———. 1960. *The firmament of time.* New York: Atheneum.
amilton, T. H. 1967. *Process and pattern in evolution.* New York: Macmillan.
well, F. 1965. *Early man.* New York: Time-Life.
ayr, E. 1970. *Populations, species and evolution.* Cambridge, Mass.: Harvard University Press.
rter, E. 1971. *Galápagos.* New York: Ballantine.
rks, M. J., and Zirkle, C. 1964. *The evolution of biology.* New York: Ronald Press.

apter 2: The Beginnings
ker, J. J. W., and Allen, G. E. 1968. *Hypothesis, prediction and implication in biology.* Reading, Mass.: Addison-Wesley.
rghoorn, E. S. 1971. The oldest fossils. *Scientific American* 224(5):30–42.
nant, J. B. 1951. *Science and common sense.* New Haven, Conn.: Yale Univ. Press.
nshaw. P. S. 1971. *This side of yesterday— extinction or utopia?* New York: Wiley.
rgensen, J. J., ed. 1970. Biology and culture in modern perspective. *Readings from Sci. Amer.* San Francisco: Freeman.
oestler, A. 1972. *The roots of coincidence.* New York: Random House.
ummel, B. 1961. *History of the earth.* San Francisco: Freeman.
iller, S. L. 1955. Production of some organic compounds under possible primitive earth conditions. *Journal of the American Chemical Society* 77:2351–2361.
onod, J. 1972. *Chance and necessity: an essay on the natural philosophy of modern biology.* New York: Knopf.
ald, G. 1955. The origin of life. In *The physics and chemistry of life,* ed. of *Scientific American.* New York: Simon and Schuster.

Chapter 3: Molecules
Asimov, I. 1954. *The chemicals of life.* New York: Abelard-Schuman.
———. 1962. *Life and energy.* N.Y.: Avon.
Baker, J. J. W., and Allen, G. E. 1965. *Matter, energy and life.* Reading, Mass.: Addison-Wesley.
Barry, J. M., and Barry, E. M. 1969. *An introduction to the structure of biological molecules.* Englewood Cliffs, N.J.: Prentice-Hall.
Bernhard, S. A. 1968. *The structure and function of enzymes.* New York: Benjamin.
Farago, P., and Lagnado, J. 1972. *Life in action.* New York: Vintage.
Neurath, H. 1964. Protein-digesting enzymes. *Scientific American* 211(6):68–79.
Roberts, J. D. 1957. Organic chemical reactions. *Scientific American* 197(5):117–126.
Steiner, R. F., and Edelhock, H. 1965. *Molecules and life.* Princeton, N.J.: Van Nostrand.

Chapter 4: Energy
Amerine, M. A. 1964. Wine. *Scientific American* 211(2):46–56.
Asimov, I. 1968. *Photosynthesis.* New York: Basic Books.
Fogg, G. E. 1968. *Photosynthesis.* New York: American Elsevier.
Goldsby, R. A. 1967. *Cells and energy.* New York: Macmillan.
Lehninger, A. L. 1961. How cells transform energy. *Scientific American* 205(3):62–82.
———. 1965. *Bioenergetics, the molecular basis of biological energy transformations.* New York: Benjamin.
Levine, R. P. 1969. The mechanism of photosynthesis. *Scientific American* 221(6):58–70.
Margaria, R. 1972. The sources of muscular energy. *Scientific American* 226(3):84–91.
Rabinowitch, E. I., and Govindjee, I. 1965. The role of chlorophyll in photosynthesis. *Scientific American* 213(1):74–83.

Chapter 5: Cells
Crick, F. H. C. 1966. The genetic code: III. *Scientific American* 215(4):55–62.
Dupraw, E. J. 1968. *Cell and molecular biology.* New York: Academic Press.
———. 1970. *DNA and chromosomes.* New York: Holt, Rinehart and Winston.
Giese, A. C. 1968. *Cell physiology.* Philadelphia: Saunders.
Goldstein, L., ed. 1966. *Cell biology.* Dubuque, Iowa: William C. Brown.
Holter, J. 1961. How things get into cells. *Scientific American* 205(9):167–180.
Ingram, V. M. 1958. How do genes act? *Scientific American* 198(1):68–74.

Kennedy, D., ed. 1965. The living cell. *Readings from Sci. Amer.* San Francisco: Freeman.
Koller, P. C. 1971. *Chromosomes and genes.* New York: Norton.
Lowey, A. G., and Siekevitz, P. 1969. *Cell structure and function.* New York: Holt, Rinehart and Winston.
McElroy, W. D. 1971. *Cell physiology and biochemistry.* Englewood Cliffs, N.J.: Prentice-Hall.
Porter, K., and Bonneville, M. A. 1968. *An introduction to the fine structure of cells and tissues.* Philadelphia: Lea and Febiger.
Robertson, J. D. 1962. The membrane of the living cell. *Scientific American* 206(4):64–82.
Swanson, C. 1964. *The cell.* Englewood Cliffs, N.J.: Prentice-Hall.
Watson, J. D. 1968. *The double helix.* New York: Atheneum.
———. 1970. *Molecular biology of the gene.* New York: Benjamin.
Woese, C. R. 1967. *The genetic code, the molecular basis for genetic expression.* New York: Harper and Row.

Chapter 6: Inheritance
Bonner, D. M., and Mills, S. E. 1964. *Heredity.* Englewood Cliffs, N.J.: Prentice-Hall.
Boyd, W. C. 1963. Genetics and the human race. *Science* 140:3571.
Crow, J. F., and Kimura, M. 1970. *An introduction to population genetics theory.* New York: Harper and Row.
Dobzhansky, T. 1963. Evolutionary and population genetics. *Science* 142:3596.
Eckhardt, R. B. 1972. Population genetics and human origins. *Scientific American* 226(1):94–102.
Gray, G. W. 1951. Sickle-cell anemia. *Scientific American* 185(2):56–59.
Johnson, W. H., and Steere, W. C., eds. 1962. *This is life.* New York: Holt, Rinehart and Winston.
Levine, R. P. 1968. *Genetics.* New York: Holt, Rinehart and Winston.
McKusick, V. A. 1965. The royal hemophilia. *Scientific American* 213(2):88–95.
Scriven, M. 1959. Explanation and prediction in evolutionary theory. *Science* 130:477–482.
Srb, A., Owen, R., and Edgar, R. 1965. *General genetics.* San Francisco: Freeman.

Chapter 7: Reproduction
Allen, R. D. 1959. The moment of fertilization. *Scientific American* 201(1):124–134.
Asdell, S. A. 1964. *Patterns of mammalian reproduction.* Ithaca, N.Y.: Cornell University Press.

Bell, P. R., and Woodcock, C. L. F. 1968. *The diversity of green plants*. Reading, Mass.: Addison-Wesley.

Bullough, W. S. 1961. *Vertebrate reproductive cycles*. New York: Barnes & Noble.

Corner, E. J. H. 1968. *The life of plants*. New York: New American Library.

Csapo, A. 1958. Progesterone. *Scientific American* 198(4):40–46.

Frazer, J. F. D. 1959. *The sexual cycles of vertebrates*. London: Hutchinson University Library.

Hardin, G. 1970. *Birth control*. New York: Pegasus.

Katchadowrian, H., and Lunde, D. T. 1972. *Fundamentals of human sexuality*. New York: Holt, Rinehart and Winston.

Masters, W., and Johnson, V. 1966. *Human sexual response*. Boston: Little, Brown.

Michelmore, S. 1964. *Sexual reproduction*. Garden City, N.Y.: Natural History Press.

———. 1965. *Sexual reproduction*. Garden City, N.Y.: Natural History Press.

Peel, J., and Potts, M. 1969. *Textbook of contraceptive practices*. New York: Cambridge University Press.

Wilmoth, J. H. 1967. *Biology of invertebrata*. Englewood Cliffs, N.J.: Prentice-Hall.

Chapter 8: Development

Bonner, J. T. 1963. *Morphogenesis*. New York: Atheneum.

Fischberg, M., and Blackler, A. W. 1961. How cells specialize. *Scientific American* 205(3):124–140.

Flickinger, R. A., ed. 1966. *Developmental biology*. Dubuque, Iowa: Brown.

Grobstein, C. 1964. Cytodifferentiation and its controls. *Science* 143:643.

Odell, W. D., and Moyer, D. L. 1971. *Physiology of reproduction*. St. Louis: Mosby.

Rugh, R., and Shettles, L. 1971. *From conception to birth: the drama of life's beginnings*. New York: Harper and Row.

Smith, C. A. 1963. The first breath. *Scientific American* 209(4):27–35.

Spratt, N. T., Jr. 1964. *Introduction to cell differentiation*. New York: Van Nostrand-Reinhold.

Sussman, M. 1960. *Animal growth and development*. Englewood Cliffs, N.J.: Prentice-Hall.

Tanner, J. M., Taylor, G. R., and the ed. of *Life*. 1965. *Growth*. New York: Time-Life.

Waddington, C. H. 1966. *Principles of development and differentiation*. New York: Macmillan.

Wilmoth, J. H. 1967. *Biology of invertebrata*. Englewood Cliffs, N.J.: Prentice-Hall.

Chapter 9: Systems and Their Control

Asimov, I. 1964. *The human body: its structure and operation*. New York: Signet.

Beck, W. S. 1971. *Human design: molecular, cellular and systematic physiology*. New York: Harcourt Brace Jovanovich.

Best, C. H., and Taylor, N. B. 1968. *The living body*. New York: Holt, Rinehart and Winston.

Currey, J. 1970. *Animal skeletons*. New York: St. Martin's Press.

Guyton, A. C. 1967. *Function of the human body*. Philadelphia: Saunders.

Hoyle, G. 1970. How is muscle turned on and off? *Scientific American* 222(4):72–82.

Huxley, H. E. 1965. The mechanism of muscular contraction. *Scientific American* 213(6):18–27.

Morton, J. E. 1967. *Guts*. New York: St. Martin's Press.

Nunnally, D., and Bogitsh, B. *Introductory zoology*. New York: Wiley.

Ramsay, J. A. 1968. *Physiological approach to the lower animals*. New York: Cambridge University Press.

Schmidt-Nielsen, K. 1964. *Desert animals*. New York: Oxford University Press.

———. 1972. *How animals work*. New York: Cambridge University Press.

Chapter 10: Nerves and Hormones

Asimov, I. 1963. *The human brain: its capacities and functions*. New York: Signet.

Baker, P. F. 1966. The nerve axon. *Scientific American* 214(3):74–82.

Case, J. 1966. *Sensory mechanisms*. New York: Macmillan.

Clegg, P. C., and Clegg, A. C. 1968. *Hormones, cells and organisms*. Stanford, Calif.: Stanford University Press.

Davidson, E. H. 1965. Hormones and genes. *Scientific American* 212(6):36–45.

Day, R. H. 1971. *Perception*. Dubuque, Iowa: Brown.

Eccles, J. C. 1963. *The physiology of synapses*. New York: Academic Press.

Frye, B. E. 1967. *Hormonal control in vertebrates*. New York: Macmillan.

Greene, R. 1970. *Human hormones*. New York: McGraw-Hill.

Griffin, D. R., and Novick, A. 1970. *Animal structure and function*. New York: Holt, Rinehart and Winston.

Heimer, L. 1971. Pathways in the brain. *Scientific American* 225(1):48–60.

McCashland, B. W. 1968. *Animal coordinating mechanisms*. Dubuque, Iowa: Brown.

Riedman, S. R. 1962. *Our hormones and ho[w] they work*. New York: Collier.

Snider, R. S. 1958. The cerebellum. *Scientif[ic] American* 199(2):84–90.

Whalen, R. E. 1967. *Hormones and behavio[r]*. New York: Van Nostrand-Reinhold.

Chapter 11: Behavior

Eibl-Ebesfeldt, I. 1970. *Ethology*. New York: Holt, Rinehart and Winston.

Hinde, R. 1970. *Animal behavior*. New York: McGraw-Hill.

Jolly, A. 1972. *The evolution of primate be[ha]vior*. New York: Macmillan.

Klopfer, P. H., and Hailman, J. P. 1967. *A[n] introduction to animal behavior: ethology['s] first century*. Englewood Cliffs, N.J[.]: Prentice-Hall.

———. 1973. *Behavioral aspects of ecolog[y]*. Englewood Cliffs, N.J.: Prentice-Hall.

Manning, A. 1972. *An introduction to anim[al] behavior*. Reading, Mass.: Addison-Wesle[y].

Marker, P., and Hamilton, W. 1966. *Mech[-]anisms of animal behavior*. New York: Wile[y].

Morris, D. 1967. *The naked ape*. New York: Dell.

Tiger, L. 1969. *Men in groups*. New York: Random House.

Tinbergen, N. 1953. *Social behavior in animal[s]*. London: Methuen.

———. 1951. *The study of instinct*. Oxfor[d]: Oxford University Press.

Wallace, R. 1973. *The ecology and evolutio[n] of animal behavior*. Pacific Palisades, Calif[.]: Goodyear.

Chapter 12: Community and Competition

Cornell, J., Mertz, D., and Murdoch, W[.], eds. 1970. *Readings in ecology and ecologic[al] genetics*. New York: Harper & Row.

Emlen, J. M. 1973. *Ecology: An evolutionar[y] approach*. Reading, Mass.: Addison-Wesle[y].

Kendeigh, S. C. 1974. *Ecology*. Englewoo[d] Cliffs, N.J.: Prentice-Hall.

Kormondy, E. 1969. *Concepts of ecolog[y]*. Englewood Cliffs, N.J.: Prentice-Hall.

Pianka, E. 1974. *Evolutionary ecology*. Ne[w] York: Harper & Row.

McNaughton, S. J., and Wolf, L. L. 197[3]. *General ecology*. New York: Holt, Rineha[rt] and Winston.

Moen, A. 1973. *Wildlife ecology*. San Fra[n]cisco: Freeman.

Ricklefs, R. 1973. *Ecology*. Newton, Mass[.]: Chiron.

Vernberg, F. J., and Vernberg, W. B. 197[0]. *The animal and the environment*. New Yor[k]:

Holt, Rinehart and Winston.

hapter 13: Populations

ppleman, B. 1965. *The silent explosion.* Boston, Mass.: Beacon.

alahan, D. 1972. Ethics and population limitation. *Science* 175:487–492.

alhoun, J. R. 1962. Population density and social pathology. *Scientific American* 206(2):139–148.

roat, T. 1972. The role of overpopulation and agricultural methods in the destruction of tropical ecosystems. *BioScience* 22:465–467.

hrlich, P. 1968. *The population bomb.* New York: Ballantine.

———, and Ehrlich, A. 1970. *Population, resources, environment.* San Francisco: Freeman.

ulett, H. 1970. Optimum world population. *BioScience* 20:160–161.

anger, W. 1972. Checks on population growth: 1750–1850. *Scientific American* 226(2):92–99.

Meadows, D., et al. 1972. *Limits to growth.* New York: Universe.

Vesting, A. 1971. Ecoside in Indochina. *Natural History* 80(3):56–61.

hapter 14: Resources, Pollution, and Values

oerma, A. 1970. A world agricultural plan. *Scientific American* 223(2):54–69.

resler, J. 1966. *Human ecology: collected readings.* Reading, Mass.: Addison-Wesley.

arson, R. 1962. *Silent spring.* Boston: Houghton Mifflin.

lement, R. 1970. Pesticide do's and don't's. *Audubon* 72(2):50–51.

ommoner, B. 1966. *Science and survival.* New York: Viking Press.

———. 1971. *The closing circle.* New York: Knopf.

ardin, G. 1968. The tragedy of the commons. *Science* 162:1243–1248.

———. 1972. *Exploring new ethics for survival.* New York: Viking.

appe, F. 1971. *Diet for a small planet.* New York: Ballantine.

Murdoch, W., ed. 1971. *Environment: resources, pollution, society.* Stamford, Conn.: Sinauer.

Rienow, R., and Rienow, L. 1967. *Moment in the sun.* New York: Ballantine.

hepard, P., and McKinley, D., eds. 1969. *The subversive science: essays toward an ecology of man.* Boston: Houghton Mifflin.

Wagar, J. A. 1970. Growth versus the quality of life. *Science* 168:1179–1184.

Glossary

Abortion: The expulsion or removal of an embryo or fetus before it can survive on its own.

Acceptor: A molecule that unites chemically with hydrogen or electrons in the oxidation-reduction reactions of cells.

Acetylcholine (as′·ə·t′l·kō·lēn): A chemical agent that transmits neutral impulses across synapses from one neuron to another.

Acid: A substance which releases hydrogen ions into a solution. It has a sour taste, unites with bases to form salts.

Actin: The moveable filament of protein found in the myofibrils of muscle.

Action potential: Changes in the electrical potential across the membrane of a nerve cell that occur during nerve impulse transmission.

Active transport: The movement of solutes against a concentration gradient, usually requiring the expenditure of cell energy.

Adaptive radiation: The diversity of species which is attributed to sub-populations adapting to a unique set of environmental conditions.

Adenine (ad′·n·ēn′): One of the nitrogenous bases that are found in nucleic acids and in ATP.

Adenosine diphosphate (ə·den′·ə·s′n dī·fäs′·fāt): ADP—A product formed by the hydrolysis of ATP, Adenosine Triphosphate, with the release of energy.

Adenosine triphosphate (ə·den′·ə·s′n trī·fŏs′·fāt): ATP—A compound that is found in all cells which serves as an energy source as it is broken down into ADP plus phosphate with the release of energy.

Adrenaline (ə·dren′·′l·in): See epinephrine.

Aerobe (er′·ōb): An organism that uses oxygen in its respiration.

Afferent nerve: A nerve which directs impulses toward the central nervous system.

Afterbirth: The placenta with its associated membranes which is expelled from the uterus at birth.

Alcohol: A carbon-hydrogen molecule which contains one or more OH groups.

Allele (ə·lēl′): One of two or more genes which produce a specific characteristic such as blood type, hair color, etc.

Alpha waves: A particular pattern of brain waves which are characteristic of a person who is awake, but at rest.

Alternation of generations: The alternation of haploid and diploid stages in the life cycle of a species.

Alveolus (al·vē′·ə·ləs): One of the tiny air sacs of the lung, in which carbon dioxide and oxygen are exchanged.

Amino acid: An organic molecule that contains at least one carboxy group, COOH, and one amino group, NH_2.

Amnion (am′·nē·ən): A fluid filled sac which surrounds the embryo in the uterus, or in the embryonic egg of egglaying animals.

Anaerobe (an·er′·ōb): An organism which does not utilize oxygen in its respiration.

Analogous: A term that describes structures which serve the same function but have different morphologies and embryonic origin.

Anaphase (an′·ə·fāz′): The third phase of mitosis in which the chromatids separate and move to opposite sides of the cell.

Androgen (an′·drə·jən): A male sex hormone which is secreted by the testicles.

Angiosperm: A flowering plant.

Animal pole: That end of an animal zygote which has relatively little yolk and experiences more rapid cell division.

Anterior: Toward the head or front.

Anther (an′·thər): A floral structure at the tip of the stamen in which male spores are formed.

Anthropomorphism: Attribution of human characteristics to other animals.

Antigen (an′·tə·jən): Any substance which triggers an immune response.

Aorta: The large artery which in humans, leaves the heart and supplies blood to all parts of the body except the lungs.

Apical meristem (ap′·ə·k′l mer′·ə·stem′): A region at the tip of a plant shoot which is marked by actively dividing cells.

Arteriole (är·tir′·ē·ōl): A small artery.

Asexual reproduction: Reproduction without the union of sex cells; the production of new individuals from one parent cell or organism.

Asymmetric: Lacking an orderly arrangement of parts.

Atomic number: The number assigned to a particular element, determined by the number of protons in its atoms.

Atomic weight: The average weight of the atom of an element. The unit of atomic weight is $\frac{1}{12}$ the mass of carbon 12.

Atrium (ā′·trē·əm): The smaller compartment

of the heart that receives venus blood from the body or lungs.

Autonomic nervous system (ôt′·ə·näm′·ik): That part of the peripheral nervous system that regulates the internal environment by the control of glands and smooth muscles.

Auxin: A general name for a plant hormone.

Avoidance behavior: That behavior which helps an organism escape or elude a stimulus.

Axon: A conducting fiber extending from a neuron that transmits impulses away from the cell body.

Bacteriophage (bak·tir′·ē·ə·fāj′): A virus that reproduces in bacterial cells.

Barbitutates: Sedative drugs that act as depressants on the central nervous system.

Bark: All the layers of woody plant stem from the vascular cambium outward.

Basal body: A bulb-like structure at the base of a cilium.

Basal metabolism: The metabolism of an organism at rest.

Base: Any chemical which releases hydroxyl (OH⁻) ions in water, or joins with free protons. Bases leave a bitter taste and form salts when they react with acids.

Behaviorism: The psychological field which uses objective observation of behavior without reference to awareness on the part of the observed subject.

Benthic: Referring to the bottom of a lake or sea.

Bio-degradable: Capable of being oxidized or broken down by living organisms.

Biological clock: A mechanism or model which explains the cyclic or rhythmic changes of behavior in animals and plants.

Biological control: Using a predator or parasite of some pest organism in order to reduce the level of its population.

Biome (bī′·ōm): A geographical area which is characterized by relatively similar plant and animal representatives, for example, a tundra.

Biosphere: That portion of the earth which can support life.

Biotic community: Populations of plants and animals, sharing a region and interacting in the exchange of materials and energy.

Bladder: An organ that stores liquid such as bile or urine.

Blastocyst (blas′·tō·sist′): A blastula; embryonic stage when the cells form a hollow sphere.

Blastulation (blas′·choo·lā′·shən): The cellular activity of a zygote that results in the formation of a hollow sphere of cells, the blastula.

Bond energy: The energy stored in a chemical bond. When the bond is broken that energy is released.

Bowman's capsule: A curved sac at the beginning of the kidney nephric unit which surrounds the glomerulus.

Brain stem: The lower portion of the brain which includes midbrain, the pons, and the medulla.

Bronchiole (brän′·kē·ōl): A small branch of the bronchi in the vertebrate lung.

Bronchus (brän′·kəs): A major branch of the trachea.

Budding: Asexual reproduction seen in yeasts and other organisms where smaller cells bud or grow from a parental cell.

Bulbourethral gland (Cowper's gland): One of a pair of pea-sized glands, located below the prostate in mammalian males, that adds its secretions to the seminal fluid.

Cecum (sē′·kəm): A pouch between the small and large intestine in man. The appendix extends from the cecum.

calorie: A unit of heat needed to raise the temperature of one gram of water one degree centigrade.

Calvin cycle: A pathway of carbon dioxide which occurs during the dark phase of photosynthesis as carbohydrate is produced.

Cambium (kam′·bē·əm): A layer of meristematic tissue between the phloem and the xylem.

Cancer (kan′·sər): An abnormal and uncontrolled cell growth occurring in plants and animals.

Cannabis (kan·na·bis): A plant of the genus cannabis; marijuana and hemp are included in this group.

Capillary: A thin-walled blood vessel branching from an artery and later joining with other capillaries to form veins.

Carbohydrate (kär′·bə·hī′·drāt): An organic substance containing carbon, hydrogen and oxygen in the ratio of CH_2O. Includes sugars, starches, cellulose, etc.

Carbonic anhydrase (kär·bän′·ik an·hī′·drās): An enzyme in red blood cells that catalyzes the changing of carbon dioxide into carbonic acid and the reverse reaction.

Carcinogenic (kär′·sə·nō·jen′ik): Cancer producing.

Carnivore: An animal that consumes other animals' flesh.

Cell (sel): The basic unit of life. The simplest unit that can exist as an independent system.

Cell plate: A structure that appears along the equatorial plane as cytoplasm is divided in telophase of plant mitosis: it will form the cell wall separating daughter cells.

Central nervous system (CNS): The brain and spinal cord.

Centriole (sen′·trē·ōl′): A paired organell near the nucleus of animal cells that functions in cell division.

Centromere (sen′·trə·mir′): A structure on the chromosome to which the spindle fibe attaches that divides when chromatid separate in meiosis and mitosis.

Centrosome (sen′·trə·sōm′): The area o specialized cytoplasm containing centriole in a cell which usually lies near the nucleus

Cerebellum (ser′·ə·bel′·əm): The distinct walnut shaped structure at the back o the vertebrate brain; involved in coordination and equilibrium.

Cerebral cortex (kôr′·teks): The external laye of the forebrain or cerebrum.

Cerebrum (ser′·ə·brəm): In humans, the largest portion of the brain; coordinate voluntary action of muscles and many othe activities.

Cervix (sur′·viks): The narrow neck of the uterus which opens into the vagina.

Chain reaction: The phenomenon in which particles emitted from atomic nuclei collide with other nuclei causing an avalanche o similar events.

Chiasma (kī·az′·mə): A cross-over area or the chromosome seen in prophase I o meiosis.

Chlorophyll (klôr′·ə·fil′): The light absorbin pigments of plants and some bacteria tha trap light energy for photosynthesis.

Chlorophyll A: A type of chlorophyll commor to all plants.

Chlorophyll B: A type of chlorophyll commor in certain land plants and in green algae

Chloroplast (klôr′·ə·plast′): A chlorophyl containing organelle that functions in photosynthesis in plant cells.

Chorion (kôr′·ē·än′): A membrane surrounding the embryo.

Chromatid (krō′·mə·tid): One of two duplicated strands in a chromosome.

Chromatin (krō′·mə·tin): Indistinct, granular chromosomal material seen at interphase.

Chromosomes (krō′·mə·sōm′): Threadlike, condensed chromatin (DNA), visible at cell division only, occurring in pairs with specific numbers for each species.

Chromosome mapping: Plotting relative positions of genes along chromosomes.

Cilium (sil′·ē·əm): Hair-like cellular processes that beat rhythmically on the cell surface.

Circadian rhythm (sər·kā′·dē·ən): Pertaining to a daily cycle.

Class: A major taxonomic sub-division of a

phylum consisting of sub-ordinate groups known as orders.

...assical conditioning: Learning in which a condition stimulus presented together with an unconditional stimulus initiate a reaction to the condition stimulus.

...ine (klīn): Small variations in genetic characters across the range of a species population.

...itoris (klit′·ər·əs): An external organ of the female genitalia considered to be homologous to the penis in the male.

...one (klōn): The descendents of a single cell, generally identical.

...odon (kō′·dän): A sequence of three nucleotide bases required to form one amino acid in a protein chain synthesized on the RNA messenger.

...elom (sē′·ləm): The cavity within the body wall and surrounding the gut in higher animals.

...itus (kō′·it·əs): Sexual intercourse.

...olchicine (kăl′·chə·sēn′): A meiotic poison which causes the doubling of chromosome numbers in plants.

...ollar cell: Certain flagellated cells in sponges which create a current and ingest food particles from the water.

...ollenchyma (kə·len′·ki·mə): A plant supportive tissue consisting of thick-walled, elongated cells.

...olloid (kăl′·oid): A liquid containing fine particles of matter in permanent suspension.

...olon: The large intestine.

...onditioned response: Any response given to a learned stimulus.

...onditioned stimulus: A stimulus that causes a response after training.

...ongenital: Any condition present at birth.

...onspecific: Members of the same species.

...onvergent evolution: A similarity in genetically different organisms resulting from their adaptation to similar habitats.

...opulation: Sexual intercourse or coitus.

...ork cambium: The perpetually growing tissue of a plant or stem that forms the bark.

...orolla (kə·rä1′·ə): All the petals of a flower.

...orpus callosum (kôr′·pəs kə·lō′·səm): A connecting strip of nerve fibers that coordinates the activities of the cerebral hemispheres.

...orpus luteum (kôr′·pəs lōo′·tē·əm): The body formed by the ovarian follicle cells after the egg has been released.

...otyledon (kăt′·′l·ēd′·′n): The seed leaf in the angiosperm embryo.

...ovalent bond: A bond formed between two atoms as they share electrons.

Creatine (krē′·ə·tēn′): The amino acid part of creatine phosphate, important in muscle contraction.

Crista (kris′·tə): Inner membranes of the mitochondria.

Crop: A food storage sac located below the esophagus in birds, earthworms and other animals.

Cross fertilization: Fertilization of one flower by another.

Cross-over: An exchange of chromosomal material between homologous chromosomes in prophase I of meiosis.

Cytochrome (sīt′·ə·krōm′): An iron containing compound which functions in electron transport during oxidative respiration.

Cytoplasm (sit′·ə·plaz′m): The living matter in a cell, outside of the nucleus.

Cytosine (sīt′·ə·sēn′): One of the nitrogenous bases that are found in nucleic acids.

Dark reactions: The stage of photosynthesis which includes carbohydrate production.

Deme (dēm): An interbreeding population.

Dendrite: The short branched fiber of the nuron which receives nerve impulses.

Density dependent factor: Any influence on the growth rate of a population that is a function of the population size.

Deoxyribonuclease (dē·äk′·sə·rī′·bō·nōō·klē·ās): An enzyme that destroys DNA.

Deoxyribonucleic acid (dē·äk′·sə·rī′·bō·nōō·klē′·ik as′·id): DNA, a nucleic acid containing deoxyribose which is found chiefly in the nucleus of cells. DNA is the genetic message, functioning in protein synthesis.

Depolarization: Reversal of the polarization of a resting neuron by the inward flow of sodium ions.

Diaphragm: The muscular dome below the lungs which functions in breathing. Also a contraceptive device which fits over the cervix.

Dicot (dī′-kät′): Abbreviation for dicotyledon which is any of the higher plants which produces two seed leaves.

Differentiation: Cellular specialization during development.

Dihybrid (dī·hī′·brid): The result of a genetic cross in which two different traits are involved.

Diploidy (dip′·loi′·dē): Having the full number of chromosomes (2N).

Directional selection: In evolution, that selection which favors one or the other of two alleles in a deme.

Distal convoluted tubule: That region of the nephric unit which empties into the collecting tubule. It regulates the solutes present

in the filtrate.

Dominant: In genetics, the allele of a pair which is expressed while the other is hidden or repressed.

Dorsal: Upward or toward the back.

Dorsal root ganglion: A cluster of receptive cell bodies that lies outside each segment of the spinal column.

Double bond: A covalent bond in which two pairs of electrons are shared between two atoms.

Double helix: A double spiral in which each strand is wound around the other.

Duodenum (dōō′·ə·dē′·nəm): The first segment of the small intestine posterior to the stomach.

Ecology: The study of the interrelationship of the organism and its environment.

Ecosystem: A community of organisms interacting together and their physical environment.

Ectoderm (ek′·tə·durm′): The outer cell layer in an animal embryo that gives rise to skin, nervous tissue, etc.

Effector: A muscle or gland that responds to impulses from an efferent neuron.

Efferent nerve fiber: A nerve fiber that carries impulses from the central nervous system to an effector.

Ejaculation: The forceful expulsion of semen from the penis.

Electrolyte: A solution of ions that exhibits electrical conductivity.

Electron: A minute, negatively charged particle of an atom, in motion about the nucleus.

Electronegative: A state of attraction to the negative electron by other atoms or molecules.

Electron shell: A theoretical band around an atomic nucleus where electrons of a certain energy level are found.

Element: A substance consisting of atoms of one kind only, such as iron, gold, sulfur, etc. There are 102 known elements.

Embryo: The developing stage of any organism.

Embryonic induction: An influence one group of embryonic cells has on another during development.

Emphysema (em′·fə·sē′·mə): Distension and rupturing of the alveoli of the lungs, interfering with respiration.

Endergonic (end·er·gon′·ik): Any chemical reaction characterized by the absorption of energy from an outside source.

Endocrine gland (en′·də·krin): A gland that releases hormones into the blood. Each hormone has specific target cells somewhere in the organism.

Endoderm (en'·də·durm'): The innermost cell layers of an embryo which differentiate into such organs as the digestive tract, liver and pancreas.

Endogenous theory (en·däj'·ə·nəs): The theory which explains biological rhythm by assuming the existence of regulatory mechanisms known as biological clocks.

Endometrium (en'·dō·mē'·trē·əm): A temporary glandular lining of the uterus, in which the fertilized egg implants and develops.

Endoplasmic reticulum: Layers of membrane folded through the cytoplasm of a cell, forming complex inner surfaces, after being covered with ribosomes.

Endosperm (en'·də·spurm'): Nutritive tissue in the seeds which nourishes the developing embryonic sporophyte.

Entropy (en'·trə·pē): The opposite of "Free Energy." The degree of randomness in a system.

Enzyme (en'·zīm): An organic catalyst, usually a protein, capable of changing the rate of chemical reactions in cells.

Epidermis (ep'·ə·dur'·mis): The outermost cell layer of an organism.

Epididymus (ep'·ə·did'·i·məs): An oblong, convoluted tube, resting on the testes which stores sperm, releasing them into the spermatic duct.

Epinephrine (ep'·ə·nef'·rin): Adrenaline; a hormone secreted by the adrenal medulla.

Epithelium (ep' ə·thē'·lē·əm): Tissue covering internal and external surfaces.

Equatorial plate: The plane of the cell in which chromosomes arrange themselves as the cell begins metaphase.

Equilibrium: The point at which a chemical reaction proceeds in both directions at the same rate.

Esophagus: The muscular tube connecting the pharynx to the stomach.

Essential amino acid: An amino acid which is necessary to the metabolism of the organism.

Estrogen (es'·trə·jən): A female hormone; one of the hormones involved in the production of secondary sex characteristics and the menstrual cycle.

Eucaryotic (yū·kar'ē·o'tic): Cells having a true nucleus and inner membrane structures.

Evolution: The continuous genetic changes occurring in populations as a result of adapting to the environmental changes.

Excitation: Absorption of energy by electrons resulting in their movement to greater distances from the nucleus. Also, may refer to stimulation of a neuron.

Exergonic (ek'ser·gon·ik): Any chemical reaction which releases energy.

Exogenous theory (ek·säj'·ə·nəs): That theory which says that biological rhythms are due to external geophysical factors.

Exoskeleton (ek'·sō·skel'·ə·t'n): An external skeleton, a hardened outside covering of arthropods such as crayfish and insects.

Exothermic reaction (ek'·sə·thur'·mik): A chemical reaction in which heat is released.

Eutrophication (yoo·träf'·i·kā·shən): The aging process whereby a body of water supports increasing numbers of organisms, often causing lakes to become marshes and then terrestrial communities.

Fat: A compound composed of three fatty acids linked to glycerol.

Fatty acids: Constituents of fats and oils. Long chains of carbon atoms with hydrogen attached, and ending with carboxyl groups.

Fermentation: Anaerobic respiration with the usual end product being alcohol and carbon dioxide.

Fertilization: The union of sperm and egg.

Fetus: The term used to describe the human embryo after eight weeks of development.

Filtrates: Substances which have passed through a filter.

Flagellum (flə·jel'·əm): A threadlike organelle which extends from the cells surface; it may beat with whiplike motions.

Follicle (fäl'·i·k'l): A layer of cells surrounding the egg cell in the ovary.

Follicle stimulating hormone: FSH; A hormone produced by the anterior pituitary which stimulates the growth and maturation of eggs and sperm.

Food chain: Transfer of food energy from producers such as green plants to consumers such as herbivores and carnivores.

Food web: A scheme showing complex feeding relationships among the organisms in a community.

Forebrain: The anterior most sub-division of the vertebrate brain consisting of the cerebral hemispheres and certain anterior structures such as the thalamus.

Foreskin: A fold of skin at the end of the penis which is removed in circumcision.

Fruit: Layers of tissues produced by the ovary of an angiosperm.

Functional group: A group of atoms in an organic molecule which are free to participate in chemical reactions.

Gamete (gam'·ēt): A male or female sex cell

with half the chromosomal material. Fertilization is the union of gametes, producing the full amount of hereditary material.

Gametogenesis (gə·mēt'·ə·jen'·ə·sis): The formation of gametes.

Gametophyte (gə·mēt'·ə·fīt'): The sexual generation of a plant in which the cells have half the normal chromosome number. It alternates with the sporophyte generation.

Ganglion (gan'·glē·ən): A group of nerve cell bodies, situated outside the central nervous system.

Gastrulation (gas'·troo·lā'·shən): The formation of the gastrula stage of an embryo. This is the stage in which the embryo acquires its three germ layers.

Gene: The portion or portions of DNA that produce a recognizable effect or trait, e.g. genes produce enzymes, color pigments, etc.

Gene frequency: The fractional distribution of alleles for a specific trait in a population, e.g. 0.3T + 0.7t — the gene frequency for tallness and shortness.

Genetic code: The biochemical specification (DNA) for producing a particular genetic trait of the entire specie's characteristics.

Genetic drift: Changes in the composition of a gene pool caused by the random loss of a particular gene through death or migration rather than by selection.

Genetic mapping: The determination of the position of genes on chromosomes, relative to each other, expressed as map units.

Genotype (jen'·ə·tīp'): The assortment of genes which make up the genetic characteristics of an organism.

Genus (jē'·nəs): A major sub-division of a family, consisting of one or more species.

Geotropism (jē·ät'·rə·piz'm): The influence of gravity upon growth or movement.

Germ cell: The sex cell; sperm or egg.

Germ layer: A layer of undifferentiated embryonic cells.

Gestation period (jes·tā'·shən): The duration of pregnancy.

Gill: The respiratory organ of certain acquatic organisms through which oxygen and carbon dioxide are exchanged with water.

Gizzard: A muscular sac in the digestive system which mechanically changes food by mashing it against sand particles; found in birds and some other animals when chewing is not possible.

Gland: A group of cells specialized for secretion.

Glands of Bartholin: Mucus glands of the vagina and labia majora whose secretions lubricate the vagina during intercourse.

Glans penis (glanz pē·nis): The end or caplike structure at the termination of the penis.

Glial cell: One of the numerous cells which fill the spaces between nerve cells.

Globular protein: A protein which has doubled back on itself several times so the structure is almost spherical.

Glomerulus (glö·mer′·yoo·ləs): A cluster of capillaries within the Bowman's capsule of the kidney.

Glucose: A six carbon sugar found in many fruits and animal tissues. The basic unit of starch and cellulose.

Glycerol (glis′·ər·ōl′): A three carbon alcohol molecule which combines with fatty acids to form fats and oils.

Glycogen (glī′·kə·jən): A complex chain of glucose units, produced by animals.

Glycolysis (glī·kăl′·ə·sis): The first phase of respiration where glucose is broken down with a small energy yield.

Golgi bodies (gōl′·jē): A secretory structure in some cells consisting of a complex system of folded membranes.

Gonad (gō′·nad): An ovary or testis.

Guanine (gwä′·nēn): One of the nitrogenous bases that are found in nucleic acids.

Guard cell: One of the two cells which surround the stoma in leaves or young stems of plants.

Gullet: The esophagus or food receiving tube in an animal.

Gyrus (jī′·rəs): A ridge formed by the convolutions of the cerebral cortex of the brain.

Habitat: The specific place where an organism lives.

Habituation (hah·bich·oo·ay-shun): A simple form of learning in which an organism's response to a stimulus diminishes.

Hallucinogenic (hə·lōō′·sə·nə·jen·ic): The quality of producing hallucinations.

Haploid (hap′·loid): Having the half number of chromosomes per cell.

Hardy-Weinberg Law: A law in genetics that states that in a stable population the overall frequency of genes will not change in the absence of migration or other unstabling factors.

Hemisphere: Either of two symmetrical halves of the brain.

Hemoglobin (hē′·mə·glō′·bin): A protein-iron complex in red blood cells that binds with oxygen.

Hemophilia (hē·mə·fil′·ē·ə): A sex linked condition in which the blood does not clot normally. It is also called bleeder's disease.

Herbivore (hur′·bə·vôr′): A plant-eating animal.

Heredity: The transmission of traits from parents to offspring.

Hermaphrodite (hər·maf′·rə·dīt′): An individual who possesses both female and male reproductive organs.

Heterotroph (het′·ər·ə·träf′): An organism which depends on an external source of organic substances for its food.

Heterozygous (het′·ər·ō·zī′·gəs): A mixed genotype. Different alleles for a particular trait.

High energy bond: A bond which releases a relatively high amount of energy when broken; for example, the phosphate bonds in ATP.

Hindbrain: The most posterior of the three subdivisions of the vertebrate brain. It includes the pons, the medulla, and the cerebellum.

Homeostasis (hō′·mē·ō·stā′·sis): The tendency of organisms to maintain stability and consistency.

Homolog (hăm′·ə·lôg′): One member of a pair of chromosomes.

Homozygous (hō·mō·zī′·gəs): Having identical alleles for a particular trait.

Hormone: A chemical messenger; a substance produced in the body which has a regulatory effect in another part of the body.

Hybrid: Offspring resulting from the cross of parents with different characteristics or parents of different species.

Hydrocarbon: Any organic compound which contains simply carbon and hydrogen.

Hydrogenation: To combine with hydrogen.

Hydrogen bond: A weak attractive force between two molecules where one contains hydrogen.

Hydrolysis: The breakdown of a compound by the addition of water molecules.

Hydrophilic (hī′·drə·fil′·ik): Water loving, or having an attraction for water.

Hydrophobic (hī′·drə·fō′·bik): Water fearing, or repelling of water.

Hymen (hī′·mən): A membrane which may partially block the opening of the vagina.

Hypnotic drugs: Drugs which act to bring about depression and sleep.

Hypothalamus (hī′·pə·thal′·ə·məs): The region which lies in front of the thalamus in the forebrain. It regulates the internal environment, pituitary secretions, and some of the basic drives.

Imprinting: A type of learning which occurs in very specific periods early in life and is generally not reversible.

Independent assortment: A law which was developed by Mendel that states that the distribution of one gene to the offspring is not affected by the distribution of another gene. This permits recombination or the mixture of genetic material to the offspring.

Induction: The influence of cell differentiation or development by neighboring cells.

Inert: Chemically showing a lack of activity.

Innate releasing mechanism (IRM): An ethological term which refers to a genetic mechanism that produces a specific behavioral event when triggered by a particular stimulus from the environment.

Inorganic: Pertaining to chemical substances not produced by cells and not containing carbon.

Instinct (in′·stinkt): Any inherent, unlearned behavioral pattern, which is functional the first time it happens and can occur in animals reared in total isolation.

Intercalated disc (in·tʉr′·kə·lā′·ted): The junction which lies between the sub-units of cardiac muscle fibers.

Interneuron (in′·tər·noor′·än): A nerve cell which links sensory cells and motor cells.

Intrauterine device (IUD) (in′·trə·yōō′·ər·in): A plastic device sometimes containing copper which is inserted into the uterus as a means of preventing conception.

Ion: An atom or molecule that has gained or lost one or more electrons.

Ionic bond: An attractive force between atoms that have different charges as a result of the transfer of electrons between them.

Ionization: The separation of a substance into its ions when in solution.

Isosmotic (ī′·säs·mät′·ik): A solution with the same osmotic properties as a cell or another solution.

Isotope: An atom with a lighter or heavier nucleus than the average; may exhibit radioactivity.

Jejunum (ji·jōō′·nəm): The area of the small intestine which lies between the duodenum and the ilium.

Kidney: The organ that removes soluble waste products thus maintaining the composition and concentration of solutes in the blood.

Kilocalorie (kil′·ə·kal′·ər·ē): One thousand calories or the amount of heat necessary to raise the temperature of one thousand grams of water one degree.

Kingdom: Any of five categories in which organisms are usually placed.

Krebs citric acid cycle: The oxidative phase of respiration in aerobic reactions where the fuel is broken down to CO_2 and H_2O.

Labia majum (lā′·bē·ə mā′·jəm): The large

conspicuous outer folds of skin forming the perimeter of the female reproductive organs.

Labia minum (lā′•bē•ə mī′•nəm): The smaller inner folds of skin which border the vagina.

Leucocyte (loo′•kə•sīt′): Blood cells which aid in resisting infections.

Leucoplast (loo′•kə•plast′): A colorless body in plant cells which stores proteins, lipids, and starch.

Lipase (lī′•pās): A pancreatic enzyme that digests lipids.

Lipid (lib′•id): A type of fatty organic compound.

Liver: A large organ in higher animals that secretes bile, stores glycogen and carries out other essential chemical activities.

Loop of Henle: The prominent U shaped loop in the renal tubule of the mammalian kidney.

Luteal phase (loot′•ē•əl): The latter part of the menstrual cycle in which the corpus luteum secretes hormones that prepare the uterus for the fertilized egg.

Luteinizing hormone: LH-A pituitary hormone that causes ovulation and stimulates hormone production in the corpus luteum.

Lymph: The clear yellowish fluid or plasma found in the lymphatic system.

Lymph node: A clump of cells in the lymphatic system which produces lymphocytes and acts as a filter for lymph.

Lymphocyte (lim′•fə•sīt′): Any of a variety of white blood cells which are formed in the lymph nodes.

Macronucleus (mak′•rō•noo′•klē•əs): The larger of two types of nuclei found in certain protozoans.

Mantle: The shell-secreting organ in mollusks.

Mass number: The total number of neutrons and protons in the nucleus of an atom.

Mechanistic theory: The theory that the processes of life are based on the same physical and chemical laws that apply to the non-living phenomena.

Medulla (mi•dul′•ə): The region of the hindbrain that appears to be an expanded portion of the spinal cord. It controls many activities of the internal organs.

Meiosis (mī•ō′sis): The process of cell division in which the chromosome number is halved during the formation of gametes.

Melanin (mel′•ə•nin): Coloring pigments in the hair and skin of man, other animals, and some plants.

Menstrual cycle: The events in the human female that produce a mature egg and prepares the uterus for receiving the zygote.

Meristem (mer′•ə•stem′): Undifferentiated embryonic tissue in plants.

Mesophyll (mes′•ə•fil′): The parenchyma tissue of a leaf.

Messenger RNA (mRNA): RNA that carries the genetic code to the ribosomes where it is translated into protein production.

Metabolism: The total chemical processes of life, including energy storage, release and transfer.

Metamorphosis (met′•ə•môr′•fə•sis): Any morphological change such as the transformation of an insect larva into an adult.

Metaphase (met′•ə•fāz′): The second phase in mitosis during which the chromosomes line up at the center of the cell along the equator of the spindle.

Micron: One thousandth of a millimeter.

Micronucleus (mī′•krō•noo′•klē•əs): The smaller of two types of nuclei in certain protozoa.

Microtubule (mī•krō): A microscopic tube or hollow cylinder of protein filaments.

Midbrain: The central of the three subdivisions of the vetebrate brain.

Millimicron: One thousandth of a micron.

Mitochondrion (mīt′•ə•kän′•drē•ən): An organelle in the cytoplasm which is sometimes called the powerhouse of the cell because it functions in energy producing processes.

Mitotic apparatus: Any cell part involved in the process of mitosis.

Monocot (män′•ə•kät′): Any plant which produces one seed leaf such as grasses or palms.

Mons veneris (mänz′ ven′•ər•is): A fatty mound of tissue covering the pubic bone in the human female.

Morula (môr′•yoo•lə): A stage of the embryo in which the cells are arranged in a solid mass.

Motor neuron: A nerve cell that carries impulses from the central nervous system to a muscle or gland.

Mutation: Any change in a gene that results in a permanent genetic change in the individual; a deletion or addition to the basic code of the DNA strand.

Mutualism (myoo′•choo•wəl•iz′m): A symbiotic relationship in which both organisms benefit in some way.

Myofibril (mī′•ə•fī′•brəl): The slender protein thread in the skeletal muscle which acts in contraction of the muscle.

Myofilament (mī′•ə•fil′•ə•mənt): A filament which makes up the myofibril in the skeletal muscle cell.

Myosin (mī′•ə•sin): The non-moving protein of the myofibril of muscle.

NADP or NADPH: Nicotinamide Adenine Dinucleotide Phosphate. A molecule which functions in the electron transport system of plant chloroplasts.

Narcotic: Any agent which produces stupor or insensibility.

Natality: The birth rate of a population.

Natural selection: The differential survival and reproduction of certain individuals in a population. In the same process, other organisms less suited to the environment are eliminated.

Nephron (nef′•rän): The functional unit of the kidney.

Nerve impulse: An electrochemical signal in which an impulse is transmitted from one neuron to another neuron or to other cells.

Neural groove (noor•al): An ingrowth of the neural plate of a vertebrate embryo; it eventually forms the spinal cord in mammals.

Neural plate: A thick plate formed by rapid cell division over the notochord on the dorsal side of the embryo; it eventually forms the spinal cord in vertebrates.

Neural tube: The early brain and spinal cord of a vertebrate.

Neuron: A nerve cell.

Neutron: An atomic nuclear particle which has about the same mass as a proton, but is electrically neutral.

Niche (nich): All aspects of the biological and physical environment that relate to the activities of an organism.

Nitrogenous base (nī•träj′•ə•nəs): The purine or pyrimidine constituent of nucleotides and nucleic acids. (adenine, guanine, cytosine, thymine, uracil.)

Nocturnal animal: An animal which is active mainly at night.

Node of Ranvier: A depression which separates Schwann cells which lie along neurons.

Notochord (nōt′•ə•kôrd′): A rod-like arrangement of cartilage cells which form the chief supporting structure in the chordate body. In vertebrates it is present only in the embryonic stage.

Nuclear membrane: The outer selectively permeable membrane of the cell nucleus.

Nuclear pore: Tiny pores in the nuclear membrane of cells.

Nuclease (noo′•klē•ās′): Any enzyme which acts to split nucleic acids.

Nucleic acids (noo•klē•ik′): Any of a number of acids which carry genetic information. There are two kinds: RNA and DNA.

ucleolus (nōō·klē′·ə·ləs): An organelle which lies within the nucleus of the cell. It is high in RNA and protein. It forms ribosomal RNA.

ucleotide (nōō′·klē·ə·tīd′): A molecule made up of phosphate, a 5-carbon sugar, and a purine or pyrimidine base.

ucleus: An organelle within a cell bounded by a double membrane which contains most of the DNA of the cell. Also, in the atom, a positively charged mass composed of neutrons and protons.

cipital lobe (äk·sip′·ə·t'l lōb): One of the four major lobes of the brain. It is involved in the reception and processing of visual information.

mnivore (äm′·nə·vôr′): An animal which eats both plant and animal matter.

ocyte (ō′·ə·sīt′): A cell which undergoes meiosis to form an egg cell.

ogenesis (ō′·ə·jen′·ə·sis): The process of egg formation.

otid (ō′·ə·tid′): One of four cells formed by meiosis. In animals, one will develop into an egg cell.

ganelle (ôr′·gə·nel′): A specific structure within a cell, usually bounded by a membrane.

erant conditioning: A form of learning in which an animal performs an act to receive rewards.

bital: The approximate and hypothetical path on which an electron moves around the nucleus.

der: A taxonomic subdivision of a class in plants and animals which may consist of several families.

ganic: Pertaining to chemical compounds that are normally synthesized by living organisms which contain carbon, hydrogen and oxygen.

gasm: The final stage or climax of sexual intercourse, usually followed by a period of relaxation.

ientation: An organism's means of determining its location in respect to some reference point in its environment.

smosis (äs·mō′·sis): The tendency of water to move across a membrane from an area of lower solute concentration to one that has a higher solute concentration.

vary: The female gonad in which the eggs are formed.

viduct (ō′·vi·dukt′): The tube opening near the ovary and ending in the uterus in most animals.

vulation (ō′·vyə·lā′·shən): The sudden release of a mature egg from the ovary.

Ovum: The egg or female reproductive cell.

Oxidation (äk′·sə·dā′·shən): The loss of electrons in chemical reactions; or the combination of oxygen with another substance.

Palisade cells: A layer of elongated, chloroplast containing cells which lie under the epidermis of the leaf in higher plants.

Pancreas (pan′·krē·əs): A gland which is located behind the stomach in mammals which secretes digestive enzymes.

Parasite: An organism that lives in or on another organism and causes some harm to its host as it derives food from it.

Parasympathetic nervous system (par′·ə·sim′·pə·thet′·ik): One of the divisions of the autonomic nervous system.

Parenchyma (pə·ren′·ki·mə): Thin walled, unspecialized plantstorage cells.

Parietal lobe (pə·rī′·ə·t'l): One of the four major lobes of the cortex. The detection of body position and sensory input are the major functions.

Passive transport: Movement of substance through the cell membrane without cellular energy (ATP) involved.

Penis: The male sex organ.

Peptide (pep′·tīd): A compound comprised of two or more amino acids.

Peripheral nervous system (pə·rif′·ər·əl): One of the two divisions of the vertebrate nervous system. It includes the nervous tissue outside the central nervous system.

Petiole (pet′·i·ōl′): The stalk by which a leaf is attached to the stem.

Pharynx (far′·inks): The organ below the mouth that is the common respiratory and digestive structure leading to the esophagus and larynx.

Phenotype (fē′·nə·tīp′): The observable genetic characteristics of any organism.

Pheromone (fer′·ə·mōn′): A substance which is secreted by an organism which changes the behavior in another organism of the same species.

Phloem (flō′·em): The vascular tissues in plants, consisting of sieve tubes and companion cells and specializing in nutrient transport.

Phospholipid (fäs′·fō·lip′·id): A fat like substance of the cell membrane, which contains phosphorus, fatty acids, glycerol, and a nitrogenous base.

Phosphorylation (fös′·fər·ə·lā′·shən): The addition of phosphate to an organic molecule such as ADP, e.g. ADP + Phosphate → ATP.

Photon: A unit of light energy.

Photoperiodism (fōt′·ō·pir′·ē·əd·iz'm): The rhythmic response of organisms to cyclic variations in light and darkness.

Photosynthesis (fōt′·ə·sin′·thə·sis): The process in which plants convert carbon dioxide and water into carbohydrates through the use of solar energy.

Phototropism (fō·tät′·rə·piz'm): A group response of plants to the presence of light.

pH scale: A scale which defines the hydrogen ion concentration of a solution. It varies from one to fourteen.

Phylum (fī′·ləm): The major primary subdivision of a kingdom. It is composed of classes.

Pigment: Any colored substance.

Pistil: The seed-bearing part of a flower.

Pituitary gland (pi·tōō′·ə·ter′·ē): A tiny endocrine gland at the base of the brain.

Placenta: Embryonic membranes in the uterus of pregnant mammals through which the embryo transports wastes and receives essential nutrients.

Plankton: Organism which drifts in a body of water; usually very small or microscopic.

Plasma: The liquid portion of the blood.

Plastids (plas′·tid): Small bodies in plant cells.

Pollen: The male microspores of flowering plants consisting of reproductive cells in a protective case.

Polypeptide (päl′·i·pep′·tīd): A group of two or more amino acids; a protein fragment.

Pons (pänz): A part or extension of the medulla; it contains nerve cells which relate to the spinal motor nerves.

Primary consumers: Animals which feed on plants; herbivores.

Primitive streak: An elongated group of cells in the early embryo which constitutes the axis of the body.

Procaryotic (prō·karee·ah·tic): Organisms that constitute the base of the food chain, mainly photosynthetic plants.

Progesterone (prō·jes′·tə·rōn′): An ovarian hormone, produced by the corpus luteum, assists in the preparation of the uterus for implantation of a fertilized egg.

Prophase (prō′·fāz): The first stage in mitosis.

Prostate gland: A gland of the male reproduction system, situated at the base of the urethra; secretes a fluid that is added to the semen.

Protein: A specific chain of amino acids. Essential in producing cell structures and a variety of enzymes.

Proton: An atomic nuclear particle which has a mass similar to a neutron, but with a positive charge.

Protoplasm: The living colloidal material

which make up cells.

Pseudopodium (sōō'·də·pō'·dē·əm): Literally, a false foot; a temporary projection from the cell used in amoeboid cells for feeding and movement.

Psychedelic (sī'·kə·del'·ik): Related to relaxation; mind-freeing, accompanied by bizarre thought patterns and behavior.

Purine: Nitrogenous bases of DNA and RNA, containing two nitrogen rings—adenine and guanine.

Pyrimidine (pi·rim'·ə·dēn'): Nitrogenous bases of DNA and RNA, containing one nitrogen ring—thymine, cytosine, and uracil.

Pyruvate (pī·rōō'·vāt): Pyruvic acid; a molecule formed at the end of glycolysis.

Radiant energy: Sunlight energy.

Radioactivity: Particles or energy emitted from radioactive elements as they decay.

Radula (raj'·oo·lə): The rough surfaced tongue or feeding organ found in many mollusks.

Receptor: A sensory structure that detects a particular form of stimulus in the environment such as light or sound.

Recessive: The allele which is not expressed in the phenotype of an organism.

Reduction: Adding electrons to an atom or molecule.

Reduction division: Meiosis.

Reflex: An involuntary response to a stimulus.

Reflex arc: A simple nervous pathway which involves a sensory cell, a connecting neuron and a muscle cell.

Refractory period: The recovery period following a nerve impulse during which the nerve cannot initiate a new impulse.

Regeneration: Regrowth of a lost part or tissue.

Reinforcement: In behavior conditioning, a process which increases the likelihood of a given response.

Renal (nē'·n'l): Referring to the kidney.

Replication: The process of duplicating DNA prior to cell division.

Respiration: Cellular transfer of energy in fuels to ATP energy.

Resting potential: The relative charges on both sides of a nerve fiber in the unstimulated state.

Reticular formation: A nerve cell mass in the medulla that functions as an arousal mechanism.

Retina: The layer of light sensitive cells in the vertebrate eye.

RH factor: A type of antigen in the blood.

Ribonucleic acid (rī'·bō·nōō·klē'·ik): A nucleic acid containing ribose sugar and the nitrogen base, uracil. Functions in protein synthesis.

Ribosomal RNA (rRNA): A type of RNA molecule found in the ribosomes; its actions are not clearly known, but it is believed to function in protein synthesis.

Ribosome (rī'·bə·sōm'): A grainy structure found in great number in cells. Important in protein synthesis.

Ribulose (rī'·bū·lōs): A five carbon sugar produced in the Calvin cycle during photosynthesis.

Ribulose diphosphate (rī'bū·lōs dī'·fos·fāt): A form of ribulose which acts as the carbon dioxide acceptor in the Calvin cycle of photosynthesis.

RNA polymerase: An enzyme which aids in the formation of RNA along the DNA template.

Root cap: A group of cells covering the growing tip of a root.

Salt: A product of the reaction between a base and an acid.

Sarcolemma (sar·koh·lem·ah): The surrounding membrane of a striated muscle fiber.

Schwann cell: A cell which helps to form the sheath around the peripheral nerve fibers.

Scrotum: The sac surrounding the testis.

Secondary consumers: Animals that feed primarily on herbivores.

Secondary sex characteristic: The physical development that occurs during puberty, specific for each sex.

Second filial generation (F2) (fil'·ē·əl): The offspring produced by the mating of two individuals of the first filial generation.

Secretion: A cellular product, usually released into a duct or organ to serve a function.

Self-fertilization: A sexual process in plants where gametes from the same plant combine.

Semen: The sperm-containing fluid ejaculated by the male.

Seminal duct: The tube that carries semen from the epididymus to the ejaculatory duct. It is also called a vas deferens.

Seminal vesicle: A sac like part of the male reproductive duct in which sperm are stored.

Senescence (sə·nes'·'ns): Gradual deterioration caused by age.

Sensory neuron: A neuron which receives impulses; an afferent neuron.

Sepal (sē'·p'l): The leaflike floral parts surrounding the base of the petals.

Sessile (ses'·il): Pertaining to organisms which do not move about.

Sex chromosome: Chromosomes associated with the determination of sex. In man, the 23rd pair.

Sickle cell anemia: A hereditary disease caused

by a form of hemoglobin which is defecti... in its nature and causes sickle shaped re... blood cells.

Sieve cell: A specialized phloem cell throug... which nutrients move in the vascular plan...

Sieve tube: A tube formed by joining sie... cells end to end.

Solar energy: Any radiant energy produced ... the sun.

Solute (säl'·yōōt): Any substance dissolved ... a solvent.

Solvent: A fluid in which substances m... dissolve forming solutions.

Soma (sō'·mə): Pertaining to the body.

Somite (sō'·mīt): One of the paired mes... dermal segments along the neural axis of ... embryo.

Species: The major subdivision of a genu... Individuals of a species are able to bre... among themselves under natural condition...

Spermatid (spur'·mə·tid): A developing sper... cell.

Spermatogenesis (spər·mat'·əjen'·ə·sis): T... formation of sperm cells.

Sphincter (sfink'·tər): A ring of muscles th... controls the movement of materials throu... passages, e.g. sphincters at either end of t... stomach.

Spinal cord: The complex band of neuro... which runs through the spinal column ... vertebrates to the brain.

Spindle: A structure formed in mitosis a... meiosis which consists of fine threads radi... ing from the centrosomes along whi... the chromosomes appear to move.

Spleen: An organ containing many blo... vessels which is situated near the stoma... in man. It forms certain white blood ce... and destroys worn out red blood cells. ... also acts as a reservoir for red blood ce...

Spontaneous generation: The notion th... living organisms are derived from nonlivi... matter.

Sporophyte (spôr'·ə·fīt'): The spore produ... ing, diploid generation in plants. It alterna... with the sexual or gametophyte generatic...

Stamen: The male organ of the flower whi... bears pollen.

Starch: A long chain of glucose molecul... formed in plants for glucose storage.

Stereotyped (ster'·ē·ə·tīpt'): Predictable ... unchanging.

Steroid (stir'·oid): The type of lipid consisti... mainly of fatty acids attached to compl... alcohols.

Stigma: The female part of a flower whi... receives the pollen.

Stoma (stō'·mə): A small opening in a le...

through which gases may pass.

tyle: The portion of the pistil that connects the stigma with the ovary in flowers.

ubphylum: A group of related organisms within a phylum.

substrate (sub′·strāt): A substance which is acted upon by an enzyme.

sulcus (sul′·kəs): A convolution of the cerebral cortex.

ubnormal releaser: An artificial stimulus which is more effective than the natural stimulus in releasing a particular behavioral pattern.

urface tension: A cohesive force among surface molecules that resists rupture.

ymbiosis (sim′·bī·ō′·sis): The association between two organisms of two different species when at least one organism benefits in the association.

ympathetic nervous system: It increases energy expenditure and prepares the body for emergency situations.

ynapse (si·naps′): The space between neurons, across which impulses are chemically transmitted.

ystole (sis′·tə·lē′): The contraction of the heart.

elophase (tel′·ə·fāz′): The final stage of mitosis which includes total separation of chromosomes, cytoplasmic division, and the return of the interphase nucleus.

emplate: A pattern or mold.

emporal lobe: One of the four lobes in the human brain; it contains auditory and visual centers.

estis: The male gonad where sperm and sex hormones are produced.

estosterone (tes·täs′·tə·rōn′): A male hormone, produced in the testes, important in the sex drive and producing secondary sex characteristics.

etanus: A state of continuous muscle contraction.

halamus (thal′·ə·məs): A cluster of neural cells in the posterior forebrain; it is involved in auditory and visual functions.

hermal energy: Energy due to the motion of molecules.

hreshold: The intensity level necessary to create a response to a stimulus.

hymine (thī′·mēn): One of the nitrogenous bases of DNA.

hymus gland (thī′·məs): An organ lying behind the breast bone which grows smaller as an individual ages; it functions in immune responses.

hyroxin (thī·räk′·sin): A thyroid hormone which functions in metabolism.

Trachea (trā′·kē·ə): The tube that carries air from the larynx to the bronchi in land vertebrates; in insects a complex system of branched tubules carrying air to the interior.

Tracheid (trā′·kē·id): A type of xylem cell in plants adapted for conduction and support.

Transfer RNA: Lightweight nucleic acid molecules which identify with specific amino acids during protein synthesis.

Trophic levels (träf′·ik): The various levels feeding levels in a food chain, including producers and consumers.

Trophoblast (träf′·ə·blast′): The thin wall side of a blastocyst that forms the chorion when implantation occurs in mammals.

Tropism: A growth which is a response to a specific external stimulus.

Tubal ligation: Female sterilization by cutting and tying the fallopian tube.

Tumescence (tōō·mes′·′ns): The swelling of an organ or part.

Turner's Syndrome: An abnormality due to the lack of one X chromosome; it is characterized by retarded growth and sexual development.

Umbilical cord: Blood vessels that transport substances between the placenta and the embryo.

Unconditioned response: A natural response to an unfamiliar stimulus.

Unconditioned stimulus: An unfamiliar stimulus that causes response.

Unit membrane: The basic structure of all cellular membranes.

Uracil (yoor′·ə·sil): One of the nitrogenous bases of RNA.

Ureter (yoo·rēt′·ər): The tube that conducts urine from the kidney to the bladder in higher animals.

Urethra (yoo·rē′·thrə): The tube which conducts urine from the bladder to the outside of the body in higher animals.

Uterus (yōōt′·ər·əs): In female animals, the reproductive tract in which the fetus is contained.

Vagina: The female copulatory organ and birth canal.

Vascular tissue: Any tissue which contains vessels through which fluids are passed.

Vasectomy (vas·ek′·tə·mē): Male sterilization by cutting and tying the seminal ducts.

Vein: A vessel that carries blood toward the heart.

Ventral: Referring to the undersurface or belly of an animal.

Villus (vil′·əs): A small protrusion of the intestinal wall, greatly increasing the absorbing surface of that organ.

Virulent: Capable of causing diseases.

Virus: An infectious, submicroscopic parasite that consists of a RNA or DNA core with a protein coat.

Viscera (vis′·ər·ə): A term which pertains to the internal abdominal organs of an animal. Sometimes called the guts.

Vitalism (vīt′·′l·iz′m): The notion that life has unique mystical properties which are distinct from those ascribed by chemical and physical laws.

Vitamin: A coenzyme. Any of a group of unrelated organic substances which are essential for normal metabolism.

Vulva: The external part of the female genitals.

White matter: The outer layer of the spinal cord in which myelinated fibers are found.

Xanthophyll (zan′·thə·fil): The yellow, red or orange pigments of green plants.

X chromosome: A sex chromosome. Females carry two X chromosomes while males have one.

Xylem (zī′·ləm): Water conducting and supporting tissue in vascular plants.

Y chromosome: A sex chromosome found only in males.

Yolk: The nutrient portion of the egg.

Yolk sac: An embryonic tissue that assists in the nourishment of the embryo in some fishes, reptiles, and birds, and in most mammals.

Z scheme: A series of photochemical events centered around photosynthesis which results in the net movement of electrons from water to reduced NADP.

Zygote (zī′·gōt): A fertilized egg in plants or animals.

Index

Note: page numbers in italics refer
to art.